Recent Advancement and Emerging Technologies in Multi-level Inverters

RIVER PUBLISHERS SERIES IN POWER

Series Editors:

MASSIMO MITOLO
Irvine Valley College
USA

VIJAY K. SOOD
Ontario Tech University
Canada

The "River Publishers Series in Power" is a series of comprehensive academic and professional books focussing on the theory and applications behind power generation and distribution. The series features content on energy engineering, systems and development of electrical power, looking specifically at current technology and applications.

The series serves to be a reference for academics, researchers, managers, engineers, and other professionals in related matters with power generation and distribution.

Topics covered in the series include, but are not limited to:

- Power generation;
- Energy services;
- Electrical power systems;
- Photovoltaics;
- Power distribution systems;
- Energy distribution engineering;
- Smart grid;
- Transmission line development.

For a list of other books in this series, visit www.riverpublishers.com

Recent Advancement and Emerging Technologies in Multi-level Inverters

Editors

Kundan Kumar
National Institute of Technology, Jamshedpur, India

Ambrish Devanshu
National Institute of Technology, Silchar, India

Subrata Banerjee
National Institute of Technology Durgapur, India

Lairenlakpam Robindro
CSIR-Indian Institute of Petroleum, India

Sheldon S. Williamson
Faculty of Engineering and Applied Sciences, at Ontario Tech University, in Oshawa, Ontario, Canada

River Publishers

Routledge
Taylor & Francis Group
NEW YORK AND LONDON

Published 2026 by River Publishers
River Publishers
Broagervej 10, 9260 Gistrup, Denmark
www.riverpublishers.com

Distributed exclusively by Routledge
605 Third Avenue, New York, NY 10017, USA
4 Park Square, Milton Park, Abingdon, Oxon OX14 4RN

Recent Advancement and Emerging Technologies in Multi-level Inverters /
by Kundan Kumar, Ambrish Devanshu, Subrata Banerjee,
Lairenlakpam Robindro, Sheldon S. Williamson.

Routledge is an imprint of the Taylor & Francis Group, an informa business

ISBN 978-87-7004-143-0 (hardback)
ISBN 978-87-4380-187-0 (paperback)
ISBN 978-87-4380-145-0 (online)
ISBN 978-87-4380-144-3 (master ebook)
DOI: 10.1201/9781003669845

Contents

Preface

The rapid advancements in power electronics have led to significant innovations in the design and application of Multi-Level Inverters (MLIs). MLIs have become a critical technology in modern power systems due to their ability to achieve higher efficiency, lower harmonic distortion, and superior power quality compared to traditional two-level inverters. Their applications span a wide range of fields, including renewable energy systems, electric vehicles (EVs), industrial automation, and smart grids. Recognizing the growing importance of MLIs, this book, "*Recent Advancement and Emerging Technologies in Multi-Level Inverters,*" provides a comprehensive exploration of the latest developments, challenges, and emerging technologies in this field.

This book aims to serve as a comprehensive resource for both researchers and practitioners, covering fundamental principles, advanced topologies, control strategies, and optimization techniques for MLIs. Bringing together contributions from leading experts, the book presents a diverse range of topics reflecting current trends and future directions. The chapters delve into innovative areas such as switched capacitor-based MLIs, modular multilevel converters, reduced switch count topologies, transformer-integrated cascaded inverters and advanced modulation techniques. Additionally, the book addresses critical aspects such as fault diagnosis, voltage balancing, and the integration of MLIs into renewable energy systems and EV powertrains.

A unique feature of this book is its emphasis on practical implementation. Along with theoretical insights, it provides step-by-step guidance on designing and optimizing MLIs using simulation tools like MATLAB. Real-world case studies further demonstrate the practical application of MLIs, offering readers a deeper understanding of their utility and performance in various systems.

The primary objective of this book is to bridge the gap between theory and practice, equipping readers with the knowledge required to design, implement, and optimize MLIs for diverse applications. Additionally, we highlight key research challenges and future opportunities in this rapidly evolving field, encouraging further innovation and exploration.

This reference book is organized into twelve chapters. The organization of the book is discussed hereunder.

Chapter 1 presents recent developments, challenges, and applications of Multi-Level Inverters. This chapter provides an overview of the most current advancements, difficulties, and uses of MLIs. A comparison of different topologies, description of different control techniques for MLI are also elaborated in this chapter.

Chapter 2 comprehensively reviews recent advances in switched capacitor based multilevel inverters (SCMLIs) within the realm of power electronics. The chapter critically examines the challenges in switched capacitor-based multilevel inverter designs, emphasizing recent research on circuit topologies and modulation strategies, including aspects like voltage balancing. Moreover, it discusses the integration of SCMLIs in electric vehicles and renewable energy systems and highlights their potential to enhance efficiency, improve power quality, and seamlessly integrate with modern power grids.

Chapter 3 explores reduced switch count multilevel inverters (RSC-MLI), a technology suited for medium-voltage, high-power applications. It provides an in-depth explanation of different RSC-MLI topologies, covering symmetrical and asymmetrical modes of operation, and discusses the challenges and future research directions for multilevel inverters (MLI). Additionally, the chapter highlights the applications of MLI in motor drives, Flexible AC Transmission Systems (FACTS) devices, and grid-integrated renewable energy systems.

Chapter 4 describes modular multilevel converters (MMCs) that have solidified their position as an advanced and futuristic technology in the field of power electronics converter, presenting distinctive advantages across diverse applications, spanning from high-voltage and high-power to medium-voltage and medium-power systems. This chapter covers various aspects of the MMC, including its overall structure, workings, mathematical models, control mechanisms, and diverse applications.

Chapter 5 provides a brief review of various types of multilevel inverters, including their classification and mathematical models. A method has been discussed for determining switching states and the inverter output voltage, allowing the estimation of the filter inductor current without the need for a current sensor.

Chapter 6 explores the optimization of switching strategies in transformer integrated cascaded multilevel inverters (CMI) employing the repeating sequence method for configurations with different levels. The chapter delves into the implementation of the repeating sequence technique and compares the performance with selective harmonic elimination PWM (SHEPWM) with respect to total harmonic distortion (THD).

Chapter 7 describes Z-source based multilevel inverter topologies for photovoltaic applications. This chapter primarily focuses on the derivation, classification, and comparison of various Z-source networks. Additionally, it provides a detailed review of Z-source-based three-level and multilevel topologies, evaluating them based on factors such as the number of components, input current patterns, advantages, and disadvantages. Finally, some of the latest proposals are presented, enabling the reader to make an informed choice of a Z-source inverter for photovoltaic (PV) applications.

Chapter 8 analyzes a seven-level binary-structured ACMLI topology using various modulation techniques of MC-PWM. To implement the seven-level inverter with multicarrier techniques, six carriers were utilized, including PD-PWM, POD-PWM, APOD-PWM, and PS-PWM. These techniques were compared in terms of Total Harmonic Distortion (THD) in both the output voltage and current, considering different modulation and frequency indices. Additionally, some hardware results are presented to demonstrate the functioning of the ACMLI.

Chapter 9 provides an in-depth exploration of multilevel converter technologies, with a specific focus on the cascaded H-bridge multilevel inverter (CHBMLI) and the modular multilevel converter (MMC), both of which are crucial for medium-voltage, high-power drives. The CHBMLI is examined in detail, covering its symmetrical and asymmetrical operation modes, comparisons with fundamental multilevel inverters (MLIs) such as the diode-clamped multilevel inverter (DCMLI) and the flying capacitor multilevel inverter (FCMLI), as well as various pulse-width modulation (PWM) techniques.

Chapter 10 proposes an asymmetrical cascaded bridge inverter utilizes three DC sources and a minimal number of devices to generate a 19-level output voltage. By employing a mathematical switching technique, this chapter presents a straightforward control and modulation strategy. The proposed bridge configuration results in a more compact design, reduced component losses, and lower installation costs. Furthermore, the recommended configuration meets unique efficiency and cost criteria by minimizing conduction and switching losses.

Chapter 11 presents a novel MMC-based configuration for hybrid energy storage, featuring an optimized control strategy that effectively manages diverse components to enhance system performance. The proposed algorithms mitigate low-frequency ripple in submodule capacitor voltage, thereby improving MMC reliability and longevity, as demonstrated through detailed simulations and experimental validations. Additionally, a comprehensive analysis of various voltage-balancing algorithms and modulation techniques for MMCs in SSTs is provided, introducing two improved algorithms that undergo thorough verification.

Chapter 12 presents the different aspects of MLI Integration in EV Powertrain and performance evaluation using drive cycle analysis. MLI integration significantly enhances EV powertrain performance by improving efficiency, reducing harmonic distortion, and optimizing battery usage. Drive cycle analysis provides an effective way to evaluate real-world performance, helping optimize control strategies and maximize EV range and efficiency.

We extend our gratitude to the authors who have contributed their expertise and research to this book. We also appreciate the reviewers for their valuable feedback, which has helped ensure the quality and relevance of this work. We hope this book will serve as an essential reference for researchers, engineers, and students in power electronics and related disciplines, inspiring new ideas and advancements in Multi-Level Inverter technology.

Acknowledgement

We are deeply grateful to all the individuals and institutions that have contributed to the successful completion of this book, "*Recent Advancement and Emerging Technologies in Multi-Level Inverters.*" This work would not have been possible without the collaborative efforts of the authors, reviewers, and experts in the field of power electronics, whose invaluable insights and contributions have significantly enhanced the quality of this publication.

We extend our sincere thanks to our contributors for their dedication and expertise in preparing the chapters. Their innovative ideas and in-depth analyses have laid the foundation for this comprehensive resource, which we hope will benefit researchers, engineers, and practitioners in the field of power electronics and electrical engineering.

We also wish to express our gratitude to the reviewers for their constructive feedback, which played a pivotal role in refining the content and ensuring the highest standards of academic rigor and relevance.

A special thanks to River Publisher for their unwavering support, professionalism, and guidance throughout the publishing process. Their commitment to disseminating cutting-edge research and knowledge has been instrumental in bringing this book to fruition. We appreciate their encouragement and efforts to ensure a smooth progression from manuscript preparation to publication.

Finally, we are grateful to our respective institutions and colleagues for providing the environment and encouragement needed to complete this work. We dedicate this book to the global academic and research community, with the hope that it will serve as a valuable reference and inspire further advancements in Multi-Level Inverter technology.

The authors cordially invite any constructive comments/suggestions from the reader about this book.

Editors

List of Figures

List of Tables

List of Contributors

Abhilash, T, *Caterpillar, India*

Agarwal, Pramod, *Electrical Engineering Department, Indian Institute of Technology Roorkee, India*

Banerjee, Subrata, *Electrical Engineering Department, National Institute of Technology Durgapur, India*

Bhushan, Ravi, *National Institute of Technology Jamshedpur, India*

Biswas, Sangita Das, *Department of Electrical Engineering, Tripura University, India*

Das, Bikram, *Department of Electrical Engineering, NIT Agartala, India*

Dash, Jyoti Ranjan, *Electrical Engineering Department, Indian Institute of Technology Roorkee, India*

Dhanraj, Wairokpam, *Electrical Engineering Department, National Institute of Technology Manipur, India*

Gottapu, Kusuma, *Department of Electrical and Electronics Engineering, Sagi Rama Krishnam Raju Engineering College, India*

Jayaram, Nakka, *Department of Electrical Engineering, National Institute of Technology Andhra Pradesh, India*

Jyothsna, T. R., *Electrical Engineering, Andhra University, India*

Kirubakaran, A., *Department of Electrical Engineering, National Institute of Technology Warangal, India*

Kishore, P. S. V., *Department of Electrical Engineering, National Institute of Technology Andhra Pradesh, India*

Kumar, B. Hemanth, *Department of Electrical Engineering, Mohan Babu University, India*

Kumar, Devanand, *National Institute of Technology Karnataka, India*

Kumar, Kuncham Sateesh, *Department of Electrical and Electronics Engineering, National Institute of Technology Trichy, India*

Kumar, Kundan, *Electrical Engineering Department, National Institute of Technology Jamshedpur, India*

Kumar, Ritankar, *School of Electrical and Computer Sciences, IIT Bhubaneswar, India*

Kumari, Deepti, *National Institute of Technology Karnataka, India*

Nandi, Champa, *Department of Electrical Engineering, Tripura University, India*

Pradhan, Rudranarayan, *School of Electrical Sciences, Odisha University of Technology and Research Bhubaneswar, India*

Prasanna, Lakshmi, *Electrical Engineering, Andhra University, India*

Rachananjali, Kunamneni, *EEE, VFSTR, India*

Rahul, Jammy Ramesh, *Department of Electrical and Electronics Engineering, Gayatri Vidya Parishad College of Engineering (A), India*

Rajesh, Jami, *Department of Electrical Engineering, National Institute of Technology Andhra Pradesh, India*

Raushan, Ravi, *National Institute of Technology Karnataka, India*

Ray, Olive, *School of Electrical and Computer Sciences, IIT Bhubaneswar, India*

Robindro, Lairenlakpam, *CSIR-Indian Institute of Petroleum, Automotive Fuels and Lubricants Application Division (AFLAD), Dehradun, India*

Shakeera, Shaik, *EEE, VFSTR, India*

Swamy, Jakkula, *Department of Electrical Engineering, Indian Institute of Technology Bombay, India*

Zhang, Lei, *School of Electrical and Electronic Engineering, Nanyang Technological University, Singapore*

List of Abbreviations

AC	Alternating current
ACMLI	Asymmetrical cascaded multilevel inverter
AI	Artificial intelligence
ANPC	Active neutral point clamped
APD	Asymmetrical phase disposition
APOD	Alternate phase opposition disposition
BESS	Battery energy storage system
BMS	Battery management systems
BSA	Bird swarm algorithm
CBPSC	Cascaded bipolar switched cells
CCT	Cross-connected topologies
CCU	Crisscross unit
CD-PWM	Carrier disposition pulse width modulation
CG-SCMLI	Common ground switched capacitor multilevel inverter
CHB	Cascaded H-bridge
CHBMLI	Cascade H-bridge multilevel inverter
CMI	Cascaded multilevel inverter
CMV	Common mode voltage
CSI	Current source inverter
DC	Direct current
DER	Distributed energy resource
EM	Electric motor
EMI	Electromagnetic interface
EP	Equal-phase
EPM	Equal phase method
ESR	Equivalent series resistance
ESS	Energy storage system
EV	Electric vehicle
FACTS	Flexible AC transmission systems
FBSM	Full-bridge submodule

FLC	Flying capacitor
FC	Fuel cell
FLCI	Flying capacitor inverter
FLCMLI	Flying capacitor multilevel inverter
FLCSM	Flying capacitor submodule
FF	Feed-forward
FOC	Field-oriented control
G2V	Grid-to-vehicle
GA	Genetic algorithm
GB	Gearbox
GHG	Greenhouse gases
HB	H-bridge
HEP	Half-equal-phase
HESS	Hybrid energy storage system
HEV	Hybrid electric vehicle
HF	High frequency
HFLC	Hybrid flying capacitor
HFT	High-frequency transformer
HH	Half-height
HPF	High pass filter
HSWC	Half switched capacitor
HSC	Hexagonal switched cell
HVDC	High voltage direct current
ICE	Internal combustion engine
IDC	Indian driving cycle
IGBT	Insulated gate bipolar transistor
IM2DC	Isolated modular multilevel DC-DC converter
IPD	Interleaved phase disposition
IS	Impedance source
LCCT	Inductor-capacitor-capacitor-transformer
LDN	Level doubling network
LF	Low frequency
LPF	Low pass filter
LSPWM	Level shifted pulse width modulation
MC-PWM	Multicarrier pulse width modulation
MI	Modulation index
MIDC	Modified Indian driving cycle
ML	Machine learning
MLDCL	Multilevel DC-link

MLDCLT	Multilevel DC link topologies
MLRI	Machine learning infrastructure
MLI	Multilevel inverter
MLMT	Multilevel module topologies
MMC	Modular multilevel converter
MOSFET	Metal-oxide-semiconductor field-effect transistor
MPC	Model predictive control
MSRC-SCMLI	Multisource-based reduced-component switched capacitor multilevel inverter
MTDC	Multiterminal direct current
MVS	Maximum voltage stress
NLC	Nearest level control
NPC	Neutral point clamped
NPCMLI	Neutral point clamped multilevel inverter
NVC	Nearest vector control
OPAL-RT	OPAL-RT technologies
PD	Phase disposition
PI	Proportional-integral
PIV	Peak inverse voltage
POD	Phase opposition disposition
PS	Phase shifted
PSO	Particle swarm optimization
PS-PWM	Phase-shifted carrier pulse width modulation
PUC	Packed U-cell
PV	Photovoltaic
PWM	Pulse width modulation
QPR	Quasi-proportional resonance
qSBI	Quasi-switched boost inverter
RES	Renewable energy source
RMS	Root mean square
RSC	Reduced switch count
RSCMLI	Reduced switch count multilevel inverter
RVT	Reverse voltage topologies
SA	Switching angle
SC	Supercapacitor
SWC	Switched capacitor
SCC	SC cell
SCMLI	Switched capacitor multilevel inverter
SCSST	Series connected switched source topologies

SDCS	Switched DC supplies
SHE	Selective harmonic elimination
SHEPWM	Selective harmonic elimination pulse width modulation
SM	Submodule
SOC	State of charge
SPWM	Sinusoidal pulse width modulation
SSPST	Switched series and parallel source topologies
SST	Solid-state transformer
ST	Square T
STATCOM	Static compensator
SVC	Space vector control
SVM	Space vector modulation
SVPWM	Space vector pulse width modulation
THD	Total harmonic distortion
THI	Third harmonic injection
TSV	Total standing voltage
UPS	Uninterruptible power supplies
V2G	Vehicle-to-grid
VSC	Voltage source converter
VSI	Voltage source inverter
WLTC	Worldwide harmonized light-duty test cycle
ZSI	Z-source inverter

1

Recent Developments, Challenges, and Applications of Multilevel Inverters

Wairokpam Dhanraj[1], Kundan Kumar[2], and Subrata Banerjee[3]

[1]Electrical Engineering Dept. NIT Manipur, Imphal, Manipur, India-795004
[2]Electrical Engineering Dept., NIT Jamshedpur, Jamshedpur, Jharkhand, India-831014
[3]Electrical Engineering Dept. NIT Durgapur, Durgapur, India-713209
E-mail: wairokpamdhanraj@gmail.com; kundan01012020@gmail.com; bansub2004@yahoo.com

Abstract

MLIs are essential for modern power electronics because they can produce voltage waveforms of excellent quality, with less harmonic distortion and enhanced electromagnetic compatibility. This chapter provides an overview of the most current advancements, difficulties, and uses of MLIs. The efficiency and fault tolerance of power systems have been greatly improved by the introduction of advanced topologies, such as the hybrid-modular MLI and single-source nine-level inverters. The dynamic performance and stability have been enhanced due to the development of advanced control techniques, such as proportional-integral (PI) and quasi-proportional resonance (QPR) controls. Methods such as selective harmonic elimination (SHE) have enhanced the removal of harmonics, while advancements in reduced switch topologies and soft-charging switched-capacitor techniques have increased efficiency. Still, MLIs encounter technological obstacles such as intricate control algorithms and thermal management, as well as economic and legal hurdles. The variety and significance of MLIs are exemplified in their applications in renewable energy systems, electric vehicles, and industrial motors. Further investigation should prioritize resolving current obstacles and

DOI: 10.1201/9781003669845-1

investigating nascent technologies to fully harness the capabilities of MLIs in diverse high-power and high-voltage applications.

Keywords: Multilevel inverters, power electronics, harmonic reduction, control strategies, renewable energy systems, electric vehicles, industrial drives, efficiency, fault tolerance, thermal management

1.1 Introduction

Multilevel inverters (MLIs) have emerged as an important device in modern power electronics due to their ability to enhance power quality, reduce harmonic distortion, and increase the efficiency of power conversion systems. They are primarily beneficial in applications requiring high power and high voltage, where traditional two-level inverters fall short. MLIs are extensively utilized in renewable energy systems, electric vehicles, and industrial drives by offering improved performance and reliability [1]. This chapter explores the latest advancements, existing challenges, and varied applications of MLIs, providing a thorough understanding of their current status and future prospects. Owing to the ever-increasing requirement of good power quality for various voltage levels, the traditional low-level inverters are phased out since these devices are not able to produce better waveforms with less harmonic distortions. These devices are also accompanied by very high-power-rated switching devices. Hence, multilevel inverters are earning more recognition since they offer appealing characteristics such as very high voltage application, better power quality, and decreased rating of switching semiconductor devices [2, 3].

1.2 Recent Developments in Multilevel Inverters

Recent advancements in multilevel inverters have significantly enhanced the performance, efficiency, and reliability of MLIs. This makes them increasingly vital to various applications such as renewable energy systems, electric vehicles, power quality improvement, and smart grids. For example, the modular multilevel converter (MMC) has seen substantial improvements, particularly in circulating current suppression. This is crucial because circulating currents in the submodules can degrade the MMC's dynamic performance and safety during the start-up and shutdown phases. A notable study by Zichen Wang [4] has explored the integration of proportional-integral (PI) control with quasi-proportional resonance (QPR) control to address these issues. By combining these control methods, the study demonstrates the effect of enhanced suppression of circulating currents. This will in turn improve the overall stability and safety of MMCs used in flexible DC transmission and distributed power regulation systems.

Another significant area of development is the comparative analysis of five-level and seven-level single-phase cascaded H-bridge multilevel inverters (CHB-MLIs). CHB-MLIs are valued for their modular structure, which allows for improved voltage waveform quality. This will help in reducing harmonic distortion and enhancing power handling capabilities. A study conducted by S. Venkateshwarlu and colleagues [5] explored the design, simulation, and control strategies of these inverters. The research highlights that seven-level inverters provide superior performance compared to five-level inverters, particularly in terms of voltage waveform quality and harmonic reduction. This makes them more suitable for high-performance industrial applications and promotes energy efficiency and sustainability in power conversion systems.

In the dominion of harmonic elimination, the implementation of selective harmonic elimination (SHE) using the bird swarm algorithm (BSA) has shown promising results. Traditional SHE techniques often involve solving complex nonlinear equations, which can be computationally intensive. However, Piyush L. Kamani and his team have introduced a BSA-based approach that simplifies this process by optimizing the switching angles. This method not only reduces the computational load but also decreases total harmonic distortion (THD) in seven-level inverters and hence helps in enhancing their overall performance. The simulation and experimental results of this study confirm the efficacy of this technique in improving the quality of the inverter output [6].

Efficiency improvements in five-level multilevel inverters (MLIs) have also been a focus of recent research. Parimalasundar Ezhilvannan and his colleagues proposed a novel reduced switch topology that significantly minimizes power losses while maintaining high performance. The proposed topology employs advanced pulse width modulation (PWM) control strategies and optimized switching patterns. This leads to improved power conversion efficiency. The study reports that this new topology reduces the total harmonic distortion (THD) of the output current to 7.12% and achieves an efficiency of 96.92%. These findings suggest that reduced switch topologies can play a critical role in advancing MLI technology for applications in renewable energy systems, electric vehicles, and industrial drives [7].

Another critical development is in the area of leakage current suppression in three-level photovoltaic (PV) inverters. Leakage current can pose significant safety risks and reduce the overall efficiency of PV systems. Lingyu Wang [8] has conducted comprehensive research on various leakage current suppression techniques, including space vector pulse width modulation (SVPWM), circuit topology modifications, and the addition of filters. These techniques help mitigate the leakage current to help in enhancing the safety and performance of three-level PV inverters. The study also provides a detailed comparison of these methods, highlighting their respective advantages and limitations. This offers insights into future research directions and technical challenges in this field.

A novel switched capacitor multilevel inverter (SCMLI) topology that aims to reduce the device count while generating high voltage levels is discussed in this research [9]. This will be a significant contribution to enhancing the efficiency and cost-effectiveness of MLIs. Since this proposed topology utilizes a cascade structure, this will not only increase the number of output levels but also decrease the voltage stress on the switches at higher levels. This will help to address some common limitations in the existing design topologies. The proposed hybrid topology incorporates a DC source and a switched capacitor unit to generate high output voltage levels. This will result in less voltage stress on the switches as compared to conventional methods. By combining an optimal DC power supply with the switched capacitor unit, it becomes possible to achieve greater voltage levels, resulting in improved performance of the inverter.

A new multilevel inverter (MLI) that demonstrates a significant reduction in the number of switches is proposed in this paper [10]. This reduced switch design achieves the same output level as conventional inverters. More precisely, it utilizes only 12 switches to provide 25 different output levels, demonstrating a significant enhancement in efficiency. The inverter circuit exhibits a remarkably low total harmonic distortion (THD) of only 4.98% at a frequency of 50 Hz. The low total THD value signifies a superior output waveform, which is crucial for utilization in renewable power networks. The design of the inverter enables an optimized arrangement and packing of the circuit, thanks to the decreased quantity of components. Implementing this optimization can result in reduced expenses and enhanced dependability in real-world scenarios.

A novel topology for a seven-level inverter that effectively decreases the number of switches needed is proposed in this paper [11]. It achieves this by utilizing only six or seven switches in comparison to the standard 12 switches usually required for a seven-level MLI design. The MATLAB/SIMULINK environment was used to build and simulate both the seven-level six-switch and seven-level seven-switch inverters. This demonstrated that the proposed topologies are effective in generating the appropriate voltage levels with reduced complexity. Furthermore, a physical implementation was showcased, which verified the accuracy of the simulated outcomes and affirmed the feasibility of the suggested inverter designs in practical scenarios. The study emphasizes the benefits of decreased switching losses and reduced costs. This will lead to a reduced number of semiconductor switches and help in enhancing the efficiency and suitability of the suggested inverters for applications like renewable energy systems. Hence, the results suggest that the modified inverter configurations offer a dependable solution for multilevel inverter applications. The typical limitations of conventional designs, such as intricacy and a large number of components are also resolved.

In summary, the recent developments in multilevel inverters encompass a broad spectrum of innovations aimed at improving their performance, efficiency, and reliability. From advanced control strategies in MMCs to novel harmonic elimination techniques and efficiency improvements in reduced switch topologies, these advancements are pushing the boundaries of what is possible in power electronics. The integration of these innovations into practical applications promises to enhance the sustainability and effectiveness of various power conversion systems, making them more adaptable to the growing demands of modern electrical grids and renewable energy sources.

1.3 Topologies of Multilevel Inverters

1.3.1 Cascaded H-bridge (CHB) inverters

CHB inverters consist of multiple H-bridge cells connected in series. This allows for higher voltage levels and reduced harmonic distortion. Normally, when "n" H-bridge cells are connected in series in a single phase, it results in $2n + 1$ voltage levels. Each H-bridge cell needs its own individual DC power source. The CHB inverter has the capability to function in both symmetric and asymmetric modes. In symmetric mode, all the DC voltage sources are identical. However, in asymmetric mode, the DC voltage sources are not equal. Two types of cases have been observed in asymmetrical mode; one where the DC ratios are in binary and the other where the DC ratios are in ternary. Recent research is focused on enhancing the modularity and scalability of

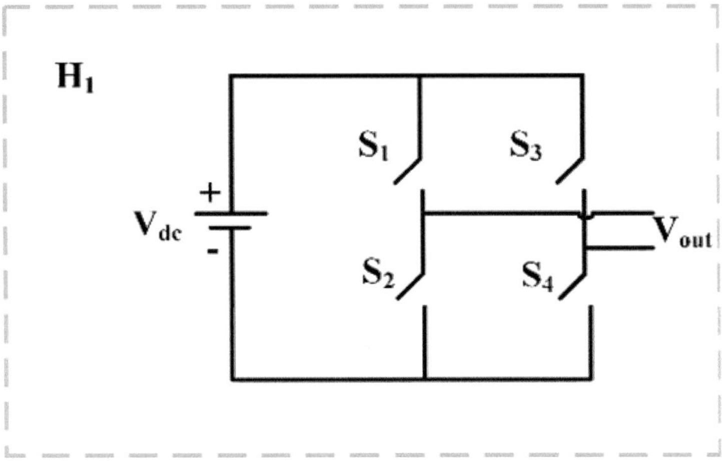

Figure 1.1 Single-cell CHB.

CHB inverters [12]. Efforts have been directed toward minimizing the number of switches required and hence reducing complexity and cost, while also improving fault tolerance to enhance reliability in large-scale applications.

Figure 1.1 shows the basic building block of a cascaded H-bridge converter, which is an H-bridge. Switches S_1, S_2, S_3, and S_4 are shown as generic switches. These switches are a combination of a transistor and an antiparallel diode, for instance, an IGBT. It also has a DC source which is represented by $V_{dc.}$ This DC source can be either fed from a rectifier, from a PV panel, or from a battery storage system. In some cases, capacitors can also be used. The H-bridge is also called a cell or a module and the CHB contains many of such cells connected in series. The DC link of each cell is isolated and hence when these cells are cascaded, there will be isolated DC links for each cell. Each cell can produce three levels of voltages (0, V_{dc}, $-V_{dc}$). If S_1 and S_3 are turned on, the voltage will be 0 (shorted DC bus). Similarly, if S_2 and S_4 are turned on, the voltage will be 0. If S_1 and S_4 are turned on, the voltage output will be +V_{dc} and if S_3 and S_4 are turned on, the voltage output will be $-V_{dc}$.

Figure 1.2 shows two cascaded modules/cells to provide a single output. With two cells, the number of levels in V_{out} is 5. With "n" cells, the number of levels in output is given by ($2n$ + 1). The individual output voltages of each cell, that is, H_1 and H_2 are independent of each other, and the overall output is the sum of these two voltages. So, the possible output voltages can

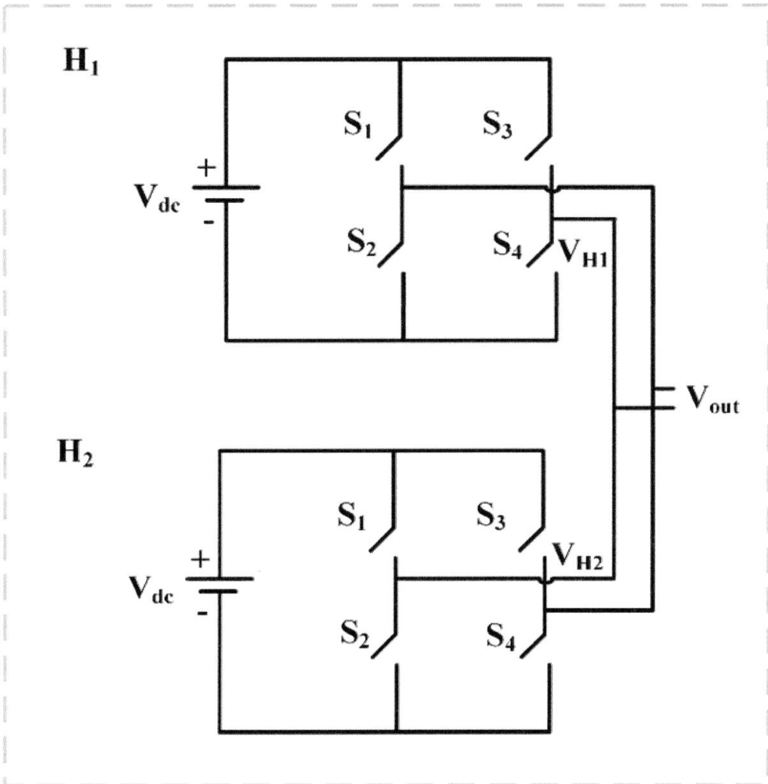

Figure 1.2 Two cells cascaded CHB.

be either of these five-levels (0, $-V_{dc}$, V_{dc}, $2V_{dc}$, and $-2V_{dc}$). These give rise to cascaded multilevel inverters.

Figure 1.3 shows a three-phase five-level cascaded H-bridge (CHB) inverter consisting of six H-bridge cells. The construction of a three-phase five-level cascaded H-bridge (CHB) is achieved by connecting two H-bridges in series per phase, as displayed in the figure. V is the DC voltage for each bridge.

Table 1.1 displays the switching logic for a five-level CHB inverter working in symmetric mode. Redundancy switching states are accessible in the CHB MLI to generate voltage levels $+E$, 0, and $-E$. These states are listed in the table. The switching combinations determine the output voltage at different levels of the inverter.

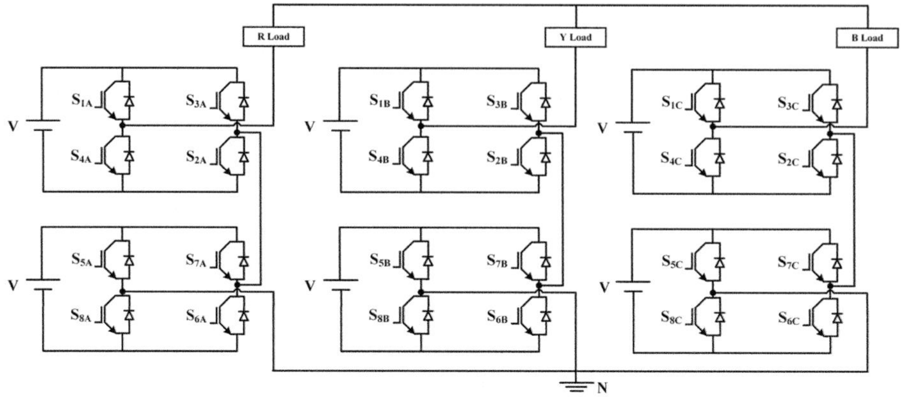

Figure 1.3 Topology of three five-level CHB inverter.

Table 1.1 Switching scheme for 5-Level CHB.

S_{1x}	S_{2x}	S_{3x}	S_{4x}	S_{5x}	S_{6x}	S_{7x}	S_{8x}	**Voltage level**
1	1	0	0	1	1	0	0	$+2E$
1	1	0	0	0	1	0	1	$+E$
0	1	0	1	1	1	0	0	$+E$
0	1	0	1	0	1	0	1	0
1	0	1	0	1	0	1	0	0
0	0	1	1	1	0	1	0	$-E$
1	0	1	0	0	0	1	1	$-E$
0	0	1	1	0	0	1	1	$-2E$

1.3.2 Flying capacitor (FLC) inverters

FC inverters use capacitors to achieve voltage levels. FLCIs operate by employing a series of capacitors to produce various voltage levels. This configuration facilitates enhanced efficiency and minimized switching losses in comparison to conventional inverters. Careful balancing of the capacitor voltages is required to ensure system stability. Innovations in this area include better capacitor voltage balancing techniques to ensure uniform voltage distribution across capacitors. Advanced control strategies have been integrated to optimize performance and efficiency, making FC inverters more viable for practical applications [13–15]. The favorable characteristics of both two and three-level topologies can be merged into a single inverter. The flying capacitor inverter has the advantage of low semiconductor expenses and delivers a multilevel output with a high output frequency and minimal

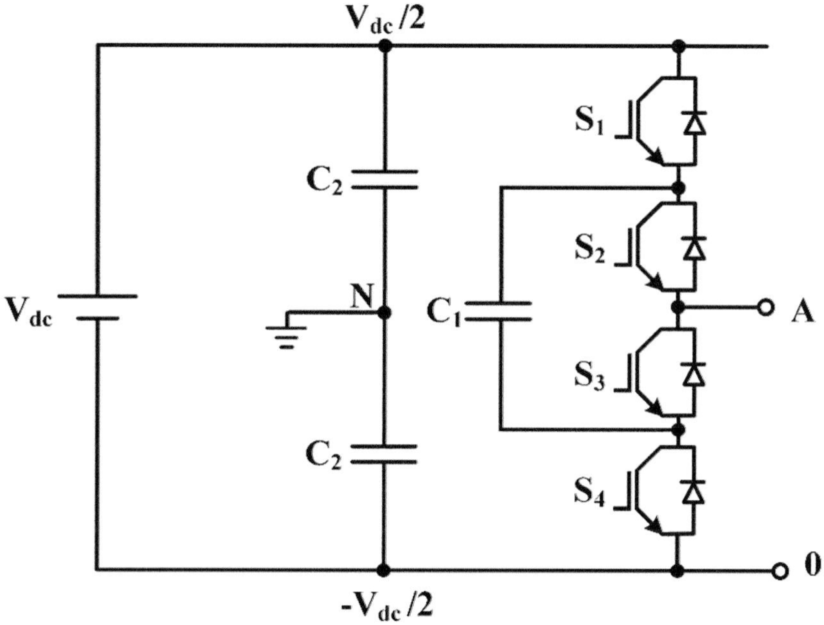

Figure 1.4 Three-level flying capacitor converter.

dynamic losses. While the input only has two levels and does not require a large DC-link capacitor bank, the output is multilevel and its frequency is a multiple of the switching frequency. Consequently, the output filter can be minimized and slower semiconductors can be employed. FC inverters are extensively utilized in applications requiring better power quality and versatility, including uninterruptible power supplies (UPS) and aerospace systems. The constraints of FC inverters emerge from the huge quantities of capacitors needed. This results in increased expenses and system sophistication. Moreover, maintaining and preserving the charge equilibrium among the flying capacitors presents an immense challenge. To address these issues, innovative control methods such as model predictive control (MPC) and capacitor voltage balancing strategies are being investigated.

Figure 1.4 shows a three-level flying capacitor inverter. The voltage rating of each capacitor is equivalent to that of the main power switch. In this type of three-level inverter topology, the possible output voltages are 0, $V_{dc}/2$, and $-V_{dc}/2$. When switches S_1 and S_2 are turned on, the output voltage equals $V_{dc}/2$. Similarly, for the negative voltage output, that is, $-V_{dc}/2$, switches

S_3 and S_4 need to be turned on. When S_1 and S_3 or S_2 and S_4 are turned on, the output voltage is 0. During this condition, the charging of capacitor C_1 occurs.

1.3.3 Neutral point clamped (NPC) inverters

Neutral point clamped (NPC) inverters were first introduced in 1981 by Baker and Nabae et al. The NPC topology, commonly referred to as diode-clamped, employs clamping diodes to facilitate the interconnection of multiple voltage levels. A sequence of capacitors and clamping diodes are used to split the DC bus voltage into smaller voltage increments. The primary benefit of NPC inverters lies in their capacity to deal with increased voltage levels, rendering them appropriate for high-power applications, including medium-voltage industrial drives and grid-connected renewable energy systems. These converters are a specific type of power converter that utilizes clamping diodes to guarantee the correct distribution of voltage among power switches [16]. NPC inverters are grid-connected, three-phase, three-wire inverters that receive power from a DC source. These inverters are often utilized in high-power applications and are the most frequently employed topology among multiple inverters. NPC inverters utilize clamping diodes to achieve multilevel voltages [17]. Recent developments have focused on reducing switching losses and improving thermal management. The adoption of advanced semiconductor materials, such as silicon carbide (SiC) and gallium nitride (GaN), has further enhanced the performance and efficiency of NPC inverters.

Figure 1.5 shows a three-level NPC arrangement which comprises 12 switches, featuring 6 diodes. It is critical to connect two capacitors in series, both of which are charged with a direct current voltage (V_{dc}). Each phase consists of four series switches, each with two diodes connected in series. The primary purpose of the diodes is to restrict the upper switches to a greater voltage relative to the zero DC-link point. A specific switching pattern will yield the output of a three-level neutral point clamped (NPC) inverter. The circuit diagram depicted in this figure provides insight into the configuration of the power circuit necessary for the three-level NPC inverter. The diodes in this circuit are arranged in a manner that produces varying voltage levels relative to the neutral point N. The voltage level is divided by capacitors C_1 and C_2, which are connected in series with a neutral point N between them. The voltage division is calculated as $V_{dc}/2$, 0, and $-V_{dc}/2$. Therefore, it is referred to as a three-level topology. With this configuration, $V_{dc}/2$ is

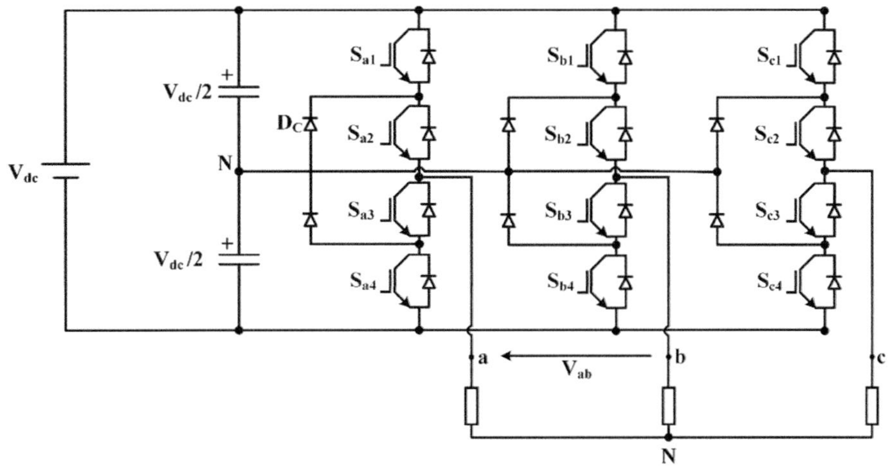

Figure 1.5 Three-level NPC inverter.

provided by the switches S_{a1} and S_{a2}. Switches S_{a3} and S_{a4} are turned on for $-V_{dc}/2$, whereas switches S_{a3} and S_{a2} are required to be turned on for 0. This power circuit will ultimately produce a three-level topology. Table 1.2 shows the comparison of different topologies.

1.4 Control Strategies

1.4.1 Model predictive control (MPC)

Model predictive control (MPC) is an advanced control approach that utilizes a dynamic model of the system to forecast its future behavior. MPC's capacity to anticipate and address possible faults before they arise improves the overall performance and stability of the system. MPC's primary benefit resides in its capacity to manage multivariable control problems while considering limits [18]. This makes it especially valuable for intricate systems like multilevel inverters (MLIs).

Notable characteristics and progressions in MPC encompass:

- Improved dynamic response: model predictive control (MPC) has the ability to rapidly adjust to system changes, resulting in a more superior dynamic response as compared to conventional control approaches. Fast and precise control is essential for applications that need high performance, such as electric automobiles and renewable energy systems.

Table 1.2 Comparison of different topologies.

Factors	**Cascaded H-bridge**	**Flying capacitor**	**Neutral point clamped**
Component count	Moderate: requires more isolated DC sources, but fewer passive components	High: requires large capacitor banks, especially for higher voltage levels	High: requires multiple diodes and capacitors to maintain voltage levels
Complexity	Moderate: each H-bridge is independent, offering modular design and simplified control	High: complex control needed for balancing multiple flying capacitors	High: requires complex diode balancing schemes and careful control of clamping diodes
Scalability	High: highly scalable due to its modular structure, easily expandable by adding more H-bridge modules	Low: difficult to scale due to high capacitor count and balancing complexity	Moderate: difficult to scale due to the increased number of diodes and complex control for higher voltage levels
Efficiency	High: modular structure with fewer components at each voltage level leads to high efficiency	Moderate: high switching losses due to large numbers of switches and capacitors	Moderate: diode losses at higher levels reduce efficiency
Harmonic distortion (THD)	Low: excellent harmonic performance due to the high number of voltage levels achievable	Low: good harmonic performance, but can vary depending on capacitor balancing	Low: good harmonic performance due to the use of multiple voltage levels
Voltage balancing	Easy: each module has independent DC sources, reducing the need for voltage balancing	Very challenging: capacitor voltage balancing is a major challenge, requiring sophisticated control methods	Challenging: requires complex control for maintaining equal voltage across capacitors
Switching losses	Low: fewer components at each level and the possibility of optimizing switching sequences	High: due to the large number of switches and capacitors involved	High: diode and switch losses increase as voltage levels increase
Applications	Widely used in renewable energy systems (solar PV, wind), electric vehicles, and distributed generation	Limited to medium-power applications such as power conditioning and uninterrupted power supplies	Suitable for industrial drives, grid-connected systems, and medium-voltage applications

- MPC demonstrates resilience to parameter variations by constantly updating its predictions using real-time data to ensure consistent and reliable performance. This is especially advantageous in settings where operational circumstances are consistently fluctuating.
- Decrease in processing complexity: conventional MPC algorithms might need a significant amount of processing resources, which restricts their use in real-time scenarios. Current studies have concentrated on creating algorithms that are more effective in reducing processing requirements. This allows for the use of MPC in real-time control situations.

1.4.2 Artificial intelligence (AI) and machine learning (ML) techniques

Artificial intelligence (AI) and machine learning (ML) are revolutionizing the domain of control techniques for MLIs. These technologies facilitate the creation of adaptive, intelligent control systems that have the ability to acquire knowledge from data and enhance their performance as time progresses.

AI and ML have various applications in MLI control, one of which is the optimization of control algorithms [19]. These techniques can dynamically adjust control parameters in real-time, hence maximizing the efficiency of the MLI. This entails employing algorithms such as genetic algorithms, neural networks, and reinforcement learning to discover the most optimal control solutions.

- Fault prediction and diagnosis: machine learning models can be taught to identify and anticipate failures in machine learning infrastructures (MLIs). This enables proactive maintenance and minimizing operational interruptions. Utilizing techniques such as anomaly detection and predictive analytics is essential for improving the dependability of a system.
- Enhancement of overall performance: through the analysis of past data and the utilization of artificial intelligence (AI) and machine learning (ML), control tactics may be continually optimized. This leads to a notable improvement in performance and also encompasses improved management of nonlinearities and disturbances, which are frequently encountered in real-world scenarios.

1.4.3 Space vector modulation (SVM)

Space vector modulation (SVM) is a powerful modulation technique used in multilevel inverters (MLIs) to generate high-quality sinusoidal waveforms.

Support vector machines (SVM) offer various advantages over traditional pulse width modulation (PWM) techniques, including better harmonic characteristics and increased utilization of the DC bus voltage [20, 21].

Significant progress in support vector machines (SVM) encompasses the following key developments:

- Decrease in computational load: conventional SVM algorithms can impose significant computational requirements, hence restricting their real-time usage. Current advancements concentrate on streamlining these algorithms, rendering them more appropriate for instantaneous management without compromising efficiency.
- Enhanced harmonic performance: SVM inherently yields reduced harmonic distortion in comparison to standard PWM techniques. The current research aims to enhance this aspect by developing hybrid SVM approaches that combine the benefits of traditional and modern methodologies.
- Hybrid SVM methods: these methods combine traditional SVM with cutting-edge techniques such as predictive control and AI-driven optimization, leading to enhanced performance. Hybrid support vector machine (SVM) techniques are specifically designed to enhance harmonic reduction, efficiency, and robustness across different operating situations.

These control mechanisms are at the cutting edge of research and development in the field of MLIs. MLIs may effectively address the increasing requirements of modern applications by providing notable enhancements in terms of efficiency, dependability, and adaptability.

1.5 Constraints in Multilevel Inverters

1.5.1 Complexity and cost

The inherent complexity and associated expenses are major barriers when it comes to multilevel inverters (MLIs) [17]. MLIs demand a substantially greater quantity of components in their design and implementation in comparison to standard inverters which results in various crucial problems:

- Increased number of components: MLIs consist of an increased number of switches, capacitors, and other passive components, resulting in elevated production expenses. This intricacy not only impacts the initial expense of the inverter but also has ramifications for its long-term upkeep and operational expenditures.

- Reliability issues: the higher number of components has a direct impact on the reliability of the system. An increased number of components leads to a greater likelihood of failures, rendering the system more susceptible to breakdowns and necessitating stronger fault detection and mitigation measures.
- Challenges in design and implementation: the complex design of MLIs requires advanced engineering and sophisticated control strategies, which can be costly and time-consuming to develop and put into action. Maximizing the synergy and effectiveness of all components contributes to the overall intricacy.
- Economic feasibility: the expense aspect might be a substantial obstacle to the broad implementation of MLIs. This is especially relevant in markets that prioritize cost considerations. The issue for academics and producers lies in finding the right balance between enhancing performance and ensuring cost-effectiveness.

1.5.2 Thermal management

Effective thermal management is essential to guarantee the dependability and durability of MLIs. The inverters' high-power densities and switching frequency result in significant heat generation, requiring efficient thermal management methods [22].

- Heat generation: The switching elements of MLIs, particularly when running at high frequencies, generate a significant amount of heat. In order to avoid overheating, it is crucial to dissipate this heat effectively. If not, this can cause harm to the components and reduce the overall performance of the inverter.
- Advanced cooling techniques: traditional cooling methods, such as heat sinks and fans, may not be sufficient for high-power applications. Researchers are exploring advanced cooling techniques, such as liquid cooling, phase change materials, and thermoelectric cooling, to enhance the process of dissipating heat.
- Material innovations: the main objective is to develop new materials that have improved thermal conductivity. Current research is underway to investigate the possibilities of enhanced ceramics, composites, and nanomaterials in improving heat management in MLIs.
- Thermal design optimization: the thermal design of multilayer insulations (MLIs) is improved using computational approaches and simulation models. This ensures that heat is managed efficiently across the

entire system. This involves enhancing the arrangement of elements and using sophisticated algorithms for managing heat.

1.5.3 Harmonic distortion

Harmonic distortion refers to the presence of additional frequencies in a signal that are not part of the original waveform. Although MLIs are intended to decrease harmonic distortion, the task of controlling and minimizing harmonics remains a considerable difficulty, particularly when dealing with fluctuating load situations [23]. Harmonic distortion can have a negative impact on the inverter's performance and efficiency, as well as the quality of power that is delivered to the load.

- Causes of harmonics: harmonics arise from the nonlinear properties of switching devices and the interplay between the inverter and the load. Fluctuating load circumstances might worsen these harmonics, hence increasing the complexity of their control.
- Effect on performance: harmonic distortion may result in heightened losses, diminished efficiency, and possible interference with other electrical devices. Furthermore, it has an impact on the overall power quality, which is of utmost importance for applications that are highly sensitive.
- Advanced techniques for filtering: advanced filtering techniques are used to reduce harmonics. Researchers are now working on the development of active and passive filters, as well as hybrid filtering methods, in order to efficiently decrease harmonic distortion.
- Strategies for control: researchers are currently studying innovative control systems, including adaptive control and real-time harmonic compensation, to effectively manage harmonic distortion in a dynamic manner. These techniques are designed to modify the operation of the inverter according to real-time conditions in order to sustain optimal performance.

1.5.4 Reliability and fault tolerance

Ensuring the dependability and resilience of MLIs is crucial for their extensive acceptance, particularly in critical applications where any period of inactivity can result in significant financial or hazardous consequences.

- Fault diagnostic and mitigation: it is of utmost importance to develop efficient fault diagnostic procedures in order to accurately identify and resolve issues before they result in system failures. Techniques like real-time monitoring, predictive maintenance, and diagnostic algorithms are being used to improve the detection of faults.

- Including redundancy in the design of MLIs can enhance their ability to tolerate faults and failures. This involves employing redundant components and parallel pathways to guarantee the continuous operation of the system, even in the presence of malfunctions.
- Advanced control algorithms are being developed to detect, isolate, and mitigate problems in real time. These algorithms utilize artificial intelligence and machine learning to improve the inverter's durability and capacity to manage unforeseen circumstances.
- Enhancing component reliability: ongoing research is dedicated to enhancing the dependability of individual components, such as switches and capacitors, by utilizing superior materials and refining production procedures. Improving the strength and efficiency of these parts can greatly increase the overall dependability of MLIs.

In order to promote and expand the use of MLIs, it is essential to address the challenges presented by complexity and cost, thermal management, harmonic distortion, and reliability. Ongoing research and innovation in these disciplines are essential for overcoming these limitations and fully exploiting the possibilities of MLIs in various applications.

1.6 Applications of Multilevel Inverters

1.6.1 Renewable energy systems and smart grid

MLIs play a crucial role in renewable energy systems, particularly in solar and wind power applications. They facilitate the integration of renewable energy sources into the grid by improving power quality and efficiency [24]. Recent developments have focused on grid-tied inverters, hybrid systems, and the integration of energy storage solutions to enhance the reliability and stability of renewable energy systems. A smart grid is the next phase in the evolution of the traditional electrical grid, using advanced digital communication, automation, and control technologies to enhance energy distribution and reliability. In these systems, MLIs are essential for integrating distributed energy resources (DERs) like solar and wind power into the grid. Their ability to provide high-quality power while minimizing harmonic distortion is critical for grid stability. MLIs enable the seamless integration of intermittent resources into the grid, thereby assuring smooth transitions between various power generation sources like wind, solar, and hydro. In energy storage applications, MLIs facilitate bidirectional power flow, enabling energy to be stored during low-demand periods and discharged when required [25]. This capability is essential for grid stabilization and peak shaving.

1.6.2 Electric vehicles (EVs)

The rapid development of the electric vehicle (EV) industry has given rise to an urgent demand for high-efficiency power electronics systems, especially inverters, which play a key role in the functioning of traction motors and battery management systems [26]. In the field of electric vehicles, MLIs are used to drive traction motors and manage battery systems. Their high efficiency and low harmonic distortion make them ideal for EV applications. Recent innovations in compact and lightweight inverter designs are contributing to the advancement of EV technology, enhancing the performance and range of electric vehicles. Besides powering traction motors, MLIs are essential in battery management systems (BMS) and charging systems. The capability of MLIs to manage bidirectional power flow facilitates effective energy conversion during both charging and discharging operations.

1.6.3 Industrial drives

MLIs are extensively used in industrial drives to control the speed and torque of electric motors. Their ability to provide high-quality power and precise control makes them suitable for various industrial applications, including robotics, conveyor systems, and manufacturing processes. In industrial drives, regulating the speed and torque of electric motors is essential for sustaining optimal performance and efficiency of diverse gear and processes. MLIs are essential in converting DC power into the AC power necessary for motor operation, facilitating smooth and precise control. Recent advancements have focused on improving the efficiency and reliability of MLIs in industrial settings.

1.6.4 High voltage direct current (HVDC) transmission

MLIs are integral to HVDC transmission systems, enabling efficient and reliable long-distance power transmission. Multilevel inverters (MLIs) play a crucial role in HVDC systems by facilitating the conversion of DC power to AC power and vice versa at both the transmission line's sending and receiving ends. MLIs are particularly advantageous in voltage source converter (VSC)-based HVDC systems, as their high efficiency, minimal harmonic distortion, and scalability facilitate the increasing integration of HVDC technology. A primary benefit of employing MLIs in VSC-based HVDC is its capacity for reactive power regulation. This functionality enables HVDC endpoints to provide voltage management and reactive power compensation, hence improving grid stability and dependability. MLIs can dynamically regulate both active and reactive power, rendering them highly suitable for grid-connected HVDC applications. Recent advancements have focused on improving the efficiency and fault tolerance of MLIs in HVDC applications. Innovations in power electronics and control strategies are contributing to the development of more robust and scalable HVDC transmission solutions [27].

1.7 Limitations of Current Technology

Even though multilevel inverter (MLI) topologies are widely used and have many benefits in high-power applications, there are several fundamental drawbacks that hinder their performance, cost-effectiveness, and feasibility. Some key challenges are the high number of components, complex control systems, voltage balancing, and harmonic distortion, among others. It is important to tackle these limitations to help advance MLI technologies, especially as they are increasingly used in renewable energy systems, electric vehicles, and industrial applications.

1.7.1 High component count and size

One of the primary challenges associated with MLIs is the high number of components. These inverters require a huge number of power semiconductor devices such as switches and diodes along with passive components like capacitors and resistors to achieve multiple voltage levels. This will lead to increased system size, cost, and complex control strategies. A major limitation of traditional MLI systems is the use of traditional transformers in high-power systems. This drastically lowers system scalability and increases its size.

One promising solution for this limitation is to reduce the component count by using hybrid topologies. For instance, combining cascaded H-bridge with neutral point clamped inverters will reduce the total number of diodes and switches required. This will result in a more efficient configuration. The adoption of solid-state transformers (SSTs) also addresses many of these concerns [28]. SSTs utilize high-frequency power electronics to deliver equivalent voltage transformation and isolation capabilities as conventional transformers at a fraction of the size and weight. Solid-state transformers eliminate the requirement for heavy magnetic transformers, enabling modular multilevel converters to be more compact, efficient, and versatile.

1.7.2 Capacitor voltage balancing

In the case of NPC topology, multiple capacitors are utilized to divide the DC bus voltage into smaller steps. However, it is challenging to ensure that all capacitors share the voltage equally [29]. In the FC topology, flying capacitors can experience voltage imbalance because of uneven charge and discharge cycles, which might influence inverter performance and reduced effectiveness.

Potential solutions include active and passive balancing techniques. In active methods, monitoring of capacitor voltages is involved and specialized circuits are used to redistribute the charges among the capacitors. In passive balancing, resistive elements are used to balance voltages but this can lead to energy losses. Advanced control algorithms like model predictive control, proportional integral control, and space vector pulse width modulation are also employed.

1.7.3 Switching losses and efficiency

Switching losses in MLIs develop because semiconductor devices are switched on and off frequently. As the number of switches increases, so do the losses, which results in reduced system efficiency. This is particularly concerning in high-power applications, as even minor efficiency losses can lead to considerable energy waste. Switching losses generate heat in the inverter, leading to thermal stress on components and ultimately shortening their lifespan.

Some promising solutions include the adoption of soft switching techniques such as zero-voltage switching (ZVS) [30] and zero-current switching (ZCS). These techniques can help in minimizing switching losses by making sure that switching happens only when the voltage or current across the switch is zero. This is very effective in reducing energy loss during switching. Nowadays, the introduction of wide-bandgap semiconductors like silicon carbide (SiC) and gallium nitride (GaN) has further minimized the switching losses. These materials can withstand higher switching frequencies which can lead to more compact and efficient inverter design. They are also capable of handling higher voltage levels and elevated operating temperatures.

1.8 Economic Factors

The economic viability is a major determining factor in the widespread adoption and popularity, particularly in industrial, commercial, and renewable energy uses. Despite the fact that MLIs offer significant technological advances such as improved power quality, reduced harmonic distortion, and enhanced efficiency, the initial cost expenses and future gains must be balanced carefully. The acceptance of MLI technologies is influenced by cost-effectiveness, return on investment (ROI), and current market trends.

1.8.1 Cost-effectiveness and return on investments

MLIs, specifically the neutral point clamped (NPC) and flying capacitor (FLC) topologies, usually incur significant upfront costs owing to the extensive number of components necessitated by their design. To generate their multilevel voltage output, these topologies require a large number of diodes, capacitors, and switches, which can increase production and material costs. A significant portion of the total system cost is represented by high-quality power semiconductor devices like IGBTs. Moreover, complex control algorithms are essential for the regulation of switching sequences and ensuring voltage balancing of the capacitors. This further maximizes the overall cost of the system.

Despite all these higher initial costs, MLIs provide long-term savings by reducing the filtering requirements. MLIs generate output voltages with reduced total harmonic distortion (THD), which reduces the necessity for costly harmonic filters. This benefit can result in significant savings on both installation and operating expenses. Hence, the return on investment for MLIs is highly dependent on these factors. Overall, MLIs provide attractive long-term ROI opportunities because of their outstanding energy efficiency and scalability.

1.8.2 Current market trends

The rapid transition toward clean energy and renewables has created significant market opportunities for MLIs. As nations and industries aim to lower their carbon emissions, there is a growing demand for efficient power electronics that can incorporate renewable energy sources into the grid. MLIs are increasingly preferred for utility-scale solar and wind power generation systems due to their capacity to manage high power levels and reduced THD.

They also have the capability to integrate smoothly with energy storage systems. The decreasing costs of solar PV modules and wind turbines, along with government incentives and regulations, have accelerated the implementation of MLIs in these areas. Recently, MLIs have become a major attraction for EV industries. The increasing popularity of electric transport systems, especially electric vehicles (EVs), has increased the need for reliable and effective inverter technologies. With the clean energy market on the rise, the need for MLIs is expected to increase swiftly. The worldwide shift toward decarbonization and the growing adoption of renewables have made MLIs a vital technology for future energy systems.

1.9 Conclusion

Multilevel inverters (MLIs) are a notable progression in power electronics, providing many advantages such as enhanced power quality, efficiency, and dependability. The numerous benefits of MLIs make them particularly appealing for a diverse array of uses, such as renewable energy systems, industrial motor drives, and electric cars. MLIs possess the potential for generating output waveforms with less harmonic distortion in comparison to conventional two-level inverters. This leads to a more reliable and consistent power supply, and this is crucial for applications that require high sensitivity. Furthermore, MLIs achieve superior efficiency by minimizing losses during switching and optimizing power conversion operations. This not only decreases energy consumption but also enhances the sustainability and cost-effectiveness of power systems. The modularity of MLIs also enables the creation of more resilient and error-tolerant systems, ensuring uninterrupted operation even in the event of component failures, and so, enhancing system reliability.

Despite these benefits, MLIs encounter certain challenges that need to be resolved in order to fully use their potential. MLIs have a natural intricacy and a greater number of components, which results in higher costs for manufacture and maintenance. This requires the use of creative design and manufacturing methods to make them more economically feasible. Efficient thermal control is essential for preventing excessive heat and ensuring the durability of MLIs, which has led to the advancement of cooling technologies and materials. Effectively controlling harmonic distortion in the presence of changing load conditions poses a significant difficulty, necessitating the use of sophisticated filtering methods and adaptable management strategies to uphold power quality. Reliability and fault tolerance are crucial for the broad deployment of MLIs in demanding applications. Ongoing research is dedicated to improving fault detection methods, redundancy, and intelligent control algorithms to achieve this. The future of machine learning interfaces (MLIs) is optimistic since ongoing progress is anticipated to tackle these obstacles and unleash novel opportunities for their implementation in diverse industries.

References

[1] V. V. Fedotov and V. V. Rozhkov, "Development and Analysis for the Optimal Structure of a Multilevel Voltage Source Inverter," 2021 International Ural Conference on Electrical Power Engineering (UralCon), Magnitogorsk, Russian Federation, 2021, pp. 121-126

[2] M. D. Siddique and S. K. Panda, "Switched-Capacitor Based Multilevel Inverter Topologies: A Design Methodology," 2022 IEEE International Conference on Power Electronics, Drives and Energy Systems (PEDES), Jaipur, India, 2022, pp. 1-5

[3] D. Sajeesh, "Simulation of Multilevel Inverter with Six Switches," 2023 5th Biennial International Conference on Nascent Technologies in Engineering (ICNTE), Navi Mumbai, India, 2023, pp. 1-5

[4] Wang, Zichen. "Recent Advances in Circulating Current Suppression Strategies for Modular Multilevel Converters." Journal of Physics: Conference Series. Vol. 2786. No. 1. IOP Publishing, 2024.

[5] V. M, P. R. G and S. Prasad. S, "Single Phase Advanced Multilevel Inverter with Reduced Devices and THD for Industrial Applications," 2023 9th International Conference on Advanced Computing and Communication Systems (ICACCS), Coimbatore, India, 2023, pp. 1006-1011

[6] L. Piyush, R. Kamani, M. K. Patel, D. P. Kathiria, J. H. Majithiya, and H. Brahmkhatri, "Implementing Selective Harmonic Elimination in Multilevel Inverters with Optimal DC Source," e-Prime, vol. 8, p. 100622, 2024

[7] P. Ezhilvannan, D. Ramasamy, S. Subramanian, and S. Krishnan, "Enhancing power conversion efficiency in five-level multilevel inverters using reduced switch topology," Bulletin of Electrical Engineering and Informatics, vol. 13, pp. 1495-1503, 2024

[8] L. Wang, "Recent advances in leakage current suppression techniques for three-level inverters," Applied and Computational Engineering, 2024

[9] S. Debata, T. Roy, S. Sahu, T. N. Sekhar and A. Dasgupta, "Development, Analysis and Simulation Study of a Novel Switched Capacitor Multilevel Inverter Structure for Different DC Source Configurations," 2018 International Conference on Recent Innovations in Electrical, Electronics & Communication Engineering (ICRIEECE), Bhubaneswar, India, 2018, pp. 3140-3145

[10] M. S. Bhaskar, K. P. Draxe and K. M. Pandav, "A novel asymmetric multilevel inverter with minimum number of switches for renewable power grid applications," 2013 International Conference on Green Computing, Communication and Conservation of Energy (ICGCE), Chennai, India, 2013, pp. 423-427

[11] S. Debata, T. Roy, S. Sahu, T. N. Sekhar and A. Dasgupta, "Development, Analysis and Simulation Study of a Novel Switched Capacitor Multilevel Inverter Structure for Different DC Source Configurations," 2018 International Conference on Recent Innovations in Electrical, Electronics & Communication Engineering (ICRIEECE), Bhubaneswar, India, 2018, pp. 3140-3145

[12] G. Deshmukh, P. Chaturvedi and V. Rajderkar, "Single phase Cascaded H-Bridge Multilevel Inverter Topology," 2022 International Conference on Futuristic Technologies (INCOFT), Belgaum, India, 2022, pp. 1-6

[13] C. B. Barth et al., "Design and Control of a GaN-Based, 13-Level, Flying Capacitor Multilevel Inverter," in IEEE Journal of Emerging and Selected Topics in Power Electronics, vol. 8, no. 3, pp. 2179-2191, Sept. 2020

[14] L. He and C. Cheng, "A Flying-Capacitor-Clamped Five-Level Inverter Based on Bridge Modular Switched-Capacitor Topology," in IEEE Transactions on Industrial Electronics, vol. 63, no. 12, pp. 7814-7822, Dec. 2016

[15] G. Kampitsis, E. I. Batzelis, P. D. Mitcheson and B. C. Pal, "A Clamping-Circuit-Based Voltage Measurement System for High-Frequency Flying Capacitor Multilevel Inverters," in IEEE Transactions on Power Electronics, vol. 37, no. 10, pp. 12301-12315, Oct. 2022

[16] C. Li et al., "A Modified Neutral Point Balancing Space Vector Modulation for Three-Level Neutral Point Clamped Converters in High-Speed Drives," in IEEE Transactions on Industrial Electronics, vol. 66, no. 2, pp. 910-921, Feb. 2019

[17] M. J. Sathik, N. Sandeep and F. Blaabjerg, "High Gain Active Neutral Point Clamped Seven-Level Self-Voltage Balancing Inverter," in IEEE

Transactions on Circuits and Systems II: Express Briefs, vol. 67, no. 11, pp. 2567-2571, Nov. 2020
[18] F. Wang, Z. Li and Z. Liu, "Model Predictive Control Methods for Three-Level Sparse Neutral Point Clamped Inverter," in IEEE Journal of Emerging and Selected Topics in Power Electronics, vol. 8, no. 4, pp. 4355-4366, Dec. 2020
[19] P. Chavan and B. R. Patil, "Harmonic Elimination with Multilevel Inverter Based Hybrid Filter Using Artificial Intelligence Approach," 2022 International Conference on Power, Energy, Control and Transmission Systems (ICPECTS), Chennai, India, 2022, pp. 1-6
[20] B. Jacob and M. R. Baiju, "A New Space Vector Modulation Scheme for Multilevel Inverters Which Directly Vector Quantize the Reference Space Vector," in IEEE Transactions on Industrial Electronics, vol. 62, no. 1, pp. 88-95, Jan. 2015
[21] B. Jacob and M. R. Baiju, "A New Space Vector Modulation Scheme for Multilevel Inverters Which Directly Vector Quantize the Reference Space Vector," in IEEE Transactions on Industrial Electronics, vol. 62, no. 1, pp. 88-95, Jan. 2015
[22] M. Aly, G. M. Dousoky and M. Shoyama, "Reliability enhancement of multilevel inverters through SVPWM-based thermal management methodology," 2015 IEEE 2nd International Future Energy Electronics Conference (IFEEC), Taipei, Taiwan, 2015, pp. 1-6
[23] S. N L and S. G, "Hybrid Multilevel Inverter Control for Harmonic Mitigation," 2023 International Conference on Control, Communication and Computing (ICCC), Thiruvananthapuram, India, 2023, pp. 1-6
[24] A. Hassan, X. Yang, W. Chen and A. I. M. Ali, "An Advanced Switched-Capacitor Multilevel Inverter Topology With a Robust PWM Controlling Technique and Voltage Boosting Property for the Grid-Connected Renewable Energy Applications," in IEEE Journal of Emerging and Selected Topics in Power Electronics, vol. 11, no. 3, pp. 3547-3561, June 2023
[25] S. Pourfarrokh, J. Adabi and F. Zare, "A New Grid-Connected Asymmetrical Multilevel Converter for PV Application," in IEEE Transactions on Power Electronics, vol. 39, no. 9, pp. 11256-11265, Sept. 2024
[26] T. A. Stanley Raja, C. Kumar and M. Premkumar, "Novel Multilevel Inverter using Switched Capacitor unit with Voltage Boosting feature for EV Applications," 2022 International Conference on Disruptive

Technologies for Multi-Disciplinary Research and Applications (CENT-CON), Bengaluru, India, 2022, pp. 157-161

[27] D. Woldegiorgis and A. Mantooth, "Investigation of a New Alternate Arm Modular Multilevel Converter Topology for HVDC Applications," 2021 IEEE Energy Conversion Congress and Exposition (ECCE), Vancouver, BC, Canada, 2021, pp. 290-295

[28] J. He, Y. Ye and X. Wang, "ZVS and Inrush Charging Current Suppression Design for Switched-Capacitor Multilevel Inverters," in IEEE Transactions on Power Electronics, vol. 38, no. 9, pp. 10611-10616, Sept. 2023

[29] J. He, Y. Ye and X. Wang, "ZVS and Inrush Charging Current Suppression Design for Switched-Capacitor Multilevel Inverters," in IEEE Transactions on Power Electronics, vol. 38, no. 9, pp. 10611-10616, Sept. 2023

[30] J. He, Y. Ye and X. Wang, "ZVS and Inrush Charging Current Suppression Design for Switched-Capacitor Multilevel Inverters," in IEEE Transactions on Power Electronics, vol. 38, no. 9, pp. 10611-10616, Sept. 2023

2

Recent Advancements in Switched Capacitor-based Multilevel Inverters

Devanand Kumar[1], Ravi Raushan[1], Deepti Kumari[1], and Ravi Bhushan[2]

[1]National Institute of Technology Karnataka, India
[2]National Institute of Technology Jamshedpur, India
E-mail: devog09kumar@gmail.com; ravi.raushan@nitk.edu.in; deepti0159@gmail.com; ravibhushan.ee@nitjsr.ac.in

Abstract

This chapter comprehensively reviews recent advances in switched capacitor-based multilevel inverters (SCMLIs) within the realm of power electronics. It outlines the fundamental principles of multilevel inverters, underscoring the role of switched capacitors in achieving higher voltage levels and refined waveform quality. The chapter critically examines the challenges in switched capacitor-based multilevel inverter designs, emphasizing recent research on circuit topologies and modulation strategies, including aspects like voltage balancing. Moreover, it discusses the integration of SCMLIs in electric vehicles and renewable energy systems and highlights their potential to enhance efficiency, improve power quality, and seamlessly integrate with modern power grids. The chapter succinctly captures the forefront of SCMLI technology, offering insights into its transformative impact on ongoing innovation to address evolving challenges in inverter topologies.

Keywords: Boost inverter, cost function, reduced component, switched capacitor, voltage balancing

2.1 Introduction

In the past few years, there have been notable progressions in power electronics, driven by an increasing need for energy conversion systems that are both

DOI: 10.1201/9781003669845-2

efficient and compact. A noteworthy development in this field is the switched capacitor multilevel inverter (SCMLI), a crucial element in applications including renewable energy systems, electric vehicles, and microgrid designs. The SCMLI stands out for its unique approach to voltage conversion, relying on capacitors' dynamic charging and discharging to generate alternating current waveforms.

The pursuit of higher efficiency, reduced size, and improved performance has led researchers and engineers to explore novel strategies and technologies in the realm of SCMLIs. These inverters exhibit significant voltage boost capabilities, and their capacitors can achieve automatic balancing without the need for additional circuits. These advancements cover a range of innovations, including high-frequency operation [1], efficiency enhancements [2], and the incorporation of wide-bandgap semiconductors alongside advanced control strategies.

Pure switched capacitor (SC)-based switching circuit consists of several capacitors and power semiconductor devices (switches, and may diodes) that transform a fixed DC-link voltage into multilevel voltage through a series-parallel switching conversion method [4, 3]. The integrated capacitors get charged directly when it is connected to another voltage source in parallel, that voltage source can be the input source or another charged capacitor. Further, these capacitors produce discrete output voltage levels by discharging in series/parallel, either with or without the input DC voltage source. An important advantage of SCMLIs is their capability to achieve high voltage levels with a fewer number of components, leading to improved reliability. The SC circuits' inherent voltage boosting feature enables the generation of multilevel waveforms with fewer power electronic components, resulting in lower costs and reduced maintenance requirements. Additionally, the use of switched capacitors facilitates dynamic voltage balancing, enhancing overall inverter performance and lifespan. These attributes make SCMLIs a valuable and intriguing solution for various novel applications.

2.2 Topology Innovation of Switched Capacitor-based Multilevel Inverter

The SCMLIs provide a promising alternative as they require substantially fewer DC sources, semiconductor devices, and other components, and at the same time, it can boost the output voltage. Consequently, SCMLIs are preferred over traditional MLIs for applications involving medium to high voltage, offering superior performance in areas such as adjustable speed drives, electric vehicles, grid-integrated renewable energy systems, and other similar applications.

In this section, SCMLIs have been taxonomized based on the type of architecture of the inverter topology, such as H-bridge-based SCMLI, half-bridge-based SCMLI, mid point SCMLI, and common ground-based MLI. Further, the developed MLIs have been distinguished based on the load-handling capability, encompassing grid integration of renewable and sustainable energy sources and domestic or local load.

2.2.1 Single source H-bridge-based SCMLI topology

The generalization of SC converters based on the H-bridge (HB) has resulted in the creation of several topologies tailored for high-voltage applications, which are discussed in this subsection. The easiest way to converter the generated voltage level(s) to a bidirectional waveform is to connect an HB at the end of the circuit. Therefore, the same voltage level can appear as positive and negative, which allows for a decrease in the component count in the inverter topology. However, it has some limitations, like total standing voltage (TSV) becoming high as the switches in the HB circuit must be able to handle the cumulative voltage. The recent development in semiconductor technology makes this task a little easier as the new type of semiconductor device (Sic power MOSFET) has the capability to block higher drain to source voltage. A few more issues arise regarding applications such as solar photovoltaics, motor drives, and batteries. These challenges include the generation of variable high-frequency common mode voltage (CMV), that causes the leakage current propagation and electromagnetic interference. Some of the H-bridge-based inverter topologies are presented in Figures 2.1–2.3. Figure 2.1 shows the generalized SCMLI topologies based on H-bridge, which can generate $(2n + 1)$-level output voltage. Wherein "n" indicates the total number of capacitors incorporated in the topology. A generalized HB-based SCMLI is reported in Figure 2.1.a, which can be further expanded to higher levels as per the need. It boosts the output voltage by switching the capacitor in series and parallel with the supply and load. Its one SC module contains two capacitors and three switches. The topology depicted in Figure 2.1.c [5] mainly consists of a boost converter at the front end, a central active SC cell (SCC), and an HB at the back end. Because of the quasi-resonant charging of the SCs, this configuration experiences lower inrush current and reduced current stress on the semiconductor devices. The capacitors in the topology in Figure 2.1.b

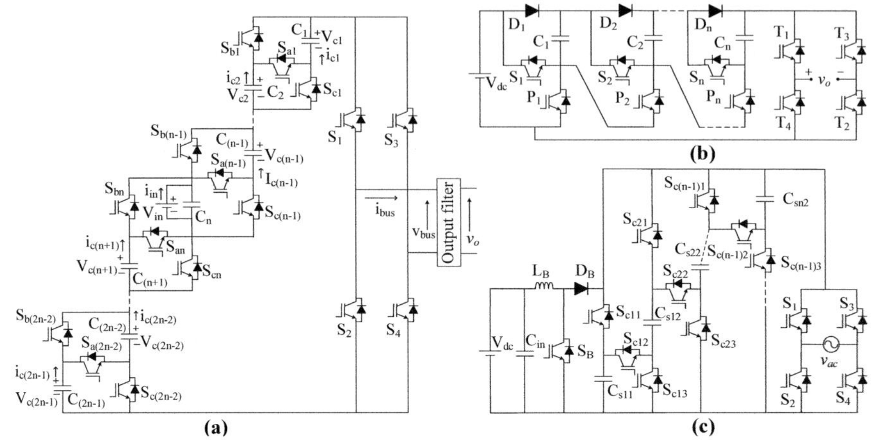

Figure 2.1 H-bridge and SP operation based $(2n + 1)$-level SCMLI. (a) [3]. (b) [4]. (c) [5].

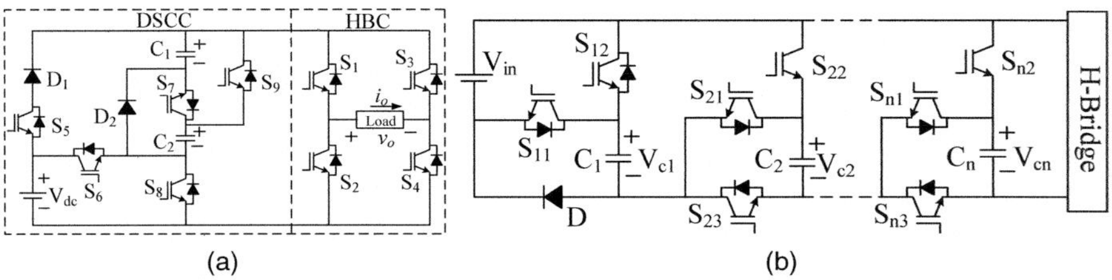

Figure 2.2 Nine-level SCMLI based on H-bridge and (a) DSCC [6] and (b) SP parallel operation [7].

get charged and discharged at the fundamental frequency, and it has a simple structure.

The MLI presented in Figure 2.2.a [6] is designed for a high-frequency power distribution system, and its devices have lesser voltage stress. It includes a front-end developed SC circuit and back-end HB. The H-bridge-based nine-level SC inverter illustrated in Figure 2.2.b [7] can be scaled to "n"-level inverter by incorporating $\log_2(n_{\text{level}} - 1) - 1$ capacitors (n_{c}), one power diode, and $3n_{\text{c}} + 3$ switches. This inverter is suitable for both low-power, as well as high-power applications, with the latter achievable through cascaded connections.

Figure 2.3 shows the seven-level SC inverter based on H-bridge. The inverter in Figure 2.3.a [8] has fewer switches as it incorporated only eight. The developed inverter in [9] has a different seven-level topology, and it is depicted in Figure 2.3.b. A closed-loop control using a proportional resonant

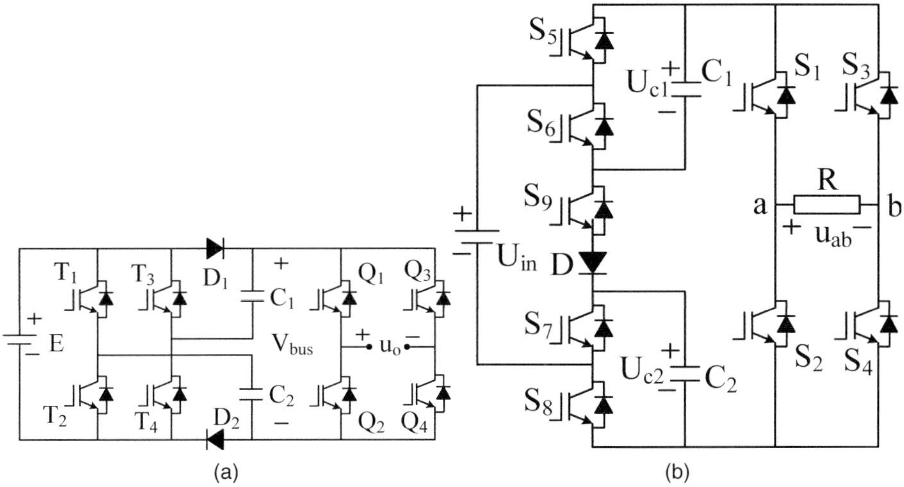

Figure 2.3 H-bridge-based seven-level SCMLI (a) [8] and (b) SP parallel operation [9].

controller has been implemented by keeping the resonant peak at a fundamental frequency to control the output voltage. However, this topology is unsuitable for high-voltage applications due to high inrush current.

2.2.1.1 Single source Half-bridge-based SCMLI topology

The SCMLI topologies based on half-bridge are presented in Figures 2.4–2.6.

Figure 2.4 depicts the five-level inverter based on the half-bridge and SCs, and it has the ability to work satisfactorily in cascaded connections also. However, the inverter in Figure 2.4.a [10] has a maximum voltage stress (MVS) equal to the peak output voltage, which poses a concern for high-voltage applications. A single-stage twice boosting five-level inverter has been developed for renewable power generation applications in [11] and depicted in Figure 2.4.b. It has a uniform MVS that equals the DC input voltage across the devices. Almost 45% of switches operate at the line frequency, which helps to reduce the losses and electromagnetic interference (EMI).

In Figure 2.5.a [12], the topology has been constructed by connecting two half-bridge modular SCs on either side of the DC supply. This design reduces the voltage stress on the inverting bridges and capacitors' voltage ripple and also lowers the output voltage's total harmonic distortion (THD) age stress of the inverting bridges, voltage ripple of the capacitors, and total harmonic distortion (THD) of the output voltage have been reduced in this inverter. Additionally, a hybrid PWM technique is employed to improve the inverter's overall performance. It can generate N_{level} output voltage level by appropriately connecting half-bridge SC modules on both sides of the DC supply, as mentioned in eqn (2.1).

$$N_{\text{level}} = 2^{(n+1)} + 2^{(m+1)} - 1, \tag{2.1}$$

here "n" indicates the total number of half-bridge SC modules connected to the left side, and "m" represents the total number of modules connected to the right side of the DC supply.

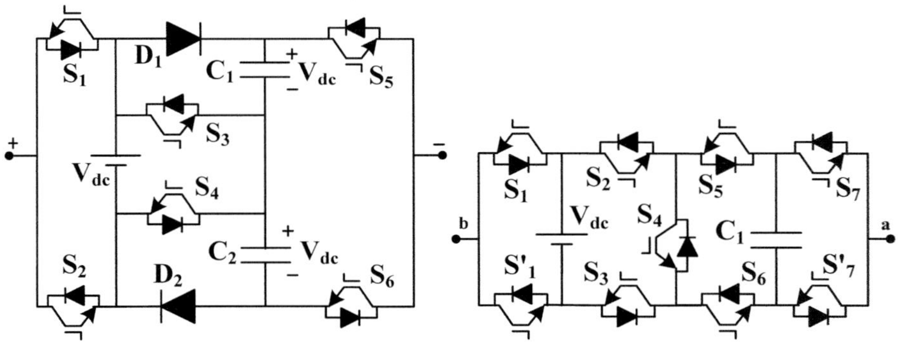

(a) (b)

Figure 2.4 Half-bridge-based five-level SCMLI (a) [10] (b) [11].

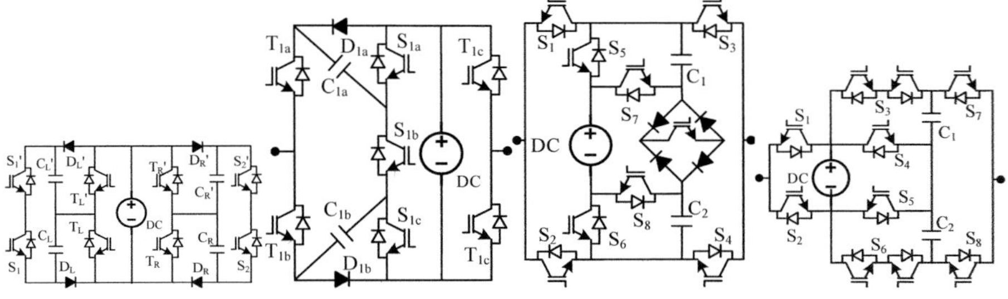

(a) (b) (c) (d)

Figure 2.5 Half bridge-based seven-level SCMLI. (a) [12]. (b) [13]. (c) [14]. (d) [15].

The presented inverter in Figure 2.5.b [13] can also generate a seven-level output voltage with a 1.5 boosting factor, and it has lesser power loss/W, TSV, and voltage stress while the inverter in Figure 2.5.c [14] has a regenerative

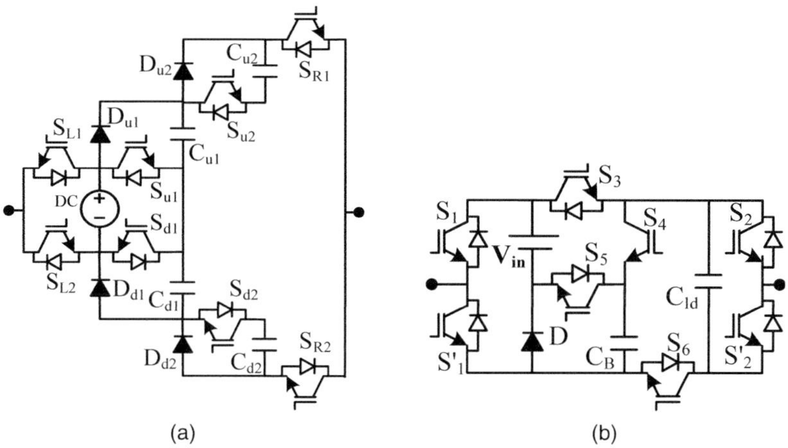

Figure 2.6 Nine-level SCMLI based on H-bridge and (a) [16] (b) [17].

performance capability, which allows the overcharge of capacitors to flow back to the DC source.

Figure 2.5.d [15] shows the seven-level inverter, which is mainly designed for distributed power generation. In this inverter, the maximum voltage stress equals the input DC source, which lowers the TSV index. It has a low inrush current and high efficiency. However, its voltage gain is limited to 1.5, and it is suitable for low-power applications.

Figure 2.6 depicts the nine-level inverters. The inverter presented in Figure 2.6.a [16] includes two half-bridge inverters, namely the left side and right side half-bridge of the converter and series connected "m" number of modified SC cells (SCCs) in the middle of the topology. The half-bridge transfers the voltages developed by the SCCs to the load. This inverter has been designed for sustainable and renewable energy applications. A hybrid nine-level inverter topology that employs an SC, flying capacitor, and H-bridge inverter is depicted in Figure 2.6.b [17]. This inverter can boost the output voltage up to twice the source voltage. However, the MVS across a few devices becomes equal to the peak output voltage. A look-up table-based simple PWM technique has been employed to generate the switching gate pulses.

2.2.1.2 Midpoint SCMLI topology

In midpoint SCMLI, the two capacitors are connected directly to the DC source, which splits the DC source voltage amplitude into two halves, and the

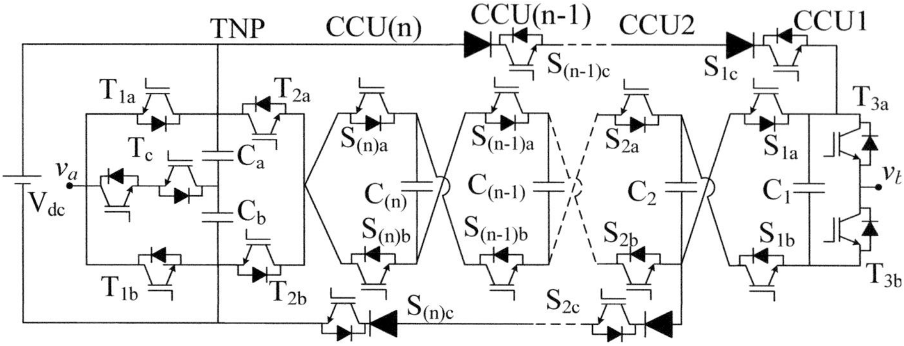

Figure 2.7 Step-up MLI using CCUs [18]

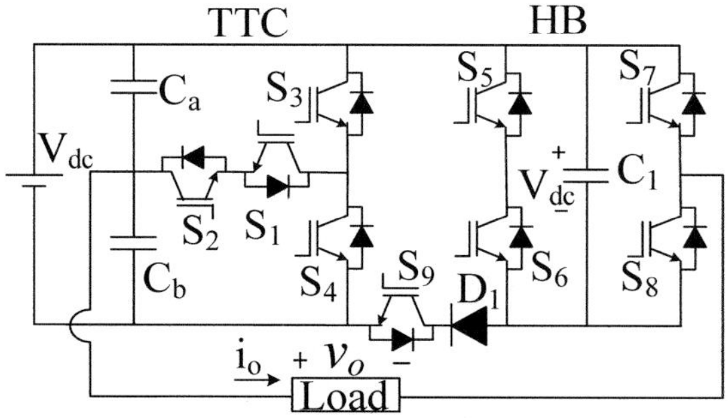

Figure 2.8 Hybrid seven-level SCI [19]

jumper/connection between these two DC-link capacitors is treated as mid-point. The load's neutral terminal is connected to this midpoint, which allows for fewer switching devices in such type of topology. Further, Figures 2.7–2.10 represent the midpoint SCMLI topology. A symmetrical step-up SCMLI topology based on crisscross units (CCUs) has been developed in [18] and depicted in Figure 2.7. In this inverter, the MVS of the devices remains lower even when the output voltage increases. As the capacitors in CCUs are getting charged up to the magnitude of the DC supply. This inverter can also operate in regenerative mode. It can generate $4n + 5$ output voltage levels with the peak amplitude of $(n + 1)$ times the supply voltage, where "n" indicates the total number of CCUs. The authors in [19] have proposed a hybrid MLI by incorporating midpoint DC link capacitors, a front-end T-type converter, a floating capacitor, and a back-end HB converter, which can generate a seven-level output voltage.

The authors have developed the nine-level active neutral point clamped (ANPC) inverter based on DC-link capacitors, floating capacitors, and half-bridge inverters, which can be observed in Figure 2.9 [20]. It is improved from the conventional ANPC MLI. However, it has various limitations: MVS becomes equal to the peak output voltage, there is no boosting factor, and a higher number of components are involved. A generalized midpoint SCMLI based on a T-type converter and "n" number of SCCs is depicted in Figure 2.10 [21]. It reduces the CMV so the leakage current also reduces. It has a lesser current spike as all the capacitors are getting charged for half of the fundamental cycle.

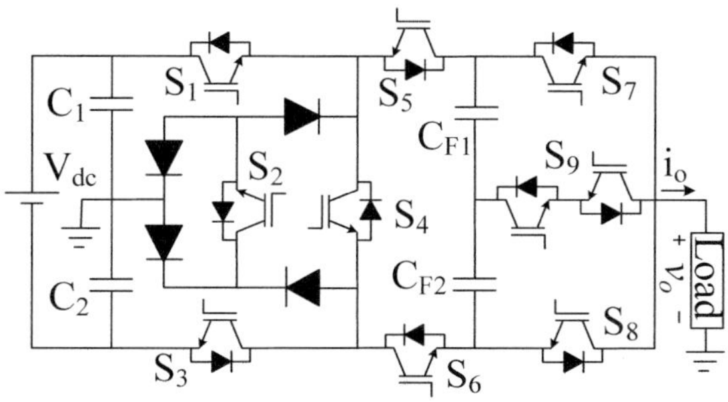

Figure 2.9 Active neutral point clamped MLI [20].

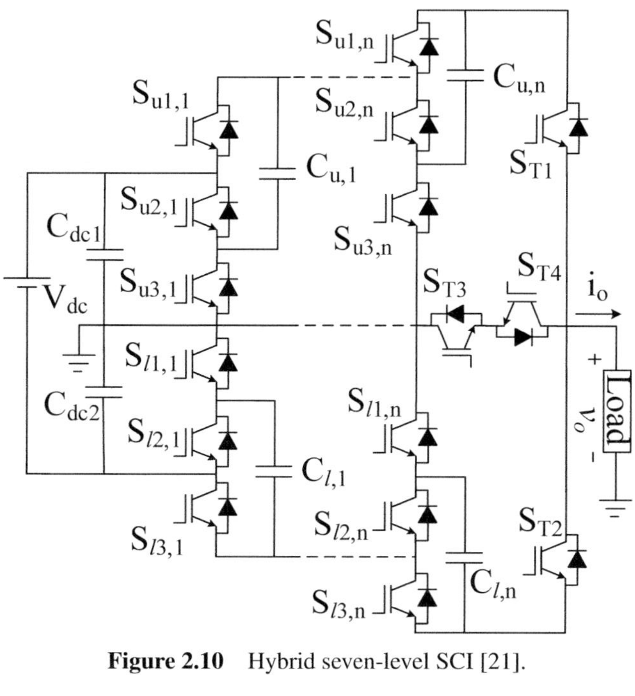

Figure 2.10 Hybrid seven-level SCI [21].

2.2.1.3 Common grounded SCMLI topology

In some of the papers, authors have claimed that the midpoint inverter reduces the leakage current in transformerless grid integration applications. However, the leakage current exists significantly even after the reduction as the high-frequency CMV existed due to the appearance of the stray capacitance between the solar photovoltaic panel's terminal and the ground. This can be mitigated by a common ground switched capacitor MLI (CG-SCMLI). As in CG-SCMLI, the grid/load's neutral terminal is connected directly to the negative terminal of the DC supply. Some of the CG-SCMLIs topologies are presented in Figures 2.11–2.12.

Figure 2.11.a [22] shows a CG-SCMLI topology based on the virtual DC bus concept. It bypasses the raised stray capacitance between the negative pole of the DC supply and the ground. Therefore, the leakage current has been eliminated, and it allows for transformerless grid integration. This inverter is suitable for low-power applications. Figure 2.11.b [23] depicts common grounded H-bridge-based MLI that works on the principle of the flying capacitor technique. It has both the common positive and common ground configuration as the inverter is connected between two grids. Therefore, it developed a common positive with one grid and common ground with the other based on the virtual DC bus concept. A CG-SC three-level inverter has been derived in [24], which is illustrated in Figure 2.11.c. For the satisfactory operation of the inverter, a peak current control strategy has been employed to generate the gating pulses and to control both active, as well as reactive power.

Figure 2.12 depicts a few five-level CG-switched capacitor inverters. The authors have applied peak current control to operate the inverter in Figure 2.12a,b [25], [26] along with the control of active and reactive power. While controlling the inverter in Figure 2.12.c [27], the hysteresis current control technique is applied.

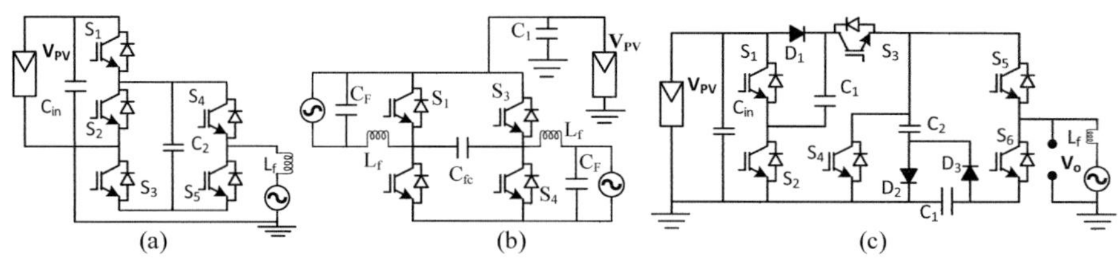

Figure 2.11 CG-SCMLI with (a) virtual ground concept [22], (b) H-bridge based transformerless topology [23], and (c) three-level inverter toplogy [24].

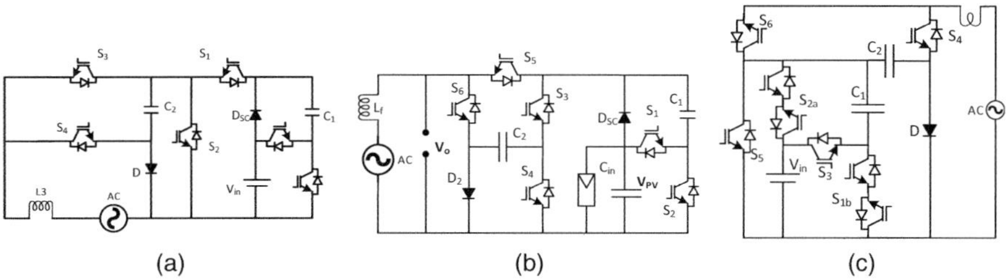

Figure 2.12 CG-SC-based five-level inverter (a) [25], (b) [26], and (c) [27].

2.3 Advanced Control Strategies

Various modulation strategies have been developed and are being utilized to control multilevel inverters. As the MLIs employ numerous switches to produce the staircase output voltage, the order of switch operation and the duration of each step hold significant importance in reducing the THD. Figure 2.13 broadly categorizes the pulse width modulation (PWM) strategies for MLIs based on the switching frequency, such as lower/fundamental, high, and variable switching frequency. In a fundamental switching frequency-based PWM strategy, a switch can be commutated once or twice per cycle, while multiple commutations occur in a high-frequency-based PWM strategy. The hybrid frequency modulation technique is the combination of the fundamental switching frequency technique and the high switching frequency technique.

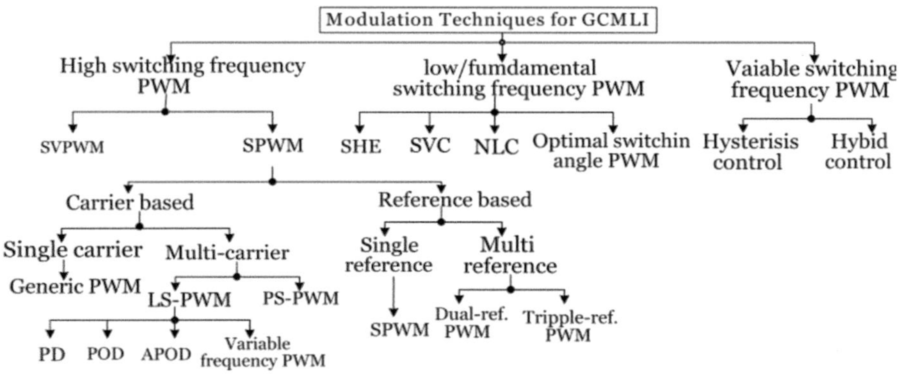

Figure 2.13 Categorization of the modulation strategies for MLIs [1].

2.4 Losses and Efficiency Considerations

2.4.1 Losses

The total losses that occur in the MLI-based system are a combination of various types of losses, which are described below.

2.4.1.1 Conduction losses (P_{cl})

The inverter's conduction losses during the fundamental cycle can be found by summing the conduction losses at each voltage level, and it is expressed by eqn (2.2).

$$P_{cl} = \sum_{lvl=-N_L}^{+N_L} \left(\sum_{n_s=1}^{N_{sw}} I_{sr}{}^2 R_{con} + \sum_{n_c=1}^{N_{cap}} I_{cr}{}^2 R_{ecap} + \sum_{n_d=1}^{N_d} I_{dav} V_{ON} \right), \quad (2.2)$$

where "N_L" represents the total number of voltage steps in the half cycle while N_{sw}, N_{cap}, and N_d indicate the total number of switches, capacitors, and diodes, respectively. The respective I_{sr}, I_{cr} are the root mean square (RMS) currents through the switches and capacitors, while I_{dav} is the average current through diodes. Whereas R_{con} is the ON state resistance of the switches, and R_{esr} is the equivalent series resistance of the capacitors.

2.4.1.2 Switching losses (P_{sl})

Switching losses occur in the switches due to their nonideal on/off behavior. During these transitions, voltage and current overlap, particularly during the sudden on/off process. For the purpose of calculating switching losses, it is assumed that voltage and current change linearly during these transitions. The average switching losses can be found using eqn (2.3).

$$P_{sl} = f_{sw} \times \left[\sum_{n_s=1}^{N_{sw}} \left\{ \sum_{T_{sw}} \frac{1}{6} \left(V_{ON} I_{ON} T_{ON} + V_{OFF} I_{OFF} T_{OFF} \right) \right\} \right], \quad (2.3)$$

where f_{sw} is the switching frequency, $T_{sw} = \frac{1}{f_{sw}}$. V_{ON} and I_{ON} represent the transition voltage and current, respectively, during switching on while V_{OFF} and I_{OFF} during switching off. T_{ON} and T_{OFF} indicate the switching on and switching off duration.

2.4.1.3 Capacitor ripple losses ($P_{c,ripp}$):

The amount of energy dissipated in the capacitor during charging of the capacitor and the charging path is excluded from the load is qualified as capacitor ripple loss. The losses mainly depend on the operating frequency and capacitance of the capacitor. It can be evaluated as

$$P_{\mathrm{c,ripp}} = (0.5) f_{\mathrm{sw}} \sum_{\mathrm{n_c}=1}^{N_{\mathrm{c}}} C_{\mathrm{n_c}} \Delta V_{\mathrm{n_c}}{}^{2}, \tag{2.4}$$

where $C_{\mathrm{n_c}}$ and $\Delta V_{\mathrm{n_c}}$ are the capacitance and voltage ripple, respectively, of the $n_{\mathrm{c}}{}^{th}$ capacitor.

2.4.2 Efficiency (η_I)

The efficiency of the MLI can be defined as the total power achieved at the inverter's output compared to the input power.

The inverter's total losses can be found by adding the losses mentioned above. Out of these losses, conduction losses and ripple losses are significant losses. Therefore, by reducing these two losses, the efficiency of the MLI can be enhanced. The efficiency of the MLI can be evaluated as

$$\eta_{\mathrm{I}} = \frac{P_{\mathrm{OI}}}{P_{\mathrm{OI}} + P_{\mathrm{cl}} + P_{\mathrm{sl}} + P_{\mathrm{c,ripp}}}, \tag{2.5}$$

where P_{OI} is the output power of the MLI.

2.5 Associated Parameters with SCMLI

2.5.1 Total standing voltage (TSV) of the semiconductor devices

It is the sum of the maximum blocking voltages of all the semiconductor devices incorporated into the inverter. The respective TSV of the switches and diodes can be found using eqn (2.6) and (2.7).

$$TSV_{\mathrm{s}}(\mathrm{p.u.}) = \frac{1}{V_{\mathrm{op}}} \sum_{\mathrm{n_s}=1}^{N_{\mathrm{s}}} V_{\mathrm{mb_{n_s}}} \tag{2.6}$$

$$TSV_{\mathrm{d}}(\mathrm{p.u.}) = \frac{1}{V_{\mathrm{op}}} \sum_{\mathrm{n_d}=1}^{N_{\mathrm{d}}} V_{\mathrm{mb_{n_d}}}, \tag{2.7}$$

where $TSV_{\mathrm{s}}(\mathrm{p.u.})$ and $TSV_{\mathrm{d}}(\mathrm{p.u.})$ indicate the per-unit TSV for switches and diodes, respectively. V_{op} denotes the MLI's peak output voltage, and $V_{\mathrm{mb_{n_s}}}$ represents the maximum blocking voltage (MBV) of the $n_{\mathrm{s}}{}^{\mathrm{th}}$ switch while $V_{\mathrm{mb_{n_d}}}$ is the MBV of $n_{\mathrm{d}}{}^{\mathrm{th}}$ diode.

2.5.2 Boosting factor (BF)

It is the ratio of the MLI's peak output voltage V_{op} to the applied source voltage V_{in}. It can be expressed as

$$BF = \frac{V_{op}}{V_{in}}. \tag{2.8}$$

2.5.3 Cost function and its analysis

The pricing of inverters plays a crucial role in various applications. Therefore, cost function (CF) has been included in this subsection and can be expressed as eqn (2.9) [2]. The overall cost linked to the various inverters can be assessed using the CF.

$$CF = N_{sw} + N_g + N_d + N_{cap} + \alpha \times TSV + \beta \times TCD_{av}, \tag{2.9}$$

where N_g indicates the total number of gate drivers while TCD_{av} represents the average value of total conducting devices, and it reflects the power loss associated with the MLIs. The cost rises as the amount of power increases. α and β are the weight coefficients that determine the respective weightiness of TSV and TCD_{av}. Higher power losses result in higher expenses. The weight of TSV and TCD_{av} is defined by α and β correspondingly. If the value of α and β are chosen higher than one, then the importance of the TSV and TCD_{av} become more than the component count. If the chosen value of α and β are less than one, then the components count becomes more important, and the cost of switches, gate drivers, and capacitors dominates over TSV and TCD_{av} values.

2.6 COMPARATIVE Analysis and Discussion

In this section, comparative analysis has been done in terms of output voltage levels (Level), the total number of switch count (N_{sw}), diode count (N_d), capacitor count (N_{cap}), voltage boosting factor (BF), TSV, maximum voltage stress (MVS), and maximum efficiency (η_m) while "Top" indicates the developed topology in the existing literature. Both TSV and MVS have been determined in p.u. by dividing the actual value by the peak output voltage.

Table 2.1 shows the comparison of SCMLIs based on H-bridge. The presented topology of the generalized MLIs in [3, 6, 7] can be expanded to the higher output voltage levels as per the requirement. For the same number of SCs, [7] generates a higher number of output voltage levels than

[3, 6]. However, it needs a higher number of switches. The topologies in [5, 8, 9] show the seven-level SC step-up inverter. Out of these, [8] has better performance as it incorporated only eight switches, and it has a boosting factor of 3. Whereas, the inverter topology in [5] has the feasibility for the extension to generate higher voltage levels. The inverter in [6] maintains the unity ratio of output voltage levels to the number of switches; nevertheless, its BF is just 2.

The boost-type SCMLIs based on half-bridge are compared in Table 2.2. [10, 11] shows the five-level inverter and [10] has a lesser number of switches. However, its MVS, capacitors' voltage rating, and number of capacitors are higher. The generalized modular SCMLIs presented in [12, 16] in which "N" indicates the total number of SC modules in Figure 2.5.a and "n" denotes the total number of steps in the half fundamental cycle.

The half-bridge-based seven-level inverters are presented in [13, 14, 15]. [15] has lesser MVS and the same BF as [13]. However, it employed a greater number of switches. The inverter in [14] has the highest BF at the tradeoff of higher TSV and diodes count. [17] shows the half-bridge-based nine-level SCMLI has remarkable enhancements as it has fewer switches than the number of voltage levels. However, its BF is limited by 2.

Midpoint-based SCMLIs are depicted in Table 2.3. [18, 21] have the generalized midpoint SCMLI in which "k" is the total number of steps in the half cycle and "n" indicates the total number of switched capacitor cells in Figure 2.10. The inverter topology in [20] can generate a greater range of output voltage levels than that of topology in [19], all while preserving the same MVS. Nevertheless, it exhibits a lower BF.

Table 2.1 Comparison of single source H-bridge based boost SCMLIs.

Top	Level	N_s	N_c	N_d	TSV_s (p.u)	TSV_d (p.u)	BF	MVS(p.u)	$V_C(N_c)$	f_s (kHz)	η_m (%)
[4]	2n + 3	2n + 4	n	n	3n + 1	2n + 1	n + 1	1	V_{dc}(n)	–	–
[3]	7	10	3	0	6	0	3	1	$V_{dc}(3)$	1	85
[5]	2n + 1	3n + 2	n	1	–	–	$n/(1\text{-}D_b)$	(n−1)/n	$V_{dc}/(1\text{-}D_b)$	20	98.4
[6]	9	9	2	2	5.5	–	2	1	$V_{dc}/2(2)$	1	94.18
[7]	$2^{n+1}+1$	3n + 3	n	1	6.5	0.25	4	1	$V_{dc}(N_/c/2)$,$2V_{dc}(N_c/2)$	5	–
[8]	7	8	2	2	5.33	–	3	1	$V_{dc}(2)$, 2.0	–	97
[9]	7	9	2	1	5.67	0.33	2	1	$V_{dc}(2)$	10	95.27

Table 2.2 Comparison of single source half-bridge based SCMLI with boost capability.

Top	Level	N_s	N_c	N_d	TSV_s (p.u)	TSV_d (p.u)	BF	MVS	$V_C(N_c)$	f_s (kHz)	η_m (%)
[10]	5	6	2	2	4	–	2	1	$2V_{dc}(2)$	-	97.63
[11]	5	9	1	0	9	0	2	0.5	$V_{dc}(1)$	2.5	97.91
[12]	2N+1	$4log_2(N+1)$	$4log_2(N+1)$-4	N_c	4N	–	N	(N+1)/2	$2^{(N-1)}V_{dc}(N_c)$	5	97.1
[13]	7	7	2	2	5	–	1.5	1	$V_{dc}/2(2)$	2	–
[14]	7	9	2	4	5.67	–	3	1	$V_{dc}(2)$	3	94.3
[15]	7	10	2	0	5.33	0	1.5	0.67	$V_{dc}/2(2)$	4	>94
[16]	2n+1	$2log_2$2n+2	$2log_2 2n-2$	N_c	4	–	n	1	$2^{(m-1)}V_{dc}(N_c)$	4	95.9
[17]	9	8	2	1	5	0.25	2	1	$V_{dc}/2(2)$	–	96.47

Table 2.3 Comparison of midpoint SCMLIs.

Top	Level	N_s	N_c	N_d	TSV_s (p.u)	TSV_d (p.u)	BF	MVS	$V_C(N_c)$	f_s (kHz)	η_m (%)
[18]	2k+1	3k+9	0.5k−1	0.5k−3	2(3k−2.5)/k	–	k/2	0.67	$V_{dc}(N_c)$	–	96.17
[19]	7	9	3	1	6.33	0.33	1.5	1	$V_{dc}/2(2), V_{dc}(1)$	–	97.29
[20]	9	10	4	4	5.5	–	1	1	$V_{dc}/2(2), V_{dc}/4(2)$	2.5	97.3
[21]	2n+3	6n+4	2n+2	0	6	0	(n+1)/2	(n+2)/(n+1)	$V_{dc}/2(N_c)$	5	–

Table 2.4 Comparison of CG-SCMLIs with boosting capability.

Top	Level	N_s	N_c	N_d	TSV_s (p.u)	TSV_d (p.u)	BF	MVS(p.u)	$V_C(N_c)$	f_s (kHz)	η_m (%)
[23]	3	4	2	0	6	–	1	1	$V_{dc}(2)$	16	97.80
[24]	3	6	3	3	5	2.5	2	1	$V_{dc}(1), 2V_{dc}(2)$	–	97.90
[25]	5	6	2	2	4.5	1.5	2	1	$V_{dc}(1), 2V_{dc}(1)$	20	98.10
[26]	5	6	3	2	3	1	2	0.5	$V_{dc}(3)$	50	98.10
[27]	5	8	2	1	5	1	2	1	$V_{dc}(1), 2V_{dc}(1)$	40	98.20
[28]	n	2n-1	(n−1)/2	0	2(2n−1)/(n−1)	0	(n−1)/2	2/(n−1)	$V_{dc}(N_c)$	10	96.70

CG-based SCMLIs are depicted in Table 2.4. The three-level inverters have been constructed in [23, 24] and they have almost the same efficiency and maximum voltage stress. [23] incorporated fewer switches than [24] but at the cost of lesser BF. [24, 25, 26] have constructed five-level CG-SCMLI with the same BF and almost the same efficiency. The inverter in [26] has lesser TSV and MVS, but the number of capacitors is more. A generalized n-level CG-SCMLI has been developed in [28] without the inclusion of a diode. In this design, all the capacitors get charged to the applied input voltage.

2.7 Case Study

Under this case study, a new multisource-based reduced-component switched capacitor multilevel inverter (MSRC-SCMLI) topology has been proposed to produce multiple output voltage levels. Figure 2.14 illustrates the circuit diagram of the proposed MSRC-SCMLI topology, which is capable of generating nine, 13, and 17 levels output voltage for the selection of two DC sources V_1 and V_2 in the ratio of 1:1, 1:2, and 1:3, respectively. The proposed inverter employs ten switches (S_1, S_2, ..., S_{10}), two diodes (D_1 and D_2) and two SCs (C_1 and C_2). It produces eight bipolar voltage levels and a zero across the load for $V_1 : V_2 = 1 : 3$. The capacitor C_1 gets charged up to V_1 whenever the switch S_6 turns on while C_2 gets charged up to V_2 through the D_2 and S_7. The switching sequence shown in Table 2.5 illustrates the complete operation of the proposed MSRC-SCMLI.

The proposed topology has a reduced number of components, including switches. Therefore, the conduction, switching, and ripple losses are lesser, leading to better efficiency. The size and cost are also reduced. Moreover, it has capacitors' self-voltage balancing capacity, so this objective does not need complicated switching patterns, and at the same time, it has a twofold voltage gain. It can be used for solar photovoltaic (PV) power generation as the two different PV panels can be used as DC sources instead of batteries.

The simulation parameters of the proposed MSRC-SCMLI are depicted in Table 2.6. For the simulation, the semiconductor devices have been selected as ideal. The inverter was tested under two different loading conditions: Z_1 (50 Ω) and Z_2 (50 Ω + 50 mH). The inverter has also been evaluated under

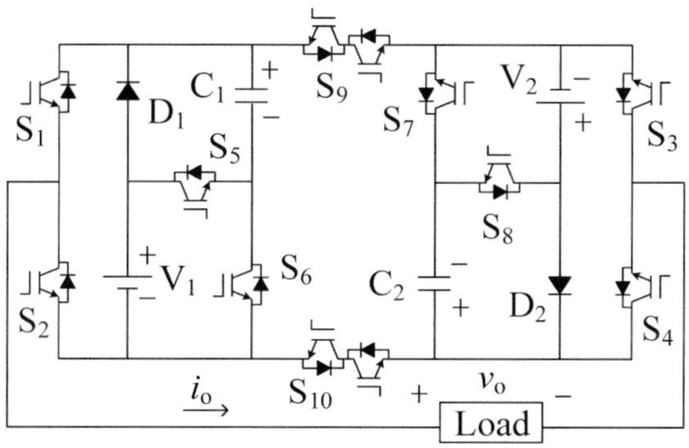

Figure 2.14 The proposed multisource based reduced component SCMLI topology.

Table 2.5 Switching sequence of the proposed SCMLI topology for $V_1 : V_2 = 1 : 3$.

S_1	S_2	S_3	S_4	S_5	S_6	S_7	S_8	S_9	S_{10}	V_o	C_1	C_2
1	0	1	0	1	0	0	1	0	1	$+8V_{dc}$	▼	▼
1	0	1	0	0	1	0	1	0	1	$+7V_{dc}$	▲	▼
0	1	1	0	0	0	0	1	0	1	$+6V_{dc}$	–	▼
1	0	1	0	1	0	1	0	0	1	$+5V_{dc}$	▼	▲
1	0	1	0	0	1	1	0	0	1	$+4V_{dc}$	▲	▲
0	1	1	0	0	0	1	0	0	1	$+3V_{dc}$	–	▲
1	0	0	1	1	0	0	0	0	1	$+2V_{dc}$	▼	–
1	0	0	1	0	1	0	0	0	1	$+V_{dc}$	▼	–
0	1	0	1	0	0	0	0	0	1	0	–	–
0	1	1	0	0	1	0	0	1	0	$-V_{dc}$	▲	–
0	1	1	0	1	0	0	0	1	0	$-2V_{dc}$	▼	–
1	0	0	1	0	0	1	0	1	0	$-3V_{dc}$	–	▲
0	1	0	1	0	1	1	0	1	0	$-4V_{dc}$	▲	▲
0	1	0	1	1	0	1	0	1	0	$-5V_{dc}$	▼	▲
1	0	0	1	0	0	0	1	1	0	$-6V_{dc}$	–	▼
0	1	0	1	0	1	0	1	1	0	$-7V_{dc}$	▲	▼
0	1	0	1	1	0	0	1	1	0	$-8V_{dc}$	▼	▼

▲ → charging and ▼ → discharging, while " – " → idle states of the capacitors.

two types of dynamic conditions: load change and change in modulation index (m_a). Figure 2.15.a shows the 17-level output voltage and current plots when the load changes from Z_1 to Z_2 at the unity modulation index. In this condition, the load voltage maintained the peak amplitude of ± 800 V while the load current peak amplitude reduced from ± 16 to ± 15.26 A. In addition, the output current lags the output voltage by $17.44°$ for Z_2 load. The corresponding capacitors' voltage is depicted in Figure 2.15.b, confirming that the capacitors' voltage is self-balance with the permissible voltage ripple limit of 5%.

Under the second dynamic condition, the output voltage and current waveforms under the change in m_a is depicted in Figure 2.16.a and corresponding capacitors voltage is represented in Figure 2.16.b. Herein, the maximum peak magnitude of the output voltage is generated as 800 V at $m_a = 1$ while the lowest peak magnitude is 100 V for $m_a = 0.1$ and the output current falls as the output voltage decreases due to the decrease in m_a under the fixed load of Z_2. The capacitor C_1 has the highest voltage ripple of 4.49 V at $m_a = 0.7$, and C_2 has the highest voltage ripple of 13.3 V at unity modulation index, and it can be seen in Figure 2.16.b.

Table 2.6 Components and their specifications for MATLAB simulation.

Components	Specifications
V_1	100 V
V_2	100 V (nine-level), 300 V (17-level)
f_{sw}	5 kHz
f_m	50 Hz
C_1 and C_2	4700 μF
R_L	50 Ω
L	50 mH

$f_m \rightarrow$modulating frequency

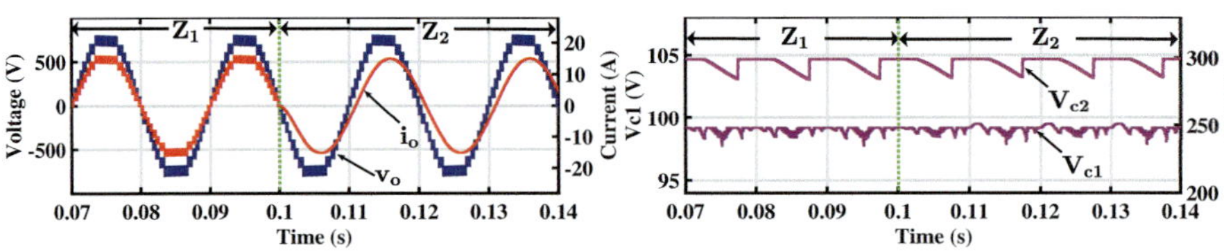

(a) (b)

Figure 2.15 Plots under load change at $m_a = 1$ and $V_1 : V_2 = 1 : 3$. (a) Output voltage and current. (b) Capacitors' voltage.

Similarly, the proposed inverter works satisfactorily for other sets of values of V_1 and V_2. The output voltage, current, and capacitors' voltage waveforms are illustrated in Figure 2.17 under the load change while in Figure 2.18 for the change in modulation index. When both V_1 and V_2 are set to 100 V, the output voltage is generated as a nine-level with a peak amplitude of ± 400 V while the output current's peak amplitude is ± 7.63 A at unity modulation index with Z_2 load. Under this condition, the maximum voltage ripple of capacitor C_1 is 8.56 and 0.65 V of C_2, which further decreases as the modulation index decreases.

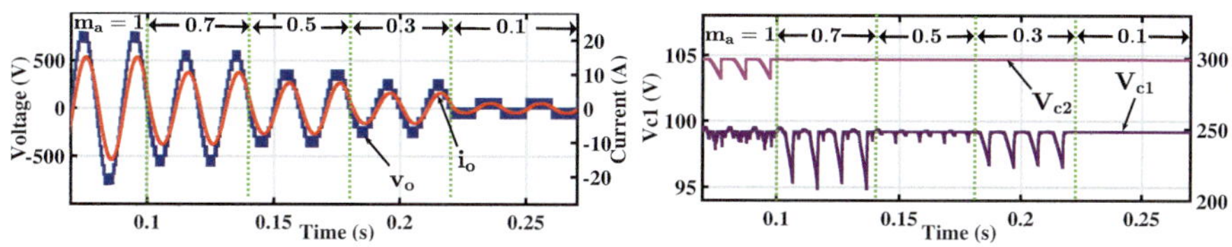

(a) (b)

Figure 2.16 Waveforms under the change in modulation index at Z_2 loading and $V_1 : V_2 = 1 : 3$. (a) Output voltage and current. (b) Capacitors' voltage.

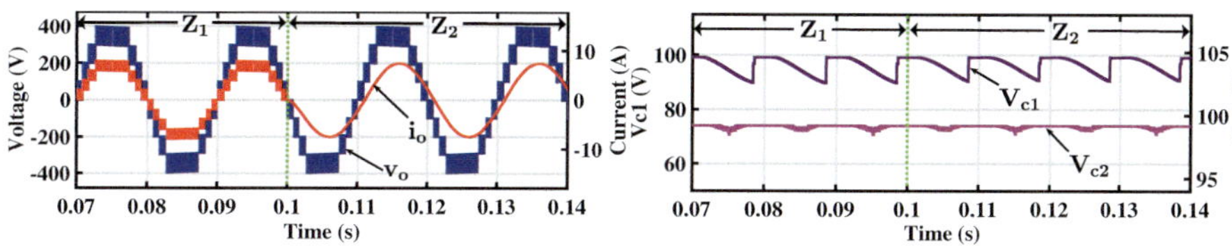

(a) (b)

Figure 2.17 Waveforms under load change at $m_a = 1$ and $V_1 : V_2 = 1 : 1$. (a) Output voltage and current. (b) The voltage across the capacitors.

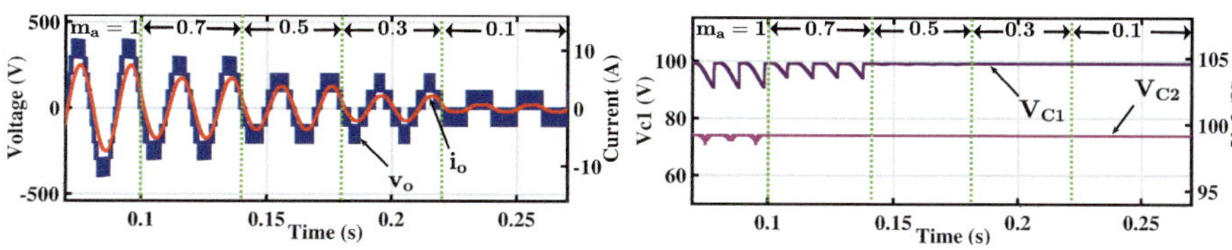

(a) (b)

Figure 2.18 Waveforms under m_a change at Z_2 loading and $V_1 : V_2 = 1 : 1$. (a) Output voltage and current. (b) The voltage across the capacitors.

2.8 Applications of SCMLI

MLIs have grown in popularity in both academic research and industrial applications due to their significant inherent advantages, catering to a range of low to high power requirements, electric vehicle power drives, smart grid systems, and integration with renewable energy systems.

2.8.1 Applications in electric vehicles

MLIs have gained significant attention in various applications, including electric vehicle (EV) infrastructure. These inverters outperform standard (traditional) two-level inverters by providing improved output waveform quality, higher voltage levels, lower harmonic distortion, and potentially better efficiency. The SCMLI with boost capability is presented for the EV application [29]. These architectures require fewer components to achieve a specific voltage level, resulting in a more compact and cost-effective inverter with higher energy density.

Multilevel inverters serve diverse functions across electric vehicle (EV) applications. In EV drivetrains, they facilitate the conversion of DC power from batteries to AC power for motors while managing torque control, thus influencing motor speed through frequency adjustment. Moreover, these inverters enable seamless integration of EV charging stations with the power grid, offering grid support functions like active power control and voltage regulation, which are crucial for stability. They also facilitate bidirectional power transfer in vehicle-to-grid (V2G) and grid-to-vehicle (G2V) systems, enhancing grid stability and enabling distributed energy storage. Additionally, multilevel inverters excel in harmonic mitigation, producing lower harmonic distortion to improve power quality. Furthermore, they operate efficiently at higher voltage levels, reducing current stresses and allowing voltage boosting when needed for charging high-voltage EVs, all while employing advanced control strategies to optimize efficiency and manage thermal performance.

2.8.2 Integration with renewable energy systems

Over the last decade, the escalating demand for energy from high-power industries and expanding user bases has placed a significant burden on the power grid. This surge in energy demand, coupled with traditional energy generation methods, has led to a noticeable increase in global warming. As a response to these environmental challenges, there has been a considerable push toward integrating renewable energy resources into the power grid, with a particular focus on PV systems, owing to their vast potential to provide sustainable and clean energy.

Researchers are increasingly drawn to MLIs over two-level inverters due to their capacity to offer benefits such as reduced EMI, improved efficiency, and larger DC link voltages. A recent proposal for a grid-integrated system designed for PV applications in [30] introduces an advanced flyback DC-DC converter and a 17-level SCMLI. This system brings advantages like balanced voltage across capacitors, a diminished number of semiconductor devices and DC voltage sources, and increased voltage levels at the output of the inverter. Strategies to address impact currents during capacitor charging, such as employing a set of inductor and a diode across it in the channel of capacitive charging current, enhancing converter efficiency, and reducing inrush current. Furthermore, the active and reactive power injection into the grid can be effectively regulated by using a current control technique. Another innovative MLI, tailored for solar PV integration, incorporates a voltage-boosting capability, generating a 13-level output voltage waveform with lesser components [31]. To mitigate challenges like harmful harmonics, leakage current, and increased voltage stress on components, a novel SCMLI topology is proposed in [32]. This topology utilizes series-connected DC sources, producing a high-quality staircase AC voltage while addressing these concerns and providing flexibility as a three-phase inverter.

2.9 Challenges and Future Directions

Despite numerous advantages, it is important to acknowledge that SCMLIs are not without challenges. Ongoing research and development efforts are essential to refine these inverters further and overcome existing limitations. Research in switched-capacitor-based multilevel inverters aims to address issues such as inrush currents, spike charging currents, harmonics, leakage currents, and increased voltage stress on components. Some general challenges and potential future directions for research and development based on the trends in power electronics applications are mentioned below.

2.9.1 Current challenges

Switched capacitor-based inverters encounter several challenges that demand attention in research and development. Voltage balancing among capacitors poses a significant hurdle, particularly in larger systems, where precise balance is imperative for reliable operation. Moreover, the complexity of modulation techniques and control strategies requires exploring advanced methods like space vector modulation and sinusoidal pulse-width modulation to optimize performance while simplifying control complexities, especially in high-voltage systems. Efficiency improvement remains a continual challenge, necessitating efforts to minimize losses, enhance thermal management, and boost overall energy conversion efficiency. Additionally, identifying and optimizing topology for specific applications without compromising performance is crucial. Scalability across various power ratings and applications is also essential for wider adoption, alongside careful management of component stress and voltage stress on capacitors, to ensure longevity and reliability.

2.9.2 Potential future directions

Research endeavors could target developing advanced control strategies aimed at simplifying the operation of switched capacitor-based inverters while enhancing performance, potentially leveraging machine learning techniques for optimization. Furthermore, exploring modular design approaches could improve the scalability of these inverters, making them adaptable to various power ratings and applications. Integration into smart grid systems presents another avenue for investigation, focusing on the role of switched capacitor-based inverters in grid support functions like voltage regulation and power factor correction. Additionally, efforts to enhance fault tolerance mechanisms and reliability are essential, alongside exploring integration with renewable energy systems to optimize compatibility with variable inputs. Finally, standardization efforts could promote interoperability and wider acceptance within the industry, streamlining integration processes.

As technology evolves and power electronics applications continue to advance, these potential future directions could help address current challenges and further enhance the capabilities of switched capacitor-based inverters.

2.10 Conclusion

Switched-capacitor multilevel inverters (SCMLIs) have revealed as a promising solution in power electronics. These inverters address challenges associated with conventional counterparts by integrating switched-capacitor (SC) basic units into multilevel circuit configurations. Single-input, DC-source

SCMLIs enhance voltage conversion gain while minimizing the total standing voltage (TSV) index and power switch count. Multi-input SCMLIs achieve flexibility in output voltage levels with fewer semiconductor devices. Challenges persist, including cost considerations, TSV index optimization, and addressing current stress. Midpoint-clamped SCMLIs enable full DC-link utilization but face setbacks in generalization and voltage conversion gain. Capacitor-grouped SCMLIs, with a virtual DC-link concept, mitigate leakage current concerns but encounter challenges in generalization and overall conversion gain. Ongoing research is crucial for refining SCMLI technologies for practical applications. Their modular structure ensures scalability and flexibility, making SCMLIs suitable for diverse applications, from renewable energy systems to motor drives, advancing power electronics with higher voltage capabilities and enhanced efficiency.

References

[1] Akanksha Sinha, Kartick Chandra Jana, Madan Kumar Das, 'An inclusive review on different multi-level inverter topologies, their modulation and control strategies for a grid connected photovoltaic system,' Sol. Energy, vol. 170, pp. 633-657, 2018.

[2] M. D. Siddique, S. Mekhilef, S. Padmanaban, M. A. Memon and C. Kumar, 'Single-Phase Step-Up Switched-Capacitor-Based Multilevel Inverter Topology With SHEPWM,' IEEE Trans. Ind. App., vol. 57, no. 3, pp. 3107-3119, May-June 2021.

[3] Y. Hinago and H. Koizumi, 'A Switched-Capacitor Inverter Using Series/Parallel Conversion With Inductive Load,' IEEE Trans. Ind. Electron., vol. 59, no. 2, pp. 878-887, Feb. 2012.

[4] E. Babaei and S. S. Gowgani, 'Hybrid Multilevel Inverter Using Switched Capacitor Units,' IEEE Trans. Ind. Electron., vol. 61, no. 9, pp. 4614-4621, Sept. 2014.

[5] M. N. H. Khan, M. Forouzesh, Y. P. Siwakoti, L. Li, and F. Blaabjerg, 'Switched capacitor integrated (2n+1)-level step-up single-phase inverter,' IEEE Trans. Power Electron., vol. 35, no. 8, pp. 8248âĂŞ8260, Aug. 2020.

[6] J. Liu, J. Wu, J. Zeng and H. Guo, 'A Novel Nine-Level Inverter Employing One Voltage Source and Reduced Components as High-Frequency AC Power Source,' IEEE Trans. Power Electron., vol. 32, no. 4, pp. 2939-2947, April 2017.

[7] Ngo, Bac-Bien, Nguyen, Minh-Khai, Kim, Jae-Hong, Zare, Firuz, 'Single-phase multilevel inverter based on switched-capacitor structure,' IET Power Electron. vol. 11, no. 11, pp. 1858-1865, 2018.
[8] W. Peng, Q. Ni, X. Qiu and Y. Ye, 'Seven-Level Inverter With Self-Balanced Switched-Capacitor and Its Cascaded Extension,' IEEE Trans. Power Electron., vol. 34, no. 12, pp. 11889-11896, Dec. 2019.
[9] J. Chen, C. Wang and J. Li, 'A Single-Phase Step-Up Seven-Level Inverter With a Simple Implementation Method for Level-Shifted Modulation Schemes,' IEEE Access, vol. 7, pp. 146552-146565, 2019.
[10] Saeedian, Meysam, Hosseini, Seyyed Mehdi, Adabi, Jafar, 'Step-up switched-capacitor module for cascaded MLI topologies,' IET Power Electron., vol. 11, no. 7, pp. 1286-1296, 2018.
[11] N. Sandeep, J. S. M. Ali, U. R. Yaragatti and K. Vijayakumar, 'A Self-Balancing Five-Level Boosting Inverter With Reduced Components,' IEEE Trans. Power Electron., vol. 34, no. 7, pp. 6020-6024, July 2019.
[12] Y. Ye, S. Chen, X. Zhang and Y. Yi, 'Half-Bridge Modular Switched-Capacitor Multilevel Inverter With Hybrid Pulsewidth Modulation,' IEEE Trans. Power Electron., vol. 35, no. 8, pp. 8237-8247, Aug. 2020.
[13] W. Lin, J. Zeng, B. Fu, Z. Yan and J. Liu, 'Switched-capacitor based seven-level boost inverter with reduced devices,' CSEE J. Power and Energy Syst. early access, doi: 10.17775/CSEEJPES.2020.02620.
[14] A. Khodaparast, M. J. Hassani, E. Azimi, M. E. Adabi, J. Adabi and E. Pouresmaeil, 'Circuit Configuration and Modulation of a Seven-Level Switched-Capacitor Inverter,' IEEE Trans. Power Electron., vol. 36, no. 6, pp. 7087-7096, June 2021.
[15] J. Zhao, Y. Chen, J. Zeng and J. Liu, 'Low-Voltage Stress Seven-Level Inverter Based on Symmetrical Capacitors,' IEEE J. Emerg. Sel. Topics Power Electron., vol. 10, no. 3, pp. 3033-3044, June 2022.
[16] M. Saeedian, M. E. Adabi, S. M. Hosseini, J. Adabi and E. Pouresmaeil, 'A Novel Step-Up Single Source Multilevel Inverter: Topology, Operating Principle, and Modulation,' IEEE Trans. Power Electron., vol. 34, no. 4, pp. 3269-3282, April 2019.
[17] B. S. Naik, Y. Suresh, J. Venkataramanaiah and A. K. Panda, 'A Hybrid Nine-Level Inverter Topology With Boosting Capability and Reduced Component Count,' IEEE Trans. Circuits and Systems II: Express Briefs, vol. 68, no. 1, pp. 316-320, Jan. 2021.
[18] W. Lin, J. Zeng, J. Liu, Z. Yan and R. Hu, 'Generalized Symmetrical Step-Up Multilevel Inverter Using Crisscross Capacitor Units,' IEEE Trans. Ind. Electron., vol. 67, no. 9, pp. 7439-7450, Sept. 2020.

[19] J. Liu, J. Wu and J. Zeng, 'Symmetric/Asymmetric Hybrid Multilevel Inverters Integrating Switched-Capacitor Techniques,' IEEE J. Emerg. Sel. Topics Power Electron., vol. 6, no. 3, pp. 1616-1626, Sept. 2018.

[20] Siddique, Marif Daula, Iqbal, Atif, Sathik Mohamed Ali, Jagabar, Mekhilef, Saad, Almakhles, Dhafer J., 'Design and implementation of a new unity gain nine-level active neutral point clamped multilevel inverter topology,' IET Power Electron., vol. 13, no. 14, pp. 3204-3208, 2020.

[21] S. S. Lee and K. -B. Lee, 'Switched-Capacitor-Based Modular T-Type Inverter,' IEEE Trans. Ind. Electron., vol. 68, no. 7, pp. 5725-5732, July 2021.

[22] Y. Gu, W. Li, Y. Zhao, B. Yang, C. Li and X. He, 'Transformerless Inverter With Virtual DC Bus Concept for Cost-Effective Grid-Connected PV Power Systems,' IEEE Trans. Power Electron., vol. 28, no. 2, pp. 793-805, Feb. 2013.

[23] Y. P. Siwakoti and F. Blaabjerg, 'Common-Ground-Type Transformerless Inverters for Single-Phase Solar Photovoltaic Systems,' IEEE Trans. Ind. Electron., vol. 65, no. 3, pp. 2100-2111, March 2018.

[24] N. V. Kurdkandi et al., 'A New Transformer-Less Common Grounded Three-Level Grid-Tied Inverter With Voltage Boosting Capability,' IEEE Trans. Energy Convers., vol. 36, no. 3, pp. 1896-1909, Sept. 2021.

[25] R. Barzegarkhoo, Y. P. Siwakoti, N. Vosoughi and F. Blaabjerg, 'Six-Switch Step-Up Common-Grounded Five-Level Inverter With Switched-Capacitor Cell for Transformerless Grid-Tied PV Applications,' IEEE Trans. Ind. Electron., vol. 68, no. 2, pp. 1374-1387, Feb. 2021.

[26] N. Vosoughi, S. H. Hosseini and M. Sabahi, 'A New Transformer-Less Five-Level Grid-Tied Inverter for Photovoltaic Applications,' IEEE Trans. Energy Convers., vol. 35, no. 1, pp. 106-118, March 2020.

[27] Barzegarkhoo, Reza, Mojallali, Hamed, Shahalami, Seyed Hamid, Siwakoti, Yam P., 'A novel common-ground switched-capacitor five-level inverter with adaptive hysteresis current control for grid-connected applications,' IET Power Electron. vol. 14, no. 12, pp. 2084-2098, 2021.

[28] M. Samizadeh, X. Yang, B. Karami, W. Chen, F. Blaabjerg and M. Kamranian, 'A New Topology of Switched-Capacitor Multilevel Inverter With Eliminating Leakage Current,' IEEE Access, vol. 8, pp. 76951-76965, 2020.

[29] A. Janabi and B. Wang, 'Switched-Capacitor Voltage Boost Converter for Electric and Hybrid Electric Vehicle Drives,' IEEE Trans. Power Electron., vol. 35, no. 6, pp. 5615-5624, June 2020.

[30] S. K. Haghighian, H. -G. Yeh, M. G. Marangalu, N. V. Kurdkandi, M. Abbasi and H. Tarzamni, 'A Seventeen-Level Step-Up Switched-Capacitor-Based Multilevel Inverter With Reduced Charging Current Stress on Capacitors for PV Applications,' IEEE Access, vol. 11, pp. 118124-118143, 2023.

[31] M. Ali, M. Tayyab, A. Sarwar and M. Khalid, 'A Low Switch Count 13-Level Switched-Capacitor Inverter With Hexad Voltage-Boosting for Renewable Energy Integration,' IEEE Access, vol. 11, pp. 36300-36308, 2023.

[32] H. K. Jahan and M. Abapour, 'Switched-Capacitor-Based Multilevel Inverter for Grid-Connected Photovoltaic Application,' IEEE Trans. Power Electron., vol. 36, no. 9, pp. 10317-10329, Sept. 2021.

3

Reduced Switch Count Multilevel Inverters and Their Applications

Jami Rajesh[1], Nakka Jayaram[2], P. S. V. Kishore[3], Jakkula Swamy[4], and Kusuma Gottapu[5]

[1,2,3]Department of Electrical Engineering, National Institute of Technology Andhra Pradesh, India
[4]Department of Electrical Engineering, Indian Institute of Technology Bombay, India
[5]Department of Electrical and Electronics Engineering, Sagi Rama Krishnam Raju Engineering College, India
E-mail: jamirajesh.sclr@nitandhra.ac.in;jayaram@nitandhra.ac.in; kishorepulavarthi.sclr@nitandhra.ac.in; swamyjakkulanitap@gmail.com; gkusuma9@gmail.com

Abstract

This chapter explores reduced switch count multilevel inverters (RSC-MLI), a technology suited for medium-voltage, high-power applications. It provides an in-depth explanation of different RSC-MLI topologies, covering symmetrical and asymmetrical modes of operation, and discusses the challenges and future research directions for multilevel inverters (MLI). Additionally, the chapter highlights the applications of MLI in motor drives, Flexible AC Transmission Systems (FACTS) devices, and grid-integrated renewable energy systems.

Keywords: Reduced switch count topologies, PWM techniques, applications of multilevel inverters

DOI: 10.1201/9781003669845-3

3.1 Introduction

Inverters are power electronic circuits that convert DC voltage to AC voltage. They are broadly categorized into two main types, which are further subdivided into four classifications, as illustrated in Figure 3.1.

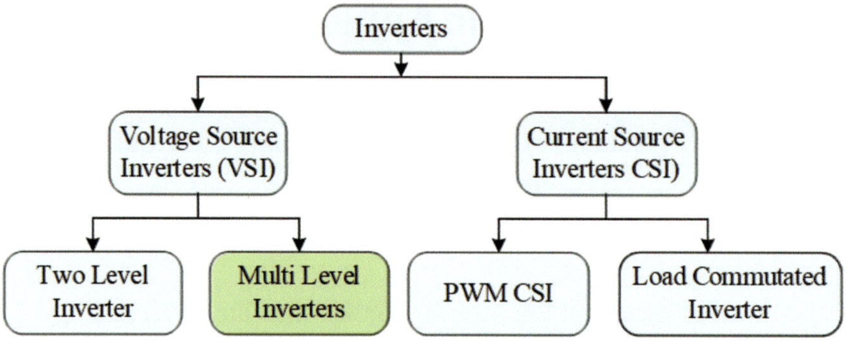

Figure 3.1 Classification of inverters.

Traditional two-level inverters produce output voltages of 0, $\pm V_{\text{in}}$, but their waveforms are rich in harmonics, making them unsuitable for AC power applications. To address this, multilevel inverters (MLIs) were developed to generate nearly sinusoidal output voltages by increasing the number of voltage levels. Total harmonic distortion (THD) decreases substantially with the rise in the number of levels. In the modern era, MLIs have transformed the field, offering significant advantages, including:

- Draws input current with fewer distortions
- Easy interface with the motor drives
- Ability to operate without transformers
- The common mode voltage is less
- They can operate at higher voltage and current levels
- Possibility of getting fault-tolerant operation
- Less dV/dt

3.2 Reduced Switch Count (RSC) MLI

The primary objective of reduced switch count multilevel inverter (RSCMLI) is to overcome the limitations of conventional multilevel inverters (MLI), specifically minimizing the number of switches, simplifying complex circuitry, and reducing costs

3.2.1 Features of RSCMLI

- The conduction loss mainly depends on the number of switches used in the topology and it will affect the efficiency of the entire topology. The RSCMLI topologies have better efficiency than conventional MLI.
- The sum of the blocking voltages of all the switches gives the total standing voltage (TSV) of the topology [1]. However, the uneven blocking voltages of the switches give the different ratings of the switches and it will reduce the cost of the topology.
- RSCMLI has better fault-tolerant ability compared to the classical MLI.
- None of the classical MLIs meet all the application requirements. The RSC-MLI can be easily developed depending on the application requirement.

3.2.2 Classifications

The RSCMLI topologies can be connected in any physical arrangement such as a ladder, staircase, or cascade structure. Thus, the RSCMLI can be categorized as below.

Modular topology structures: topologies that enhance the number of levels without affecting the device's rating are known as modular topologies.

Unit-based configurations: the topologies designed only for some fixed number of voltage levels are called unit-based configurations.

Half-bridge module-based structures: the topologies that are specifically designed only for two levels and considered these as the basic building units are called half-bridge module topologies.

Configurations with H-bridge structure: to utilize the benefits of traditional HB structure, multiple configurations are used by integrating the H-bridge module in any way.

Structures with bidirectional devices: most of the RSCMLI topologies use bidirectional switches to clamp the midpoint of the DC supplies.

Symmetrical/asymmetrical topologies: the topologies having the same level of input DC voltages are called symmetrical topologies otherwise those are called asymmetrical topologies.

Topologies with power distribution: to achieve this, the topologies must maintain the average current flowing through each source is equal, which makes the average power even.

Topologies with equal blocking voltages: in RSCMLI, the topologies employ different types of connection patterns for switches, input DC sources, and other auxiliary components. This impacts the voltage rating/stress across the switching devices. The topologies can be categorized as even and uneven blocking voltages.

3.2.3 Generalized RSCMLI topology structures

3.2.3.1 Structures with different voltage levels and polarity generators

Mostly, all the RSCMLI topologies generate an odd number of voltage levels at the load side. Among these, the popular generalized RSCMLI topologies are multilevel DC link topologies (MLDCLT) [2, 3], switched series and parallel source topologies (SSPST) [4, 5], reverse voltage topologies (RVT) [6, 7], series connected switched source topologies (SCSST) [8, 9], multilevel module topologies (MLMT) [10], and cross-connected topologies (CCT) [11]. Per phase connection of these topologies with three DC supplies is depicted in Figure 3.2. To meet the level generator, the basic units are series connected.

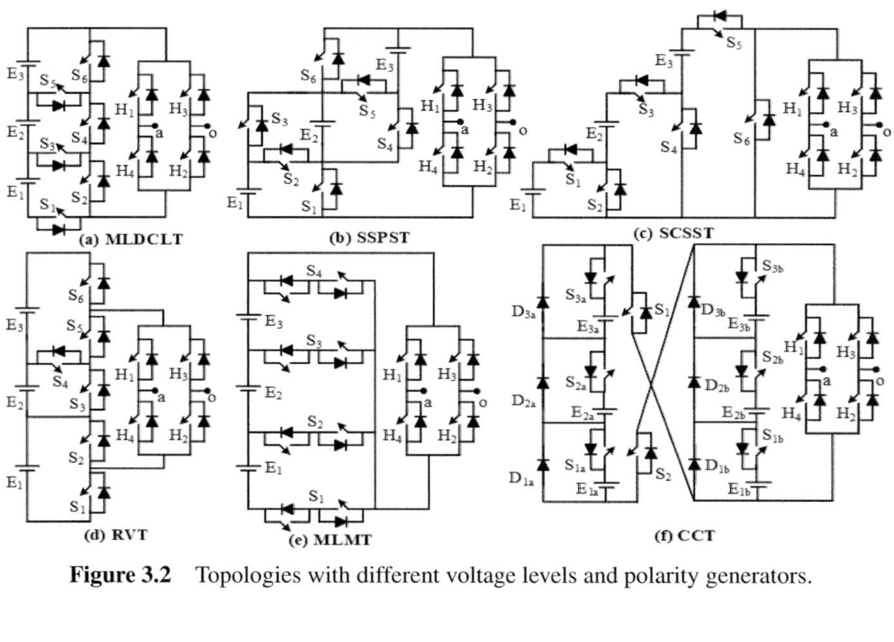

Figure 3.2 Topologies with different voltage levels and polarity generators.

3.2.3.2 T-type structured topologies

The most appropriate configurations [12–17] are depicted in Figure 3.3.

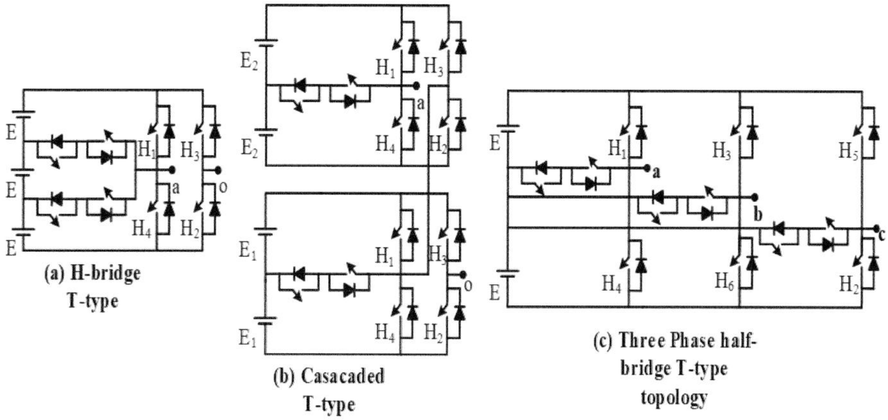

Figure 3.3 T-type structure topologies.

3.2.3.3 HSC structured topologies

In these topologies, unidirectional switches are used to create the HSC, and bidirectional devices are used to integrate HSC to the DC link.

The topological structure depicted in Figure 3.4.a features two DC supplies, E_S and E_R, on either side of the HSC [18, 19]. The bidirectional device on both sides of the HSC structure is similar to the bidirectional device on one side of HSC but connects both sides of HSC and DC links through bidirectional devices [18, 19] as depicted in Figure 3.4.b. The structure shown in Figure 3.4.c is the extension of Figure 3.4.b [20] The structure of these extended HSC structures is depicted in Figure 3.4.d and 3.4.e.

3.2.3.4 Ladder structured topologies

Few topologies have ladder-based structures with unidirectional or bidirectional devices connected on both ends of the DC link. The most often used topologies belonging to this category are cascaded bipolar switched cells (CBPSC), switched DC supplies (SDCS), and packed U-cell (PUC). These topologies are depicted in Figure 3.5. Here, PUC and SDCS topologies have unidirectional switches, and CBPSC topology has bidirectional switches.

3.2.4 Stacked structures

These topologies are created by connecting two or more inverter configurations. The most popular stacked topologies are categorized in this section.

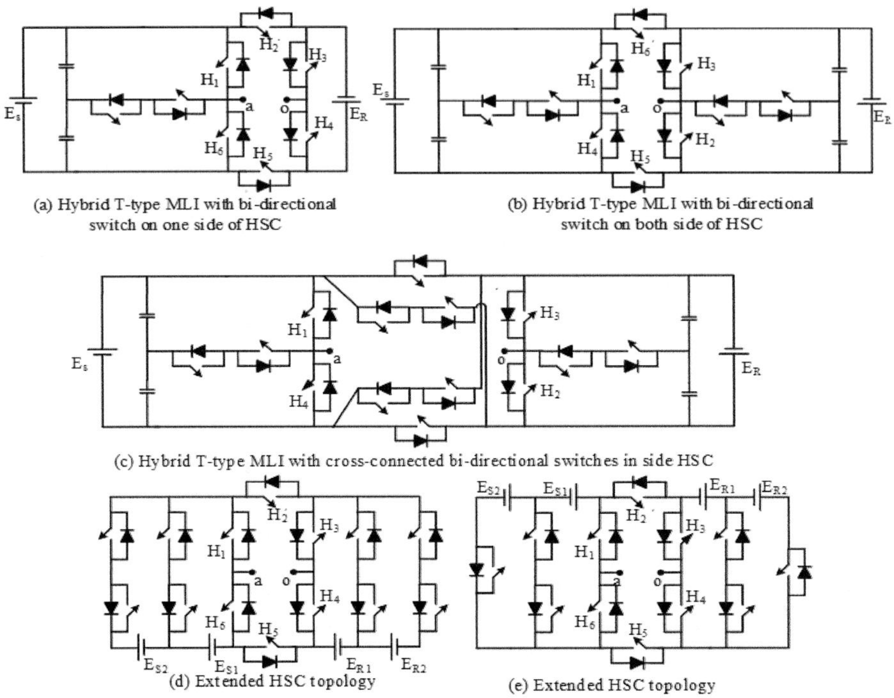

Figure 3.4 Hybrid T-type structures and extended HSC structures.

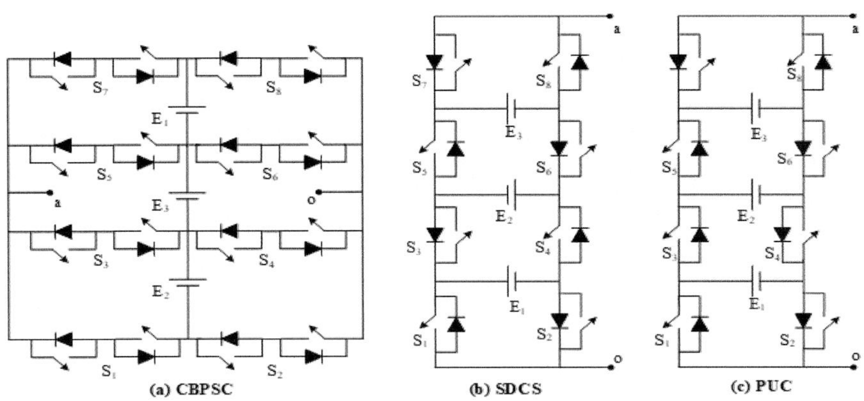

Figure 3.5 Ladder structured topologies.

3.2.4.1 Stacked topology with level doubling network (LDN)

Level doubling network (LDN) for symmetrical seven-level cascaded H-bridge MLI (CHBMLI) structures to get an equivalent 13-level asymmetric configuration and it is depicted in Figure 3.6.

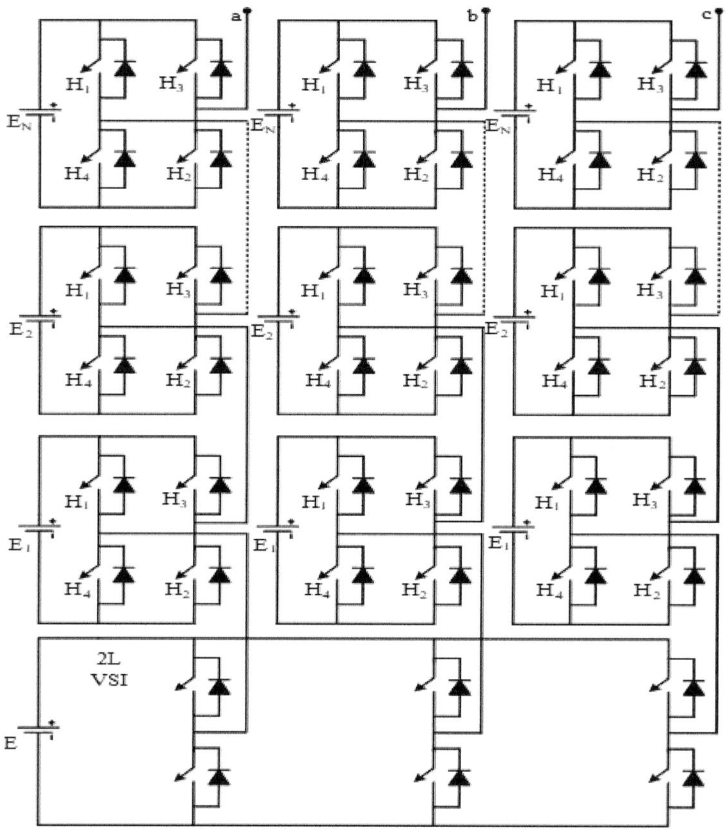

Figure 3.6 Stacking of CHB with two-level inverter.

3.2.4.2 Stacked topology without level doubling network

The stacked topology of MLI without LDN is explained in this section. Here, the topologies are configured with the stacking connection of hybrid topologies.

A hybrid 11-level RDC-MLI topology was created by stacking a five-level ANPC converter module and FC module and it is depicted in Figure 3.7. A stacked topology with an LDN network is formed by a hybrid connection of the ANPC module and FC module.

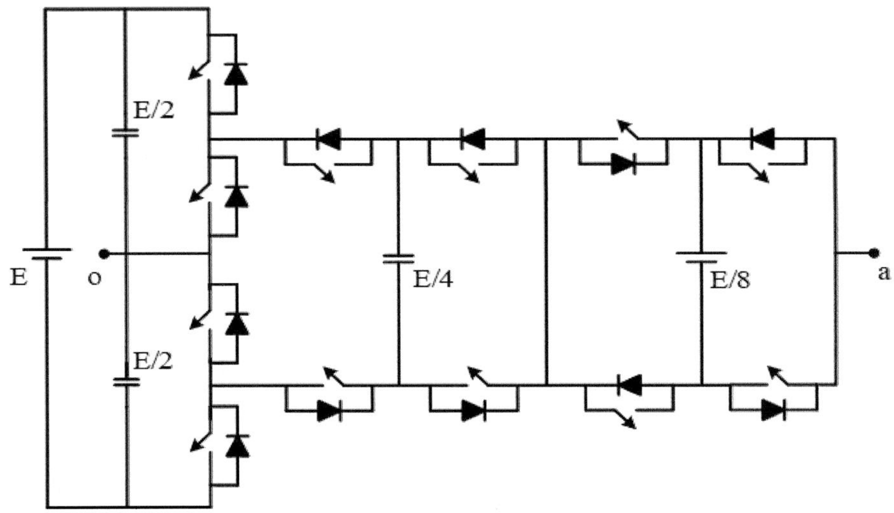

Figure 3.7 A hybrid structure by stacking a five-level ANPC module and FC module.

Figure 3.8 configures a five-level hybrid flying capacitor (HFC) module created by stacking a three-level FC unit and two two-level converter units.

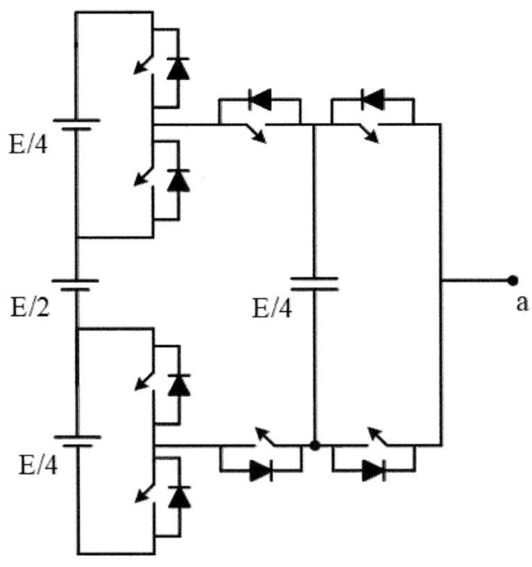

Figure 3.8 A five-level hybrid FC module inverter.

3.2.5 Unit-based RSCMLI

Unit-based RSCMLI is used to simplify the topological size and complexity. However, these configurations can produce finite voltage levels. The topologies under RSCMLI are discussed in this section.

3.2.5.1 Basic unit RSCMLI

The basic RSCMLI topological structure to obtain five-level unipolar voltage is depicted in Figure 3.9. The RSCMLI structure has two parts and these are a three-cell structure and a one-cell structure. Three-cell structures have three voltage sources connected through five unidirectional devices and single-cell structures have one voltage source and two unidirectional devices.

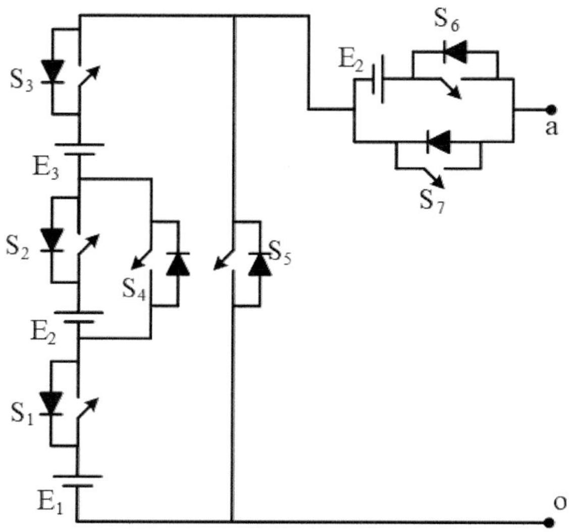

Figure 3.9 Basic unit RSCMLI topology.

3.2.5.2 Symmetrical unit-based topologies

The symmetrical unit-based topologies are discussed in this section. However, two reduced switch symmetrical topologies are considered here.

- *Five-level configuration*: each unit depicted in Figure 3.10.a depicts the reduced switch symmetrical five-level inverter. The cascading of two units results in nine-level inverters.
- *Nine-level configuration*: The nine-level symmetrical topology seems to be a T-type configuration and it is shown in Figure 3.10.b.

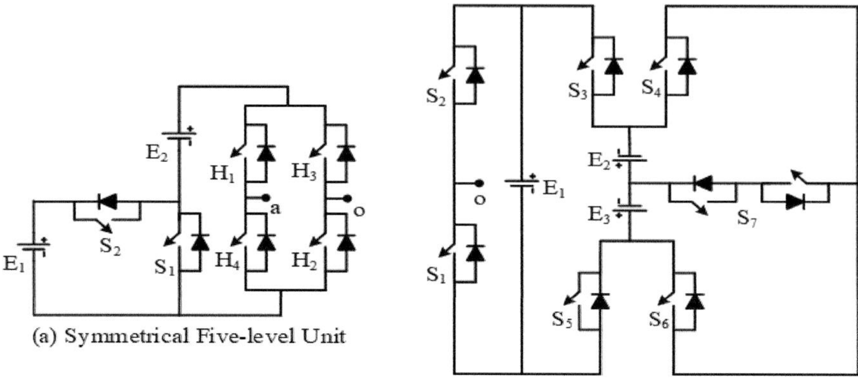

(a) Symmetrical Five-level Unit

(b) Symmetrical Nine-level Unit

Figure 3.10 Symmetrical unit-based configurations.

3.2.5.3 Asymmetrical unit-based configurations

These are E-type and Square T (ST)-type as shown in Figure 3.11. The E-type structure depicted in Figure 3.11.a has four DC supplies, two switches, and six unidirectional switches configured in an envelope structure. ST-type topology is depicted in Figure 3.11.b.

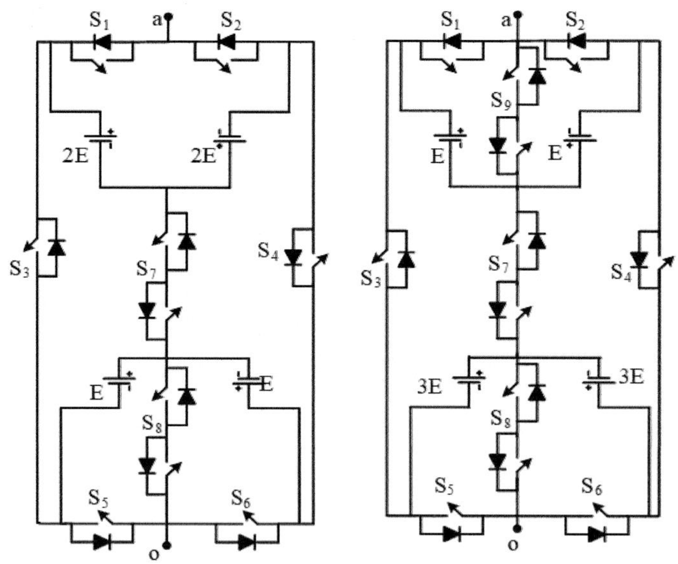

(a) E-type Unit

(b) ST-type Unit

Figure 3.11 Asymmetrical unit-based topologies.

3.2.5.4 Hexagonal switched cell topologies

The basic unit of the hexagonal switched cell (HSC) topology is depicted in Figure 3.12.

3.2.5.5 Extended HSC topologies

These configurations are created by augmenting the basic HSC units with extra DC sources.

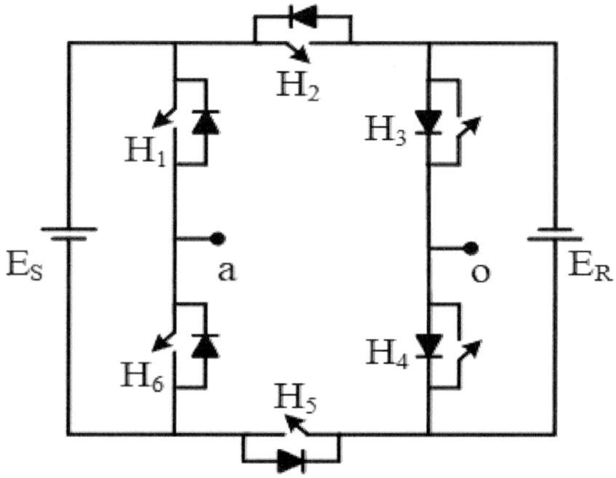

Figure 3.12 HSC topologies.

Topology 1: a nine-level basic HSC unit is depicted in Figure 3.13 and it is extended by cascading several units and supports asymmetry.

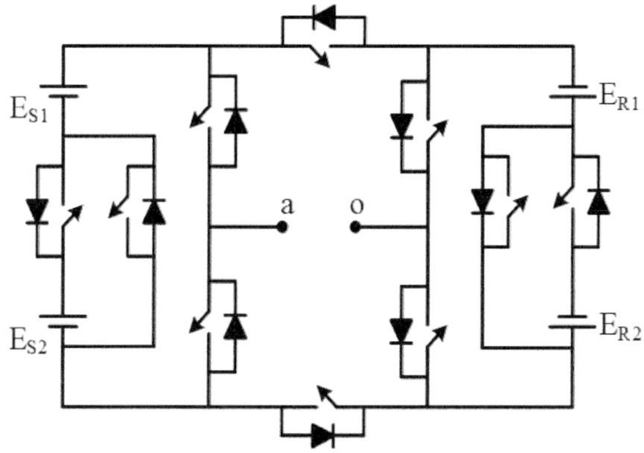

Figure 3.13 A nine-level stretched HSC unit.

Topology 2: another extended basic HSC unit with a 31-level output voltage is depicted in Figure 3.14. By choosing $E_{S1} = E, E_{S2} = 5E, E_{R1} = 2E$, and $E_{R2} = 10E$. This topology has four DC sources.

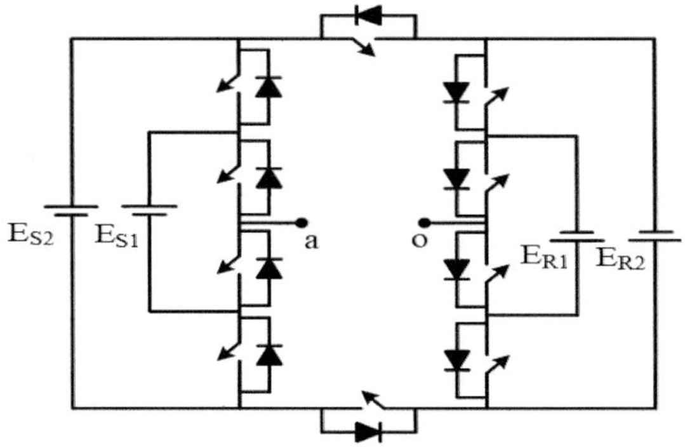

Figure 3.14 Stretched HSC unit.

Topology 3: a basic structure with three DC supplies and eight unidirectional devices is depicted in Figure 3.15.

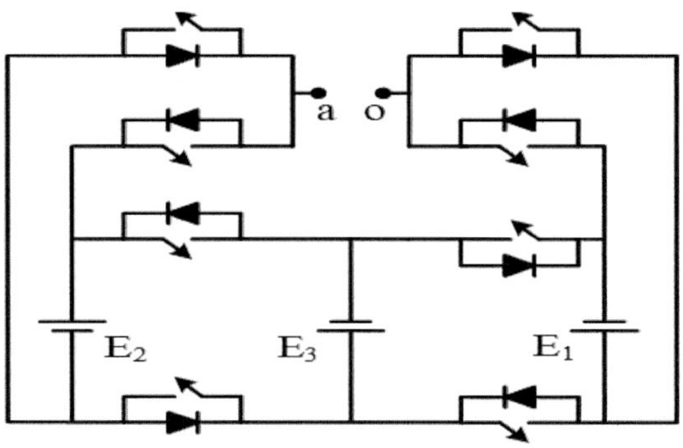

Figure 3.15 Extended HSC unit with three DC supplies.

3.2.5.6 Asymmetrical T-type structures

These units are created from a T-type HB structure.

Structure 1: this configuration was created from HB constructed T-type structure to encourage the asymmetrical operation of DC supplies. The configuration is depicted in Figure 3.16.

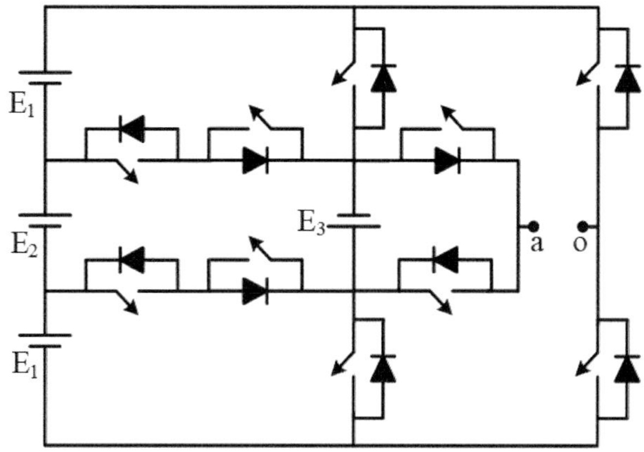

Figure 3.16 A stretched HB-constructed T-type structure.

Structure 2: a T-type structure is depicted in Figure 3.17. The maximum voltage stress occurs across the changing switches S_1 and S_2 and works on the fundamental frequency.

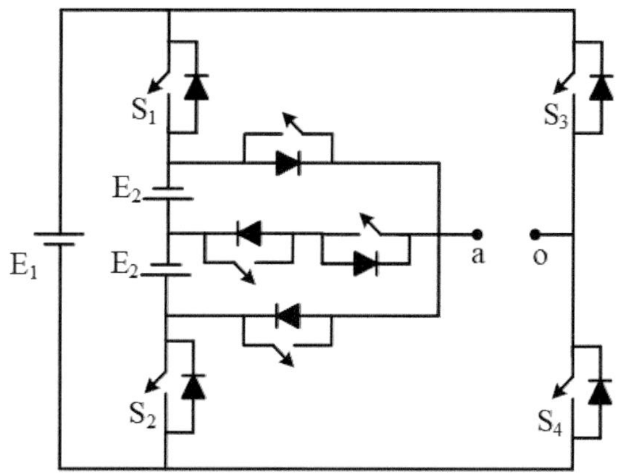

Figure 3.17 An extended T-type H-bridge topology.

3.2.6 Switched capacitor (SC) structures

To discuss issues like DC link stabilizing the voltage of the capacitor and reducing the number of input DC supplies, SC-based configurations are developed. These configurations have segmental structures through SC units coupled in different ways like series, parallel, or cascade. By switching the DC link capacitors in series or parallel with the input DC sources, these configurations produce the necessary voltage levels and ensure the charge balance of the capacitors. As stated, SC topologies are discussed in this section.

3.2.6.1 Series/Parallel based SC structures

(i) Topology-1: basic SC structure: A basic SC unit is depicted in Figure 3.18.a. to enhance this configuration, authors suggested three possible configurations of SC units.

Configuration 1: in Figure 3.18.b shows a topology having series connection of *n* SC units with a common DC voltage source.

Configuration 2: incorporating polarity generator in the topology, SC units are connected in cascaded and it is shown in Figure 3.18.c. Now the topology looks like five level inverters.

Configuration 3: target is to reduce the capacitor voltage and device count; an updated topology is proposed with basic SC units as shown in Figure 3.18.d.

The detailed comparison between the above configurations is presented in Table 3.1.

Table 3.1 The device comparative and significant features. **m*: number of SC modules.

Topology	H-bridge /other	Levels in phase voltages	Number of DC supplies	Number of switches	Number of capacitors	Symmetrical /asymmetrical /both	Voltage stress (equal /unequal)
Series-parallel SC topologies: basic unit	**Config-1**	$2m+3$	1	$2m+4$	m	Symmetrical	Equal
	Config-2	$4m+1$	m	$6m$	m	Both	Equal
	Config-3	$4m+1$	m	$3m+4$	m	Both	Unequal

Topology-2: modified SC unit: in this topology, as shown in Figure 3.19, each unit has two modules. The topology produces nine levels on the output side. The overall gain of the topology is four.

3.2.6.2 FBC with inherent SC unit

The configurations under this category are described in this section.

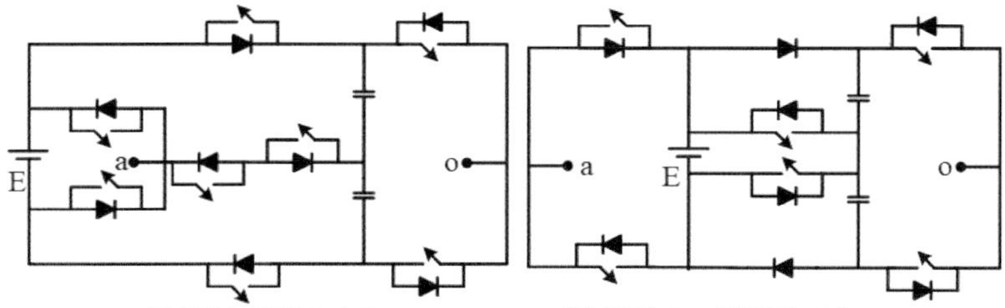

(a) With SC Module (b) With two HB Modules

Figure 3.21 Five-level SC units.

3.2.6.4 SC doubler structures

The proposed configuration is depicted in Figure 3.22.a. This topology has two different DC sources, three unidirectional switches, one capacitor, and two power diodes. To enhance the output voltage levels, more voltage supplies, capacitors, and switches are to be connected, and it is shown in Figure 3.22.b.

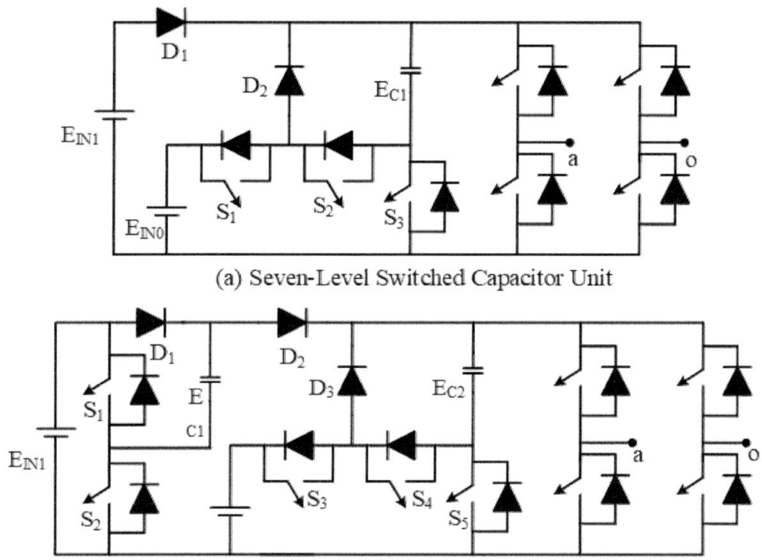

(b) SC Cell with voltage doubler circuit

Figure 3.22 SC doubler and HBC topologies.

3.2.6.5 Modified HSC-based SC structure

The SC configuration depicted in Figure 3.23 is created from half switched capacitor (HSC) unit.

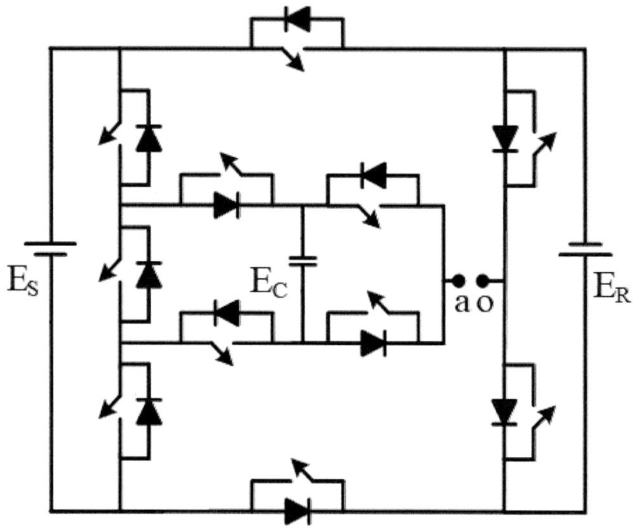

Figure 3.23 Modified HSC-based SC structure.

3.2.6.6 Modified T-type based SC structures

All the configurations are discussed in this section.

Structure-1 (five-level T-type with one DC supply): the configuration is deployed in Figure 3.24.a. The gain of the topology is one and can be extendable to higher levels by cascading the units.

Structure-2 (dual T-type): a seven-level dual T-type structure is depicted in Figure 3.24.b. The gain of this topology is 1.5.

Structure-3 (seven-level boost topology): A SC RDC-MLI with a gain of three is depicted in Figure 3.24.c.

Structure-4 (nine-level cross-connected boost structure): A nine-level boost structure is depicted in Figure 3.24.d. Here, each capacitor is charged to *E*/2, and capacitor voltages are self-balanced irrespective of modulation index and load power factor.

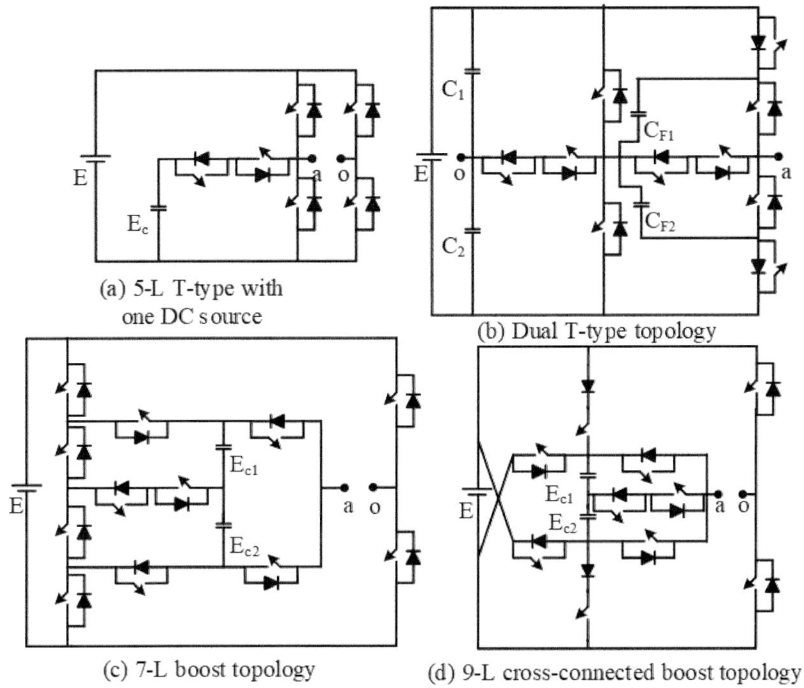

Figure 3.24 Modified T-type-based SC topologies.

3.2.6.7 Single-stage SC unit

The single-stage SC unit topology is shown in Figure 3.25. The gain of the topology is two.

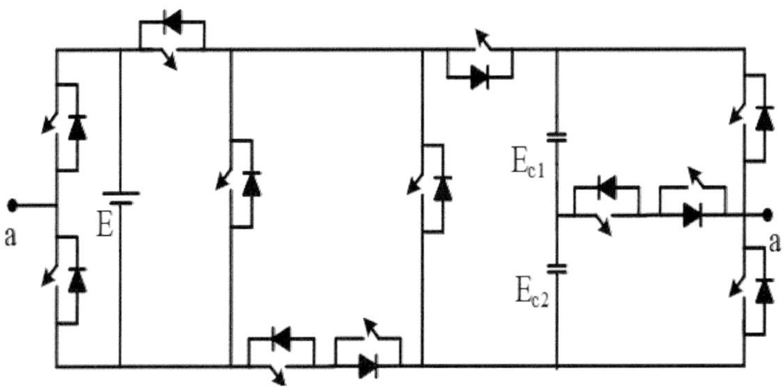

Figure 3.25 Single-stage SC unit configuration.

3.2.6.8 Seven-level SC structure

The seven-level SC module depicted in Figure 3.26 has three capacitors C_1, C_2, and C_3. Among them, C_1 and C_2 act as DC link capacitors and C_3 acts as SC.

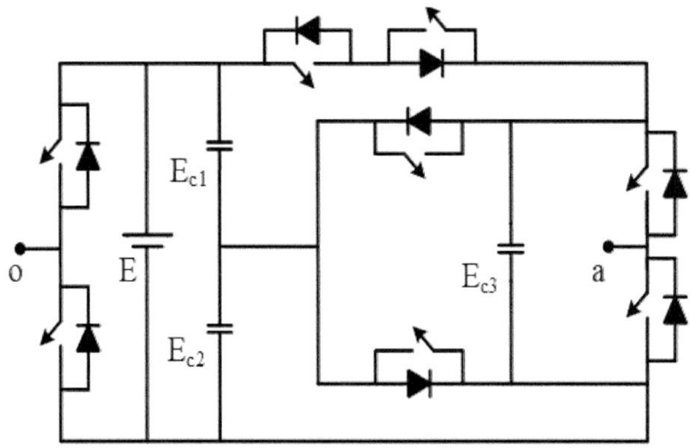

Figure 3.26 Seven-level SC module

3.2.6.9 Nine-level quadruple boost configuration

The boost inverter topology is depicted in Figure 3.27. For positive voltage level, C_1 series with C_2, and for negative voltage level, C_1 series with C_3.

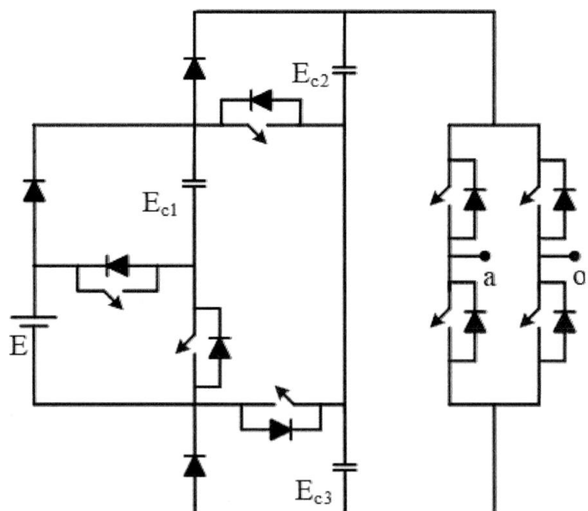

Figure 3.27 Nine-level quadruple boost inverter.

Table 3.2 The components comparison and other significant features. *m: number of SC units.

Topology	Name	H-bridge/Other	Levels in phase voltages	Number of DC sources	Number of switches	Number of capacitors	Sym/Asym/Both	Blocking voltages (Same/Different)
FBC unit with SC	Seven-level SC Tripler	Other	$6m + 1$	m	$8m$	$2m$	Both	Different
	13-level SC tripler +Doubler	Other	$12m + 1$	m	$10m$	$4m$	Both	Different
Modified T-type SC	Topology-1	HB based	$4m + 1$	m	$4m$	m	Both	Different
	Topology-2	HB based	7	1	6	4	Symmetrical	Different
	Topology-3	HB based	$6m + 1$	m	10	$2m$	Both	Different
	Topology-4	HB based	$8m + 1$	m	$8m$	$2m$	Both	Different

3.2.6.10 HSC: stacked structure

A hybrid topology was introduced by using the SC concept in stacked configuration as depicted in Figure 3.28.

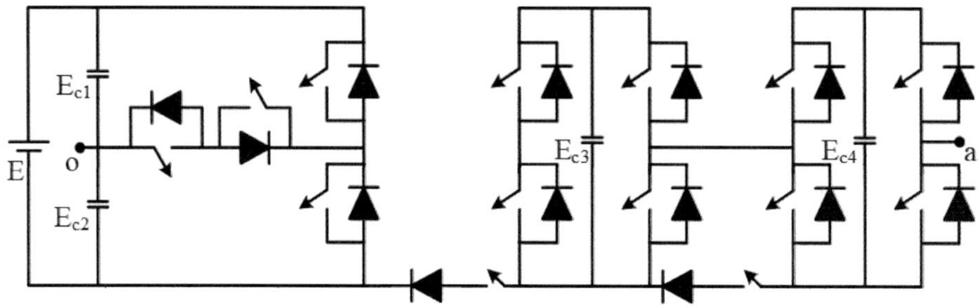

Figure 3.28 Hybrid SC with stacked structures.

The comparison between the configurations is depicted in Table 3.2.

3.2.7 Transformer-based configurations

Single-phase topologies with coupled inductors generate less number of levels. Further, to enhance the levels of voltage by reducing the switch count, incorporating a transformer at the load end is presented. Transformer-based topologies are explained in this section.

3.2.7.1 One-phase transformer-based reduced device RSCMLI

To decrease isolated DC source requirements, a configuration is proposed by concerning multiple H-bridges through the same DC source by its input side, and isolation transformers are connected at the output side as depicted in Figure 3.29. All the secondary windings are cascaded and the output is

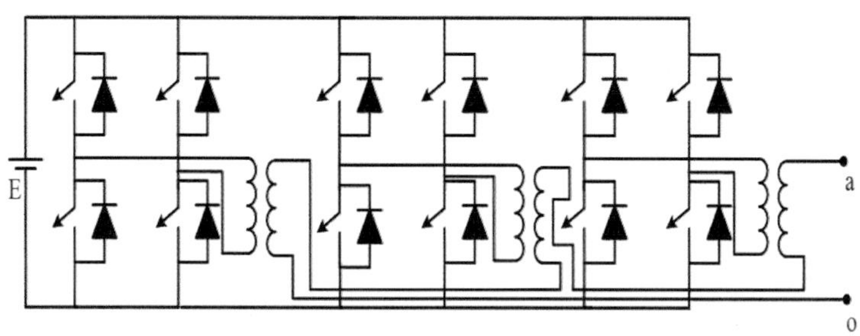

Figure 3.29 Transformer-based reduced device topology.

similar to a conventional cascaded H-bridge (CHB). Similar configurations having cascading of secondary windings are depicted in Figure 3.30.

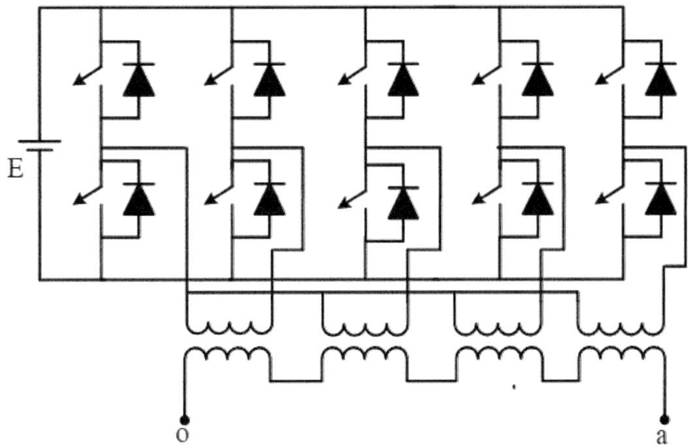

Figure 3.30 Transformer-based leg shared topology.

3.2.7.2 Two-phase five-level converter with Scott transformer

The configuration is depicted in Figure 3.31. The best feature of these topologies is the incidence of a single DC supply in common to two phases.

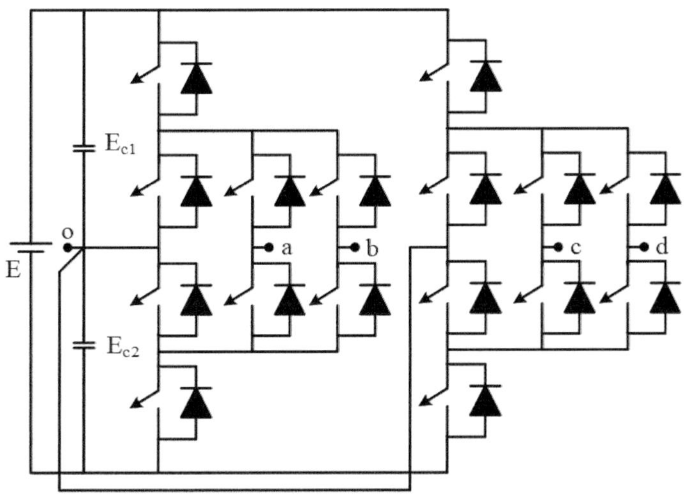

Figure 3.31 Transformer-based two-phase topology.

3.2.8 Three-phase topologies

In this section, the hybrid connection of a conventional two-level inverter is presented.

3.2.8.1 Two-level inverter-based configurations

A three-phase three-level is depicted in Figure 3.32. Among this configuration's most important features are the absence of two isolated power supplies and fast recovery diodes.

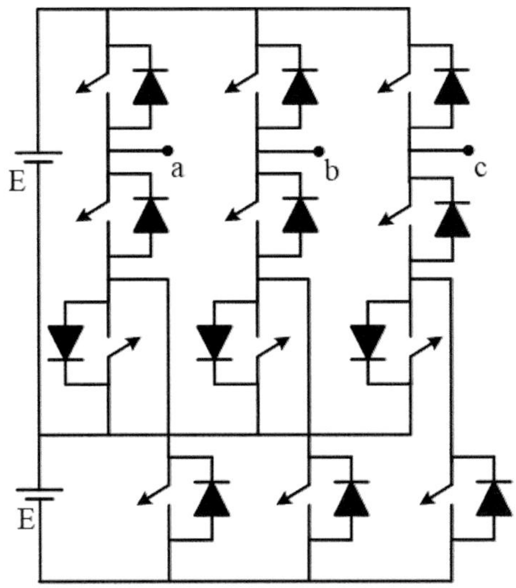

Figure 3.32 A three-phase three-level inverter.

3.3 Applications of MLIs

MLIs are employed in many arenas such as the integration of renewable energy sources into the grid, electric motor drives, and FACTS devices.

3.3.1 Renewable energy sources integrated into the grid

MLIs are used extensively in the integration of renewable sources to the grid. The CHBMLI is used to transfer the power from the two solar panels into the single-phase grid network as shown in Figure 3.33. Different types of multilevel inverters are used in grid-connected PV systems. A new MLI is developed for the grid integration of PV systems to include MPPT, and to improve the THD and the efficiency.

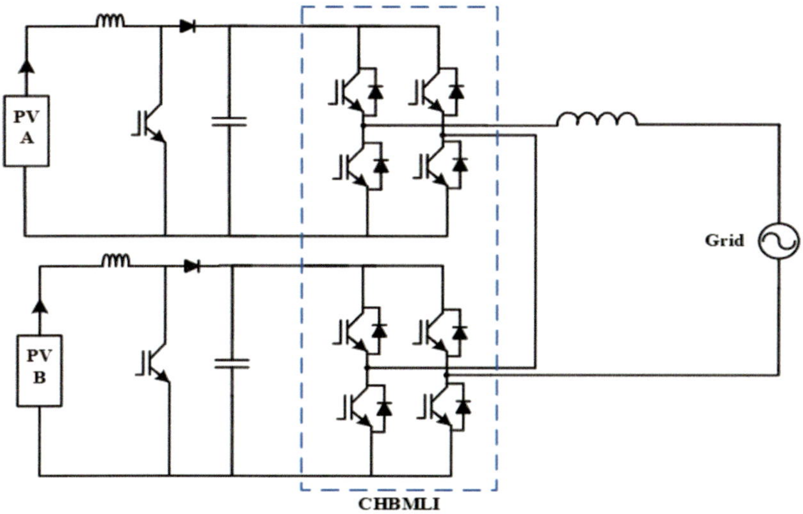

Figure 3.33 Single phase grid connected system using CHBMLI.

3.3.2 FACTS devices

Among the several types of FACTS devices available, STATCOM is regarded as a critical controller. Figure 3.34 represents the basic block diagram of STATCOM using MLI.

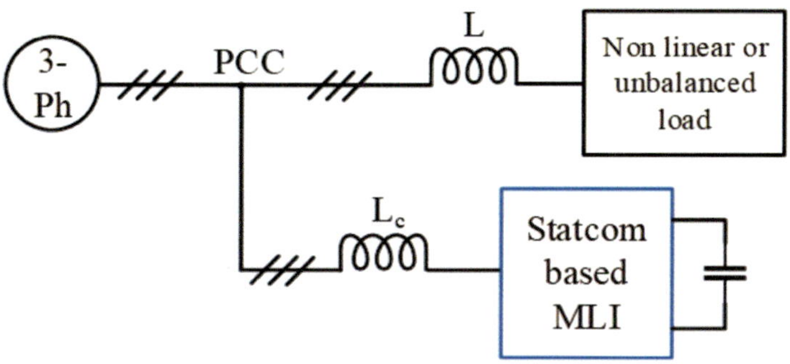

Figure 3.34 Block diagram of STATCOM using MLI.

3.3.3 Motor drives

A five-level reduced switch MLI is used in the DTFC-SVM-based control of interior type PMSM drive as shown in Figure 3.35. Analysis of the single-phase induction motor drive was done using a step-up converter integrated with a voltage multiplier circuit with a different number of levels.

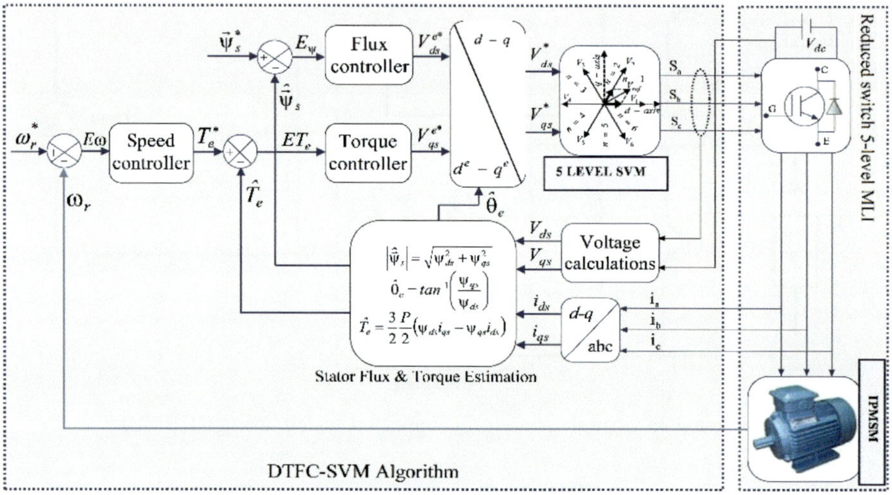

Figure 3.35 Block diagram of PMSM drive.

3.4 Conclusions

With the advancement of self-commutated switching devices, MLIs have developed into a viable choice for high-power medium-voltage applications. The operation, features, drawbacks, and applications of all traditional MLIs are covered in this chapter. MLIs are working effectively in manufacturing applications, but their size, cost, and complexity of control have prevented them from being widely adopted. Various RSC-MLI topologies have been developed in the literature to address the above-mentioned issues. In this chapter, various RSC-MLI topologies are thoroughly reviewed and categorized according to their features, structure, applicability, and selection for specific purposes. Lastly, a discussion of MLI applications in motor drives, facts devices, and grid-connected RES follows.

References

[1] K. K. Gupta, A. Ranjan, P. Bhatnagar, L. K. Sahu, and S. Jain, "Multilevel inverter topologies with reduced device count: A review," *IEEE Trans. Power Electron.*, vol. 31, no. 1, pp. 135–151, Jan. 2016

[2] J. S. Lai and F. Z. Peng, "Multilevel converters—a new breed of power converters," in Conf. Rec. 1995 *IEEE-IAS Annu. Meeting*, pp. 2348–2356.

[3] F. Z. Peng, J. S. Lai, J. W. McKeever, and J. Van Coevering, "A multilevel voltage-source converter system with balanced DC voltages," in Conf. Rec. 1995 IEEE PESC, vol. 2, pp. 1144–1150.

[4] C. Rech and J. R. Pinheiro, "Hybrid multilevel converters: Unified analysis and design considerations," *IEEE Trans. Ind. Electron.*, vol. 54, no. 2, pp. 1092–1104, Apr. 2007

[5] L. G. Franquelo, J. Rodriguez, J. I. Leon, S. Kouro, R. Portillo, and M. A. M. Prats, "The age of multilevel converters arrives," *IEEE Ind. Electron. Mag.*, vol. 2, no. 2, pp. 28–39, Jun. 2008

[6] S. Daher, J. Schmid, and F. L. M. Antunes, "Multilevel inverter topologies for stand-alone PV systems," *IEEE Trans. Ind. Electron.*, vol. 55, no. 7, pp. 2703–2712, Jul. 2008

[7] E. Najafi, A. Yatim, and A. Samosir, "A new topology-reversing voltage (RV)-for multi-level inverters," in Proc. IEEE 2nd Int., Power Energy Conf., pp. 604–608, 2008.

[8] J. Rodriguez, J. S. Lai, and F. Z. Peng, "Multilevel Inverters: A Survey of Topologies, Controls, and Applications," *IEEE Trans. Ind. Electron.*, vol. 49, no. 4, pp. 724-738, Aug. 2002.

[9] S. Lai, and F. Z. Peng, "Multilevel Converters-A New Breed of Power Converters," IEEE Trans. Ind. Applicat., vol. 32, no. 3, pp. 509-517, May/June, 1996.

[10] E. Babaei and S. H. Hosseini, "New cascaded multilevel inverter topology with minimum number of switches," *J. Energy Convers. Manag.*, vol. 50, no. 11, pp. 2761–2767, Nov. 2009.

[11] N. Arun and M. M. Noel, "Crisscross switched multilevel inverter using cascaded semi-half-bridge cells," IET Power Electron., vol. 11, no. 1, pp. 23–32, Dec. 2018.

[12] J. Rodriguez, J.-S. Lai, and F. Z. Peng, "Multi-level inverter: a survey of topologies, controls, and applications," *IEEE Trans. Ind. Electron.*, vol. 49, no. 4, pp. 724–738, Aug. 2002.

[13] N. Vosoughi, S. H. Hosseini, M. Sabahi, "A new transformer-less five-level grid-tied inverter for photovoltaic applications", *IEEE Trans. On Energy Convers.*, vol. 35, no. 1, pp. 106-118, Mar. 2020.

[14] M. Calais, L. J. Borle, and V. G. Agelidis, "Analysis of multicarrier PWM methods for a single-phase five level inverter," in Proc. *IEEE PESC'01*, Vancouver, BC, Canada, 2001, pp. 1351–1356.

[15] S. Daher, J. Schmid, and F. L. M. Antunes, "Multilevel inverter topologies for stand-alone PV systems," *IEEE Trans. Ind. Electron.*, vol. 55, no. 7, pp. 2703–2712, Jul. 2008.

[16] J. M. Carrasco, L. G. Franquelo, J. T. Bialasiewicz, E. Galvan, R. C. PortilloGuisado, M. A. M. Prats, J. I. Leon, and N. Moreno-Alfonso, "Power-electronic systems for the grid integration of renewable energy sources: A survey," *IEEE Trans. Ind. Electron.*, vol. 53, no. 4, pp. 1002–1016, Aug. 2006.

[17] B. Ge, F. Z. Peng, A. T. de Almeida, and H. Abu-Rub, "An effective control technique for medium-voltage high-power induction motor fed by cascaded neutral-point-clamped inverter," *IEEE Trans. Ind. Electron.*, vol. 57, no. 8, pp. 2659–2668, Aug. 2010.

[18] S. P. Gautam, L. Kumar, and S. Gupta, "Hybrid topology of symmetrical multilevel inverter using less number of devices," IET Power Electron., vol. 8, no. 11, pp. 2125−2135, Nov. 2015

[19] S. P. Gautam, L. K. Sahu, and S. Gupta, "Reduction in number of devices for symmetrical and asymmetrical multilevel inverters," IET Power Electron., vol. 9, no. 4, pp. 698−709, Mar. 2016

[20] K. Ding, K. W. E. Cheng, and Y. P. Zou, "Analysis of an asymmetric modulation method for cascaded multilevel inverters," *IET Power Electron.*, vol. 5, no. 1, pp. 74–85, Jan. 2012.

4

Modular Multilevel Converter

Jyoti Ranjan Dash[1], Rudranarayan Pradhan[2], and Pramod Agarwal[3]

[1,3]Electrical Engineering Department, Indian Institute of Technology Roorkee, India
[2]School of Electrical Sciences, Odisha University of Technology and Research Bhubaneswar, India
E-mail: jranjandash@ee.iitr.ac.in; rudranarayan@outr.ac.in; pramod.agarwal@ee.iitr.ac.in

Abstract

Modular multilevel converters (MMCs) have solidified their position as an advanced and futuristic technology in the field of power electronics converter, presenting distinctive advantages across diverse applications, spanning from high-voltage and high-power to medium-voltage and medium-power systems. The broad acceptance of MMCs is rooted in their fundamental modularity, scalability, high efficiency, enhanced harmonic performance, and better output characteristics. The modular topology allows attaining any desired power level by incorporating additional submodules in series. Another key advantage of the MMC is its ability to operate without a phase-shifted transformer. This significantly reduces the converter's overall cost and complexity. These combined benefits make the MMC a highly reliable and versatile technology for various claims, incorporating high-voltage direct current (HVDC) transmission, static compensators (STATCOMs), solar power integration, and high-power motor drives. However, the unique characteristics of MMCs, including the substantial number of submodules, floating capacitor voltages, and circulating currents, necessitate a sophisticated control system. This control system must manage internal and external factors effectively to ensure high dynamic performance. The unique design of MMCs demands superior precision in control to channel their full potential, providing seamless integration and operation across different power applications. This chapter

DOI: 10.1201/9781003669845-4

comprehensively addresses the fundamental operations, submodule topology, pulse width modulation, voltage balancing of submodule capacitors, control of circulating current, and output power control of MMCs. It serves as a valuable resource for understanding the complex dynamics and control strategies essential for controlling the full potential of modular multilevel converters in diverse applications.

Keywords: Modular multilevel converter, voltage balancing of submodule capacitors, circulating current, submodule, pulse width modulation, high voltage direct current

4.1 Introduction

Marquardt conceived the modular multilevel converter (MMC) in 2003 [1], a landmark development in multilevel converter technology later brought to commercial fruition by Siemens in San Francisco circa 2010 for high-voltage direct current (HVDC) transmission systems. Its design is based on a modular architecture and utilizes low-voltage insulated gate bipolar transistor (IGBT) switches for high-voltage operations, allowing direct connection to medium- and high-voltage grids without the need for a step-up transformer [2]. These distinctive features bestow upon the MMC superior attributes compared to traditional two-level and multilevel converters, including enhanced flexibility, heightened reliability, and improved cost efficiency [3].

Across various moderate and high-voltage power adaptation systems, the MMC finds application. These systems encompass HVDC transmission systems [4], motor drives operating at medium voltage [5–7], green energy systems [8, 9], systems for charging batteries [10], static synchronous compensators [11], and electric vehicle charging and drives [12, 13].

This chapter covers various aspects of the MMC, including its overall structure, workings, mathematical models, control mechanisms, and diverse applications [3, 14]. Section 4.2 delves into the circuit configurations of MMC, exploring different submodule structures and their unique features. The mathematical modeling of MMC is detailed in Section 4.3, while Section 4.4 focuses on the control strategies applied to MMC. The chapter concludes with a summary and explores potential developments in Section 4.5.

4.2 Architecture and Performance

The modular multilevel converter (MMC) is a pivotal expertise for modern applications necessitating high voltage and power-handling capabilities.

This section delves into the technical specifics of the MMC, encompassing its architecture, operational mechanisms, submodule topologies, and their unique electrical characteristics.

An MMC with three phases is shown in Figure 4.1, consisting of three phases represented as $(a, b, \text{and } c)$ and a split DC source of voltage $V_{\mathrm{d}}/2$. Each phase, also called leg, has two arms: one is the upper arm, and the other is the lower arm. These arms contain several submodules (SMs) units and an arm inductor, represented by L_{a} with an internal resistance value of r_{a}. The MMC is built in a modular way, allowing for easy adjustment of voltage and power levels. It demonstrates a capability to generate output voltage and current waveforms characterized by diminished temporal variations and minimal ripples. Typically, its output voltage exhibits exceptionally low total harmonic distortion (THD). The MMC facilitates operation at a reduced switching frequency by leveraging a multitude of levels in the output voltage. Moreover,

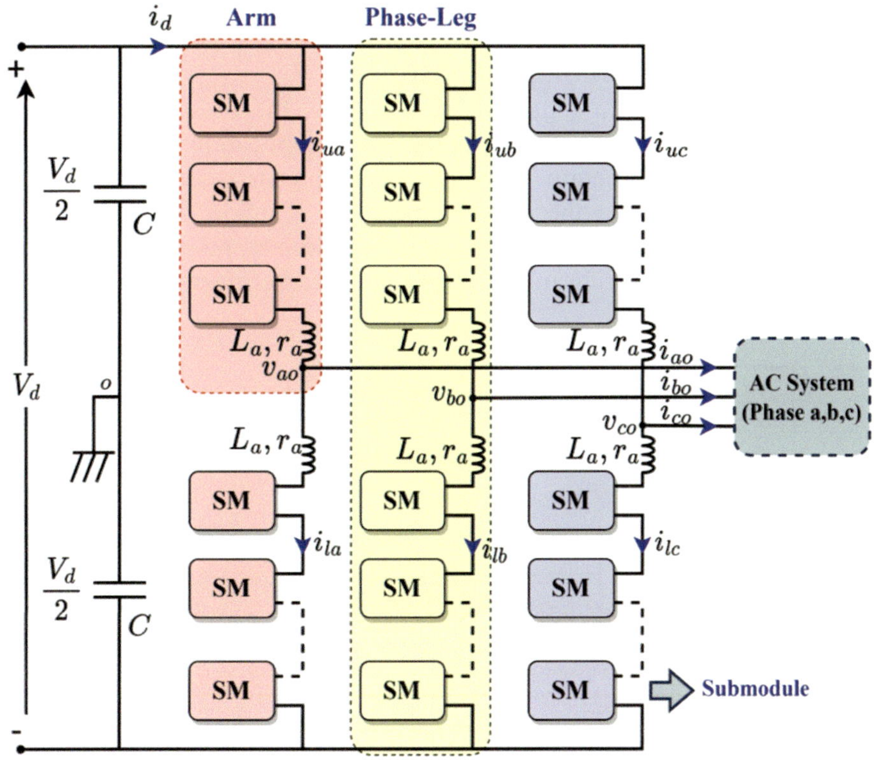

Figure 4.1 The circuit configuration of MMC.

it integrates redundant submodules within each arm, thereby guaranteeing uninterrupted operation even amidst fault conditions.

4.2.1 Submodule configuration

A submodule (SM) functions as the primary DC to AC conversion circuit, typically constructed using low voltage insulated gate bipolar transistor (IGBT) devices in the 1200–1700 V span and DC capacitors. A submodule configuration involves integrating these components, playing a necessary part in the complete operation of the MMC. Various submodule configurations are commonly employed within the MMC framework to ensure efficient DC to AC-conversion; these are:

- Half-bridge submodule
- Full-bridge submodule
- Flying capacitor submodule
- Cascaded half-bridge submodule

4.2.1.1 Half-bridge submodule (HBSM)

The HBSM featured in Figure 4.2 emerges as the most favored choice, enjoying widespread adoption in commercial applications. The simplicity inherent in its construction facilitates uncomplicated control and design procedures. Throughout the regular operation, a single switch within the HBSM conducts current at a time. This minimizes power losses due to heat generation, indicating improvement in overall efficiency for the MMC.

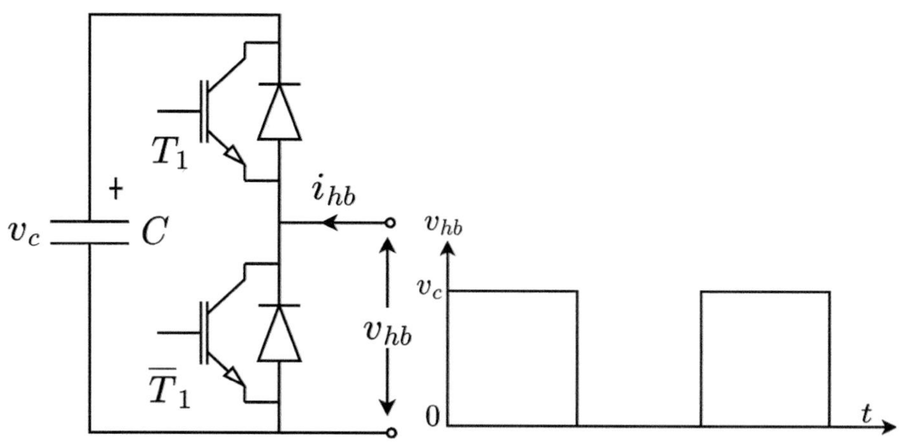

Figure 4.2 HBSM is characterized by its voltage waveform.

However, a fundamental limitation of the HBSM design is its inability to generate negative voltage outputs. This restricts its use in applications requiring a bipolar operation, such as some grid-connected converters, and limits its capacity to block DC faults directly.

4.2.1.2 Full-bridge submodule (FBSM)

The FBSM depicted in Figure 4.3 comprises four semiconductor devices and one DC capacitor. Compared to the HBSM with the same voltage rating, the FBSM incorporates twice the number of semiconductors. Despite this difference, its control and design complexities remain the same as those of the HBSM. During regular operation, two devices within the FBSM are tasked with carrying the current, which consequently leads to heightened power losses and reduced efficiency. Negative voltage levels within the FBSM restrict current flow during DC-side faults.

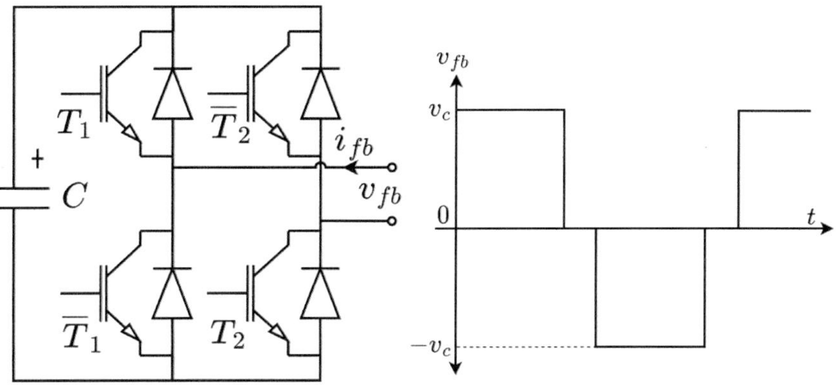

Figure 4.3 FBSM is characterized by its voltage waveform.

4.2.1.3 Flying capacitor submodule (FLCSM)

The FCSM shares similarities with the FBSM, featuring four semiconductor devices as shown in Figure 4.4. However, it necessitates the inclusion of two DC capacitors. For the three-level function of the FCSM, it is imperative to preserve the outer capacitor voltage v_{c2} at twice the voltage level of the inner capacitor voltage v_{c1}. During regular operation of the flying capacitor submodule, a maximum of two devices are responsible for carrying the current. Efficiency and power losses within the FCSM are comparable to those observed in the FBSM. Nevertheless, it is noteworthy that the FCSM exclusively generates positive voltage levels and lacks the capability to restrict current flow during DC-side faults.

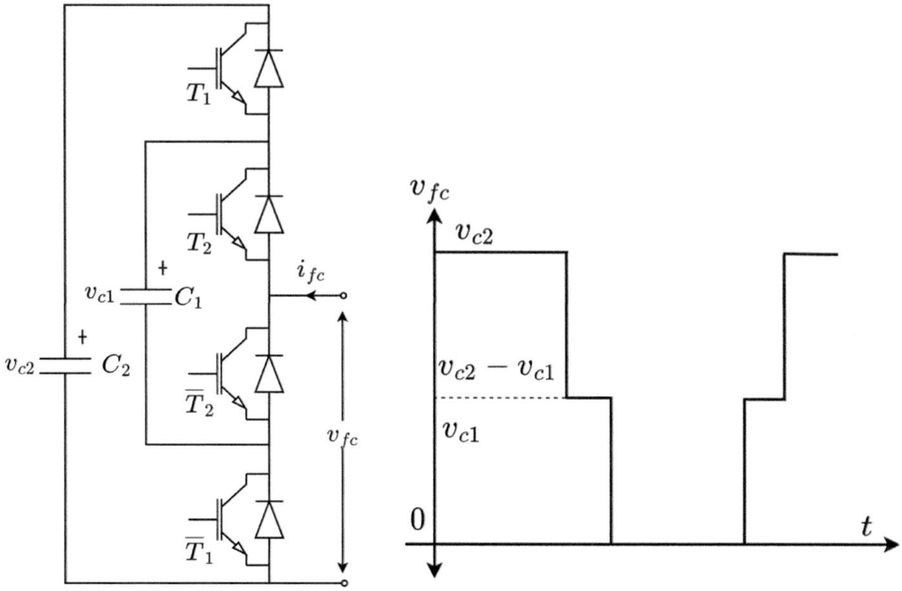

Figure 4.4 FCSM is characterized by its voltage waveform.

4.2.1.4 Cascaded half-bridge submodule (CHBSM)

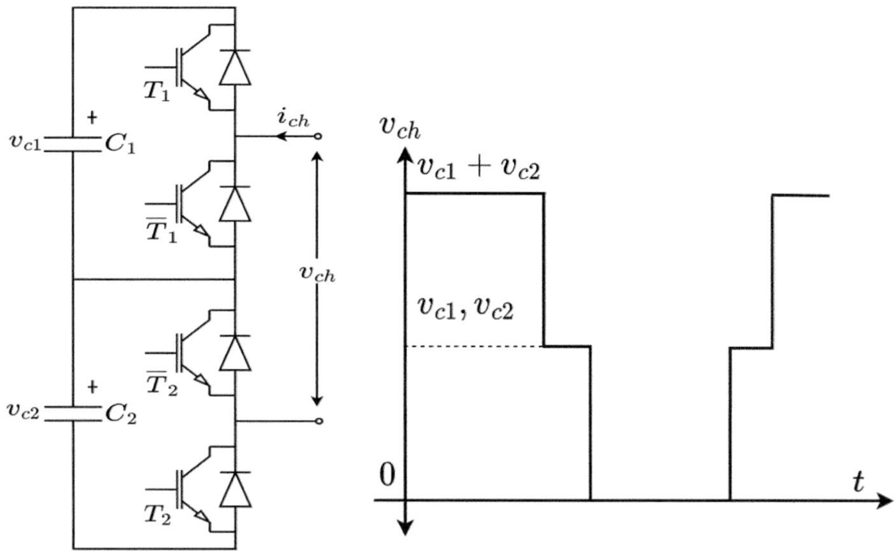

Figure 4.5 CHBSM is characterized by its voltage waveform.

The CHBSM is crafted by interconnecting two HBSM in sequences, thereby enabling three-level operation, as shown in Figure 4.5. Renowned for its straightforward control mechanism, low design complexity, and minimal power losses, the CHBSM boasts high efficiency, mirroring the characteristics of the HBSM. However, like its counterpart, the CHBSM lacks the capability to block DC-side faults and is not conducive to bipolar operation.

4.2.2 Principles of MMC operation

The operation of an MMC employing the widely used HBSM is depicted. This configuration utilizes multiple submodules (SMs) arranged in a cascaded fashion to produce a staircase waveform output voltage featuring multiple voltage levels [3, 15].

The three-phase configuration of the MMC is schematically illustrated in Figure 4.6.a. The DC side is represented by a bifurcated DC supply with a voltage of $V_\mathrm{d}/2$. The three-phase AC structure is interfaced with a passive load, which is modeled as a series combination of inductance (L_out) and resistance (R_out).

For a comprehensive grasp of the operational principles of the MMC, Figure 4.6 provides a detailed focus on the upper arm of phase-a. This illustration considers four HBSMs, designated as $(\mathrm{SM}_1 - \mathrm{SM}_4)$. It is supposed that all submodule capacitor voltages are uniform and denoted as v_c, while the output voltages of the submodules $(\mathrm{SM}_1 - \mathrm{SM}_4)$ are represented as v_h1, v_h2, v_h3, and v_h4 correspondingly.

The upper arm voltage waveform for phase-a across one fundamental period is illustrated in Figure 4.6. This waveform is the addition of the individual output voltages generated by all submodules connected in series within the upper arm of phase-a. This visual representation effectively demonstrates how the combined effect of each submodule voltage contributes to the overall output voltage waveform generated by the MMC, as shown in Figure 4.6.b.

Additionally, Table 4.1 illustrates the switching functions of the four submodules alongside their corresponding output voltages. The switching functions for $\mathrm{SM}_1 - \mathrm{SM}_4$ $(\mathrm{SM}_1{-}\mathrm{SM}_4)$ are denoted as g_ua1, g_ua2, g_ua3 and g_ua4, respectively. The arm voltage v_ua is delineated with five distinct levels, including 0, v_c, $2v_\mathrm{c}$, $3v_\mathrm{c}$, and $4v_\mathrm{c}$ expressed accordingly. This tabulated data offers a comprehensive overview of the submodule switching actions that contribute to the varied levels of the arm voltage v_ua. The equation governing

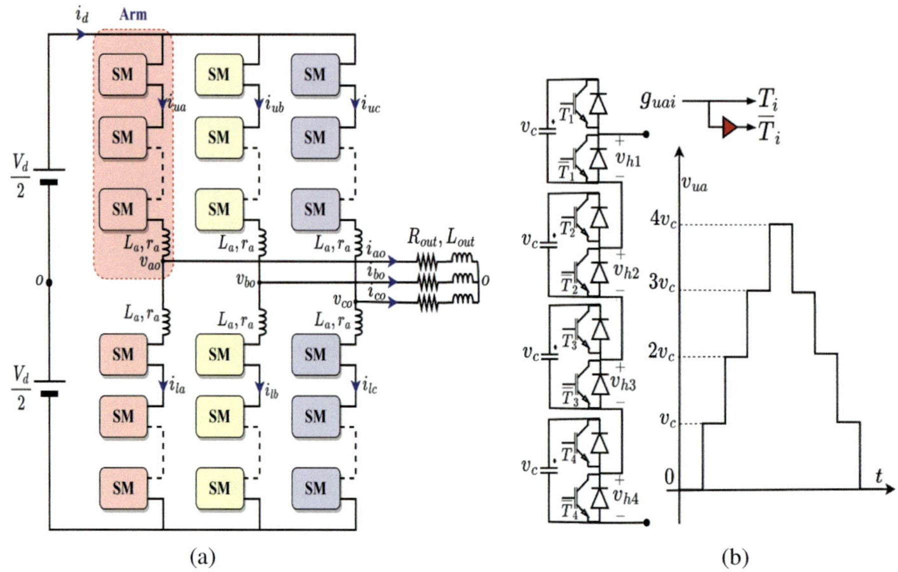

Figure 4.6 Three-phase MMC and its arm design. (a) Three-phase MMC with RL-load. (b) HBSM-based upper arm structure of phase-a and its output voltage.

the arm voltage, depicting the operation of an arm, is as follows:

$$v_{ua} = \sum_{i=1}^{n} v_{hi} = \sum_{i=1}^{n} g_{\text{uai}}\, v_{uci}. \tag{4.1}$$

In eqn (4.1), n = number of submodules = "4" and value of $v_{\text{uci}} = v_{\text{c}}$, so the expression of upper arm voltage is:

$$\begin{aligned} v_{ua} &= v_{h1} + v_{h2} + v_{h3} + v_{h4} \\ &= g_{ua1}v_c + g_{ua2}v_c + g_{ua3}v_c + g_{ua4}v_c. \end{aligned} \tag{4.2}$$

The arm voltage of a MMC is determined by the cumulative summation of the output voltages of its submodules. The output voltage of each SM is calculated as the product of its capacitor voltage and switching status. This principle is inherently scalable, accommodating any quantity of submodules per arm. Within an MMC, the strategic modulation of submodule switching in both the upper and lower arms synthesizes a multilevel voltage waveform on the AC side. By precisely coordinating the switching states of the submodules, the MMC engenders the requisite voltage levels for AC system operation. This control methodology enables the MMC to

Table 4.1 The switching state of MMC and its arm voltage level.

g_{ua1}	g_{ua2}	g_{ua3}	g_{ua4}	v_{h1}	v_{h2}	v_{h3}	v_{h4}	v_{ua}	Arm voltage level
on	on	on	on	v_{c}	v_{c}	v_{c}	v_{c}	$4v_{\mathrm{c}}$	4
on	on	off	on	v_{c}	v_{c}	0	v_{c}	$3v_{\mathrm{c}}$	
on	off	on	on	v_{c}	0	v_{c}	v_{c}	$3v_{\mathrm{c}}$	3
on	on	on	off	v_{c}	v_{c}	v_{c}	0	$3v_{\mathrm{c}}$	
off	on	on	on	0	v_{c}	v_{c}	v_{c}	$3v_{\mathrm{c}}$	
on	on	off	off	v_{c}	v_{c}	0	0	$2v_{\mathrm{c}}$	
on	off	on	off	v_{c}	0	v_{c}	0	$2v_{\mathrm{c}}$	
on	off	off	on	v_{c}	0	0	v_{c}	$2v_{\mathrm{c}}$	2
off	on	off	on	0	v_{c}	0	v_{c}	$2v_{\mathrm{c}}$	
off	off	on	on	0	0	v_{c}	v_{c}	$2v_{\mathrm{c}}$	
off	on	on	off	0	v_{c}	v_{c}	0	$2v_{\mathrm{c}}$	
on	off	off	off	v_{c}	0	0	0	v_{c}	
off	on	off	off	0	v_{c}	0	0	v_{c}	1
off	off	on	off	0	0	v_{c}	0	v_{c}	
off	off	off	on	0	0	0	v_{c}	v_{c}	
off	off	off	off	0	0	0	0	0	0

achieve high-voltage, high-power transformation while enhancing efficiency and mitigating harmonic distortion.

4.3 Mathematical Framework for MMC

Comprehending the mathematical dynamics of an MMC is imperative for exhaustive analysis, controller synthesis, and optimization. This section elucidates a continuous-time model for a three-phase MMC, as delineated in Figure 4.7.

The three-phase model of the MMC incorporates the split DC bus voltage and current, denoted as $V_{\mathrm{d}}/2$ and i_{d}, respectively. The converter's output phase currents are represented by i_{ao}, i_{bo}, and i_{co}, while the corresponding output voltages are denoted as v_{ao}, v_{bo}, and v_{co}. Furthermore, the voltages within the upper and lower arms are expressed as (v_{ua}, v_{ub}, and v_{uc}) and (v_{la}, v_{lb}, and v_{lc}) correspondingly. The upper and lower arm currents of the three-phase MMC are represented as (i_{ua}, i_{ub}, and i_{uc}) and (i_{la}, i_{lb}, and i_{lc}), respectively.

This analysis uses the symbols i_{ca}, i_{cb} and i_{cc} to represent the circulating current flow surrounded by every individual phase. The upper and lower arm currents in the MMC are determined based on the relationships established in the analogous model presented in Figure 4.7.

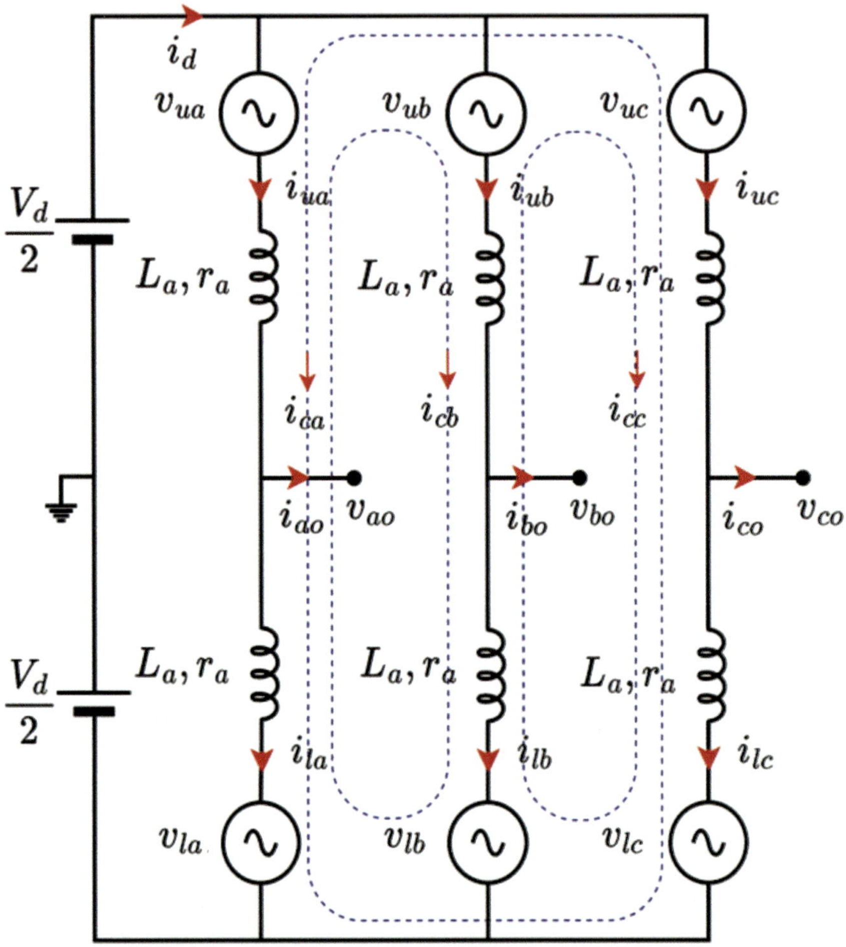

Figure 4.7 Comparable model of three-phase MMC.

$$i_{uj} = \frac{1}{3}i_d + i_{cj} + \frac{1}{2}i_{jo}, \tag{4.3}$$

$$i_{lj} = \frac{1}{3}i_d + i_{cj} - \frac{1}{2}i_{jo}. \tag{4.4}$$

In the above equations, j denotes the (a, b, c) phase of the three-phase MMC. By employing Kirchhoff's voltage law to analyze the corresponding model of the MMC, the voltages of the upper and lower arms are articulated as follows:

$$v_{uj} = \frac{V_d}{2} - L_a \frac{di_{uj}}{dt} - r_a i_{uj} - v_{jo}, \tag{4.5}$$

$$v_{lj} = \frac{V_d}{2} - L_a \frac{di_{lj}}{dt} - r_a i_{lj} + v_{jo}. \tag{4.6}$$

By combining and deducting eqn (4.5) and (4.6), the input DC and output AC side voltages may be derived as:

$$V_d = v_{uj} + v_{lj} + L_a \left(\frac{di_{uj}}{dt} + \frac{di_{lj}}{dt} \right) + r_a \left(i_{uj} + i_{lj} \right), \tag{4.7}$$

$$v_{jo} = \frac{v_{uj} - v_{lj}}{2} - \frac{L_a}{2} \left(\frac{di_{uj}}{dt} - \frac{di_{lj}}{dt} \right) - \frac{r_a}{2} \left(i_{uj} - i_{lj} \right). \tag{4.8}$$

The utilization of the above equations enables the demonstration of a three-phase MMC. The subsequent section examines circulating current and output current mathematical models employing fundamental equations.

4.3.1 Circulating currents mathematical expressions

The voltage difference concerning the upper and lower arms causes current to flow through the converter legs. Circulating current, which is mainly made up of even-order components of harmonic that correspond to multiples of $(2,\ 4,\ 8\ldots)$ with the fundamental frequency, is a defining feature of converter functioning. The expression for circulating currents is computed by summing the upper and lower arm currents. This mathematical method captures the phenomena, allowing for a thorough understanding of how circulating current emerges in the MMC system.

$$i_{cj} = \frac{i_{uj} + i_{lj}}{2} - \frac{i_d}{3}, \tag{4.9}$$

$$i_{dj} = i_{cj} + \frac{i_d}{3} = \frac{i_{uj} + i_{lj}}{2}. \tag{4.10}$$

The parameter i_{dj} denotes the $j = (a, b, c)$ phase inner difference current, representing the combined value of both circulating and direct current. The regulation of the unbalanced voltage is denoted as v_{dj}, which enables control over the inner difference current i_{dj}. Substituting eqn (4.10) in (4.7), the expression of three-phase unbalance voltage v_{dj} will be:

$$v_{dj} = L_a \frac{di_{dj}}{dt} + r_a i_{dj} = \frac{V_d}{2} - \frac{v_{uj} + v_{lj}}{2}. \tag{4.11}$$

4.3.2 Output currents mathematical expressions

The current flow in output delivered to the load travels through the converter leg. The distinction involving the upper and lower arm currents determines the current that flows into the load. From the eqn (4.3) and (4.4), the expression of output current is:

$$i_{jo} = i_{uj} - i_{lj}. \tag{4.12}$$

Substituting the eqn (4.12) in (4.8), the output voltage expression will be:

$$\begin{aligned} v_{jo} &= \frac{v_{lj} - v_{uj}}{2} - \frac{L_a}{2}\frac{di_{jo}}{dt} - \frac{r_a}{2} i_{jo}, \\ &= e_j - \frac{L_a}{2}\frac{di_{jo}}{dt} - \frac{r_a}{2} i_{jo}. \end{aligned} \tag{4.13}$$

In eqn (4.13), where e_j symbolizes the internal emf produced in phase-j. Through the parameter regulation, e_j, we can control the output voltage and current on the AC side. Considering the eqn (4.11) and (4.13), the reference voltages are

$$v_{uj} = \frac{V_d}{2} - e_j - v_{dj}, \tag{4.14}$$

$$v_{lj} = \frac{V_d}{2} + e_j - v_{dj}. \tag{4.15}$$

Using the above equation allows us to control external voltage and current. Simultaneously, it facilitates the regulation of imbalances in the inner circulating current. In the next section, we provide detailed information about MMC control.

4.4 MMC Control Strategies and Techniques

The exemplary performance of the MMC is attributed to its sophisticated control strategies, which enhance safety, reliability, and efficiency. These control purposes are classified into principal and subordinate categories. The essential objectives encompass pulse width modulation (PWM) technology, SM capacitor voltage regulation, and output current control, all of which are vital for MMC operation. In contrast, circulating current control is considered a resulting objective, influencing the MMC system's dimensions, reliability, and efficiency. This section delves into the comprehensive control schema of the MMC. Figure 4.8 depicts the drawing of the complete controller architecture of the MMC.

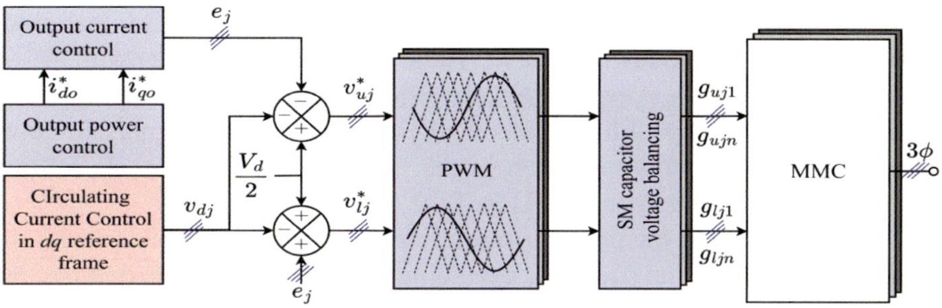

Figure 4.8 Block diagram delineating the control methodology employed in the MMC.

4.4.1 Carrier-based PWM schemes for MMC

The PWM principle controls the AC output voltage in power converters. Engineers tailor PWM schemes to decrease harmonic distortion in the output voltage while increasing its magnitude at a predetermined switching frequency. Furthermore, engineers design these modulation schemes to accomplish additional control objectives, including voltage balancing, voltage reduction due to common mode, minimization of device switching frequency, power loss mitigation, and output current ripple reduction. Among various modulation techniques employed in MMCs, phase-shifted carrier pulse width modulation (PS-PWM) stands out as a popular choice [16–18].

The PS-PWM is particularly appropriate for MMC configurations alongside smaller submodule units per arm. In instances where there are n SMs in each arm, this modulation method is executed by employing n identical triangular carriers, $c_{\text{u1}} - c_{\text{u4}}$ evenly arranged in the horizontal direction for the MMC upper arm, and n identical triangular carriers $c_{\text{l1}} - c_{\text{l4}}$ for the lower arm. The PS-PWM is accomplished by comparing these triangular carriers and the required signal.

PS-PWM for MMCs utilizes a set of n triangular carrier signals, where n signifies the number of SM units per arm. These triangular carrier signals are strategically placed with a specific phase shift denoted by the symbol α. The subsequent methods can be employed to determine the applicable phase shift between these carrier signals.

$$\alpha = 2\pi/n. \tag{4.16}$$

The triangular carrier signals for both upper and lower arms can be offset with an interleaving angle α_{in}, which is specified by:

$$\alpha_{in} = \pi/n. \tag{4.17}$$

The MMC employs an interleaving angle-based PS-PWM technique between the upper and lower arms to synthesize an output voltage with $2n+1$ distinct voltage levels. Alternatively, the MMC can achieve an output voltage with $n + 1$ levels by utilizing identical triangular carrier signals for both arms. Nevertheless, this reduction in the number of voltage levels follows a concomitant improvement in THD of the output voltage.

Figure 4.9 shows the upper arm reference signal v_{uar} associated with the upper arm carrier signal $c_{\mathrm{u}1} - c_{\mathrm{u}4}$ generating the required PWM for the upper arm of MMC. Similarly, Figure 4.10 presents the lower arm reference v_{lar} compared with the $c_{\mathrm{l}1} - c_{\mathrm{l}4}$ generated PWM for the lower arm of MMC. These pulses, along with the switching frequency, are evenly distributed among $g_{\mathrm{ua}1} - g_{\mathrm{ua}4}$ and $g_{\mathrm{la}1} - g_{\mathrm{la}4}$. The variable n_{ua} and n_{la} varies within the range of zero to four during one fundamental period, leading to the synthesis of a multilevel waveform in both arms. The interleaved angle between $g_{\mathrm{ua}1}$ and $g_{\mathrm{la}1}$ results in the generation of an $2n+1$ output voltage. Figure 4.11 indicates the phase voltage and its THD using interleaved angle PS-PWM.

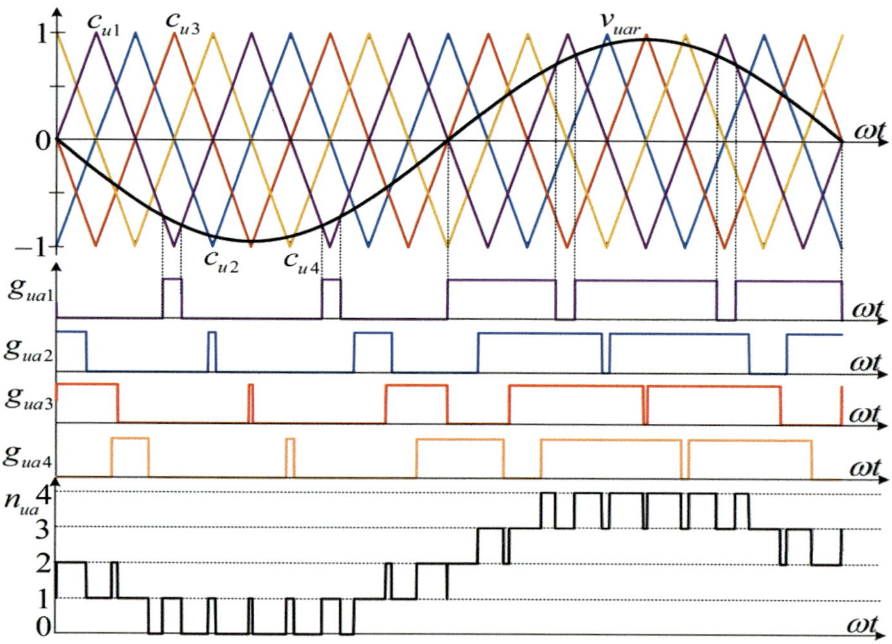

Figure 4.9 The output of PS-PWM for the upper part of the a-phase in a 3Φ MMC.

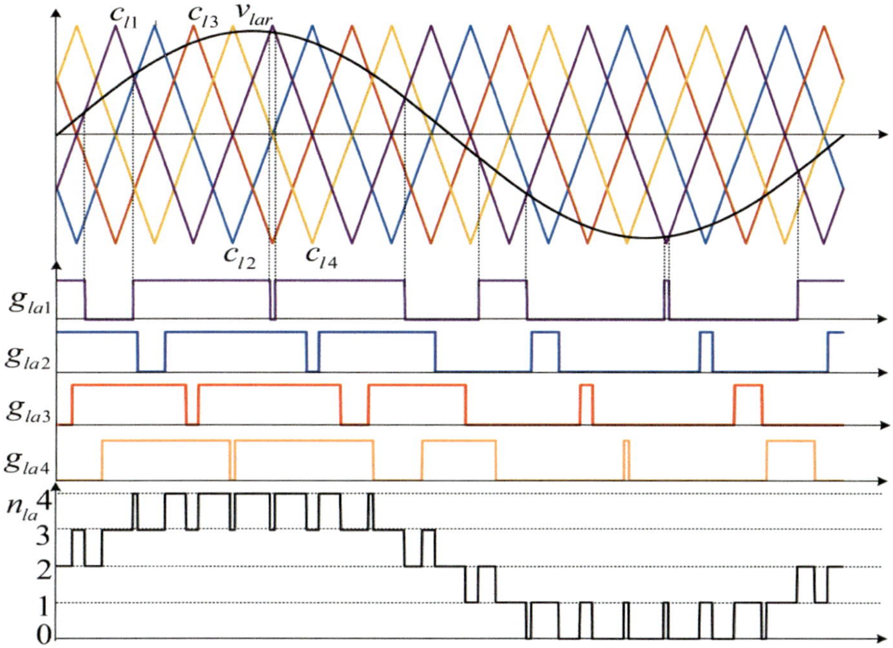

Figure 4.10 The output of PS-PWM for the lower part of the a-phase in a 3Φ MMC.

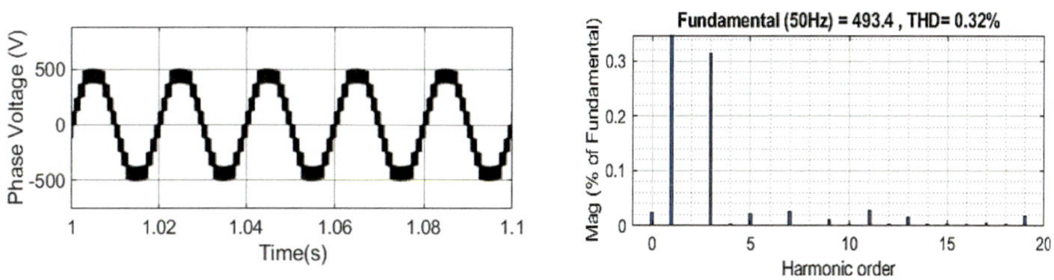

(a) (b)

Figure 4.11 (a) Phase voltage of MMC and (b) % THD of voltage ($f_{cr} = 2000$ Hz).

4.4.2 Capacitor voltage balancing

Maintaining a balanced voltage across submodule capacitors is critical for the reliable functioning of MMC [19–22]. A core principle for maintaining balanced capacitor voltages involves directly controlling their charging and discharging. This control strategy relies on two key factors: the direction

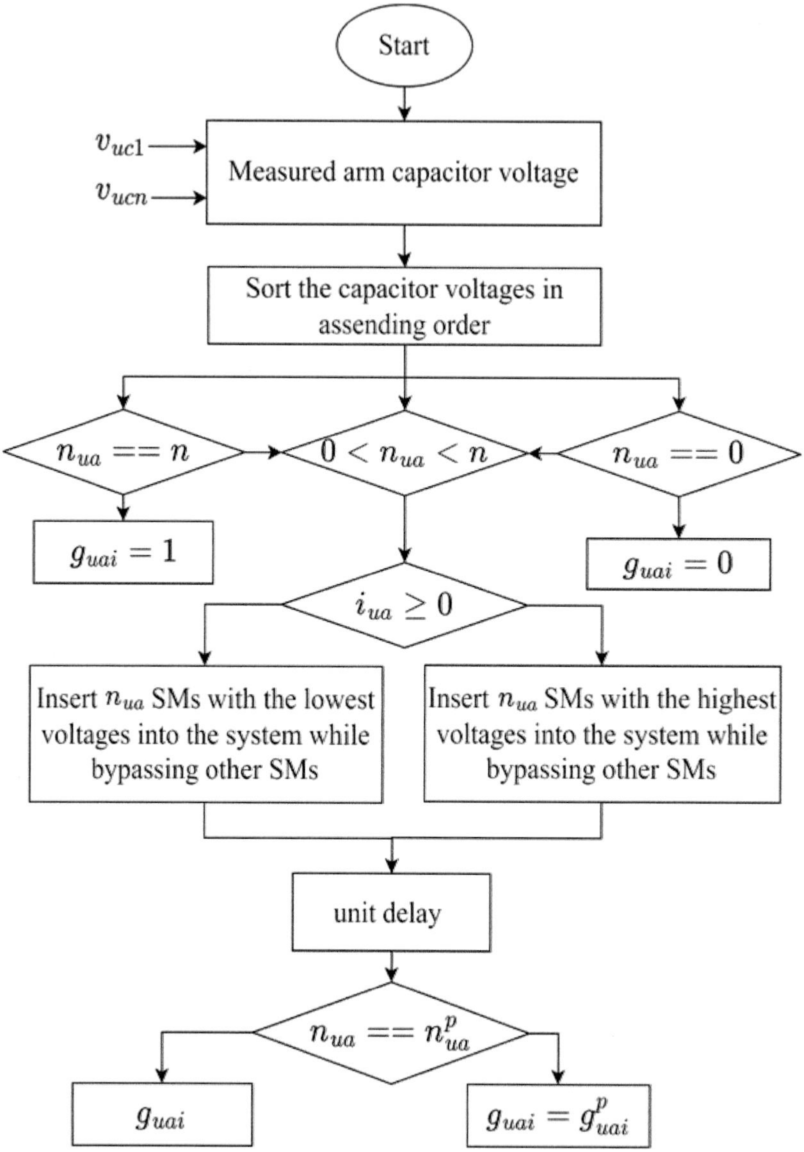

Figure 4.12 Capacitor voltage balancing algorithm for MMC.

of current streaming through the arm and the real-time voltage of every capacitor. When current flows in the positive direction, we insert capacitors with minimal voltage into the arm, promoting voltage balance. Conversely, when the current direction is reversed, capacitors with higher voltage are introduced to achieve the required voltage balancing. This method ensures an even distribution of voltage among submodule capacitors, enhancing the consistency and overall performance of the MMC system.

Achieving voltage balance is attainable during the modulation stage by employing logical functions guided by a specific algorithm. The algorithm-based method involves collective control of all submodules within each arm. Consequently, it is adaptable for implementation with various PWM schemes. The flow chart in Figure 4.12 outlines a straightforward algorithm for capacitor voltage control. The method can also be divided into the following steps, each explained in detail.

- Evaluate the capacitor voltages v_{uc1}, v_{uc2}, v_{uc2}, and v_{uc3} for individual arms, arrange them in ascending order.
- In an arm, insert all SMs when the number of SMs placed equals n, and bypass all SMs if it equals zero.
- If n other than 0 and n measures each instant's upper arm current i_{ua}. When the arm current is positive (or negative), the submodule with the minimum (or maximum) capacitor voltage is engaged; otherwise, it is disregarded.
- We evaluate the number of upper-arm submodules during the ongoing control cycle n_{ua} against that of the preceding control cycle $n_{\mathrm{ua}}^{\mathrm{p}}$. Should the ongoing control cycle be identical to the previous one, the latest gating signal g_{uai} is deployed. If not, the previous gating signal is preserved.

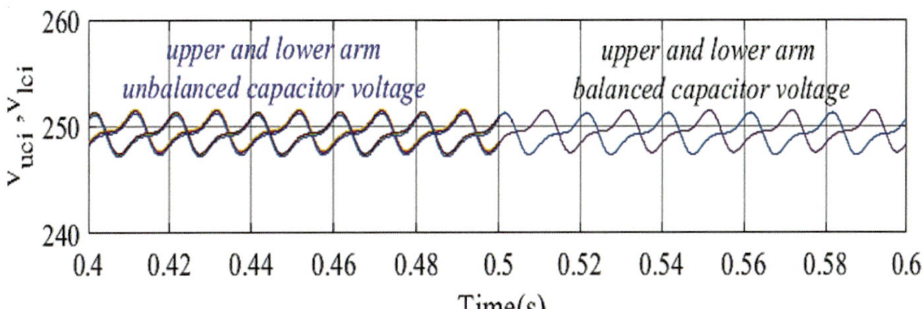

Figure 4.13 Capacitor voltage balancing using the above procedure (balancing applied at 0.5 seconds)

The principal merit of this algorithm resides in its efficacy to attenuate switching frequency, thereby mitigating switching losses. For a more granular elucidation of the algorithm's implementation in a three-phase MMC with four submodules per arm, consider an input voltage of 1 kV DC. Figure 4.13 delineates the asymmetric capacitor voltages across the upper and lower arms.

4.4.3 Strategies for circulating current regulation

The voltage disparity among the upper and lower parts of the MMC engenders circulating currents that flow internally within the converter. These currents are isolated from the AC side, exerting no impact on the AC side's current or voltage. However, they contain components at a frequency that even multiplies the AC supply frequency, known as negative sequence components. Notably, these second-harmonic currents increase the arm current's root mean square (RMS) value, resulting in significant power losses within the converter. To interpret this, the internal circulating current can be represented by two components: a constant direct current value denoted by $i_{\mathrm{d}}/3$ and a sinusoidal component oscillating at twice the fundamental frequency, as shown below.

$$\begin{cases} i_{da} = \frac{i_d}{3} + I_{ca}\sin(2\omega + \phi) \\ i_{db} = \frac{i_d}{3} + I_{ca}\sin\left(2\omega + \phi + \frac{2\pi}{3}\right) \\ i_{dc} = \frac{i_d}{3} + I_{ca}\sin\left(2\omega + \phi - \frac{2\pi}{3}\right) \end{cases}, \tag{4.18}$$

where i_{dj} indicates the inner difference current from the eqn (4.10). The peak of the circulating current is I_{cj}. The inner unbalanced voltage is defined as

$$\begin{bmatrix} v_{da} \\ v_{db} \\ v_{dc} \end{bmatrix} = L_a \frac{d}{dt} \begin{bmatrix} i_{da} \\ i_{db} \\ i_{dc} \end{bmatrix} + r_a \begin{bmatrix} i_{da} \\ i_{db} \\ i_{dc} \end{bmatrix}. \tag{4.19}$$

Implementing of $abc - dq$ transformation is necessary to regulate the circulating current efficiently with a reduced number of controllers. The resulting d-axis and q-axis voltages can be determined by applying the $abc - dq$ transformation.

$$\begin{bmatrix} v_{dd} \\ v_{dq} \end{bmatrix} = L_a \frac{d}{dt} \begin{bmatrix} i_{dd} \\ i_{dq} \end{bmatrix} + \begin{bmatrix} r_a & -2\omega L_a \\ 2\omega L_a & r_a \end{bmatrix} \begin{bmatrix} i_{dd} \\ i_{dq} \end{bmatrix}. \tag{4.20}$$

The mathematical model defined in eqn (4.20) is used to formulate circulating current regulation within the synchronous dq reference outline. Figure 4.14 exhibits the structure of the circulating current manipulate. We calculate the i_{dd} and i_{dq} current components by transforming circulating currents from measured arm currents into the synchronous dq reference frame. By adjusting the reference current components to zero, we compare reference and actual currents to determine current errors, which we then

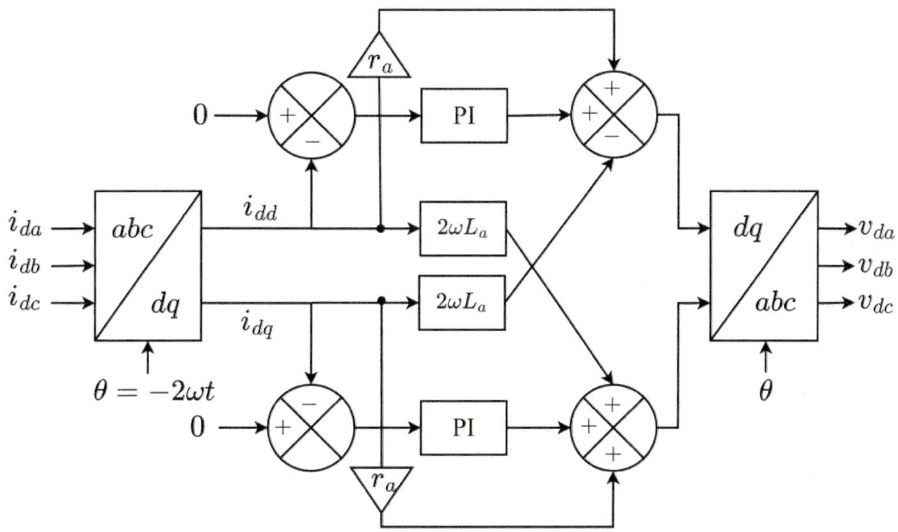

Figure 4.14 Controlling circulating current in the reference frame.

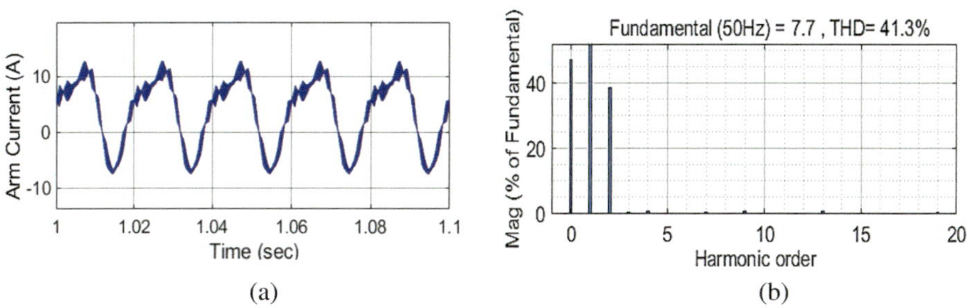

Figure 4.15 THD performance of MMC arm current absence of circulating current management (a) circulating current (b) % THD.

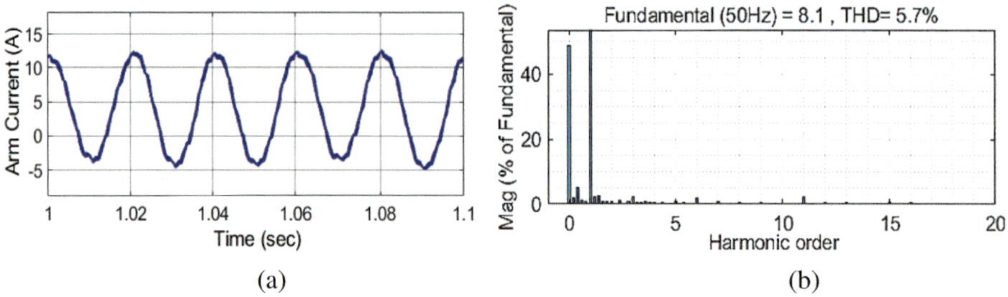

Figure 4.16 MMC arm current with circulating current control and its THD (a) circulating current (b) % THD.

reduce using PI mechanisms. To isolate the d and q axis current control loops, we integrate generated speed voltages in the inductor. The PI-regulator generates reference dq voltage instructions, which, when combined with submodule capacitor voltages and output current controllers, enhance the productivity and reliability of the MMC. We transform these reference voltages to the abc frame employing the $dq{-}abc$ alteration matrix. We reduce THD in the arm current by applying the circulating current control technique, as demonstrated in Figures 4.15 and 4.16.

4.4.4 Output power control

This section examines how to control the output of MMCs. The current regulation mechanism actively adjusts the current on the AC side, typically functioning as the inner loop for MMC output. Drawing on the MMC's structure and grid characteristics, the current controller sets and guides the goals of the outer loop control. When the MMC is connected to the AC power grid, the control system meticulously regulates the power exchange between the MMC and the grid by focusing on parameters such as DC-link voltage, active power, and reactive power. Conversely, when the MMC is linked to passive loads, the control system dynamically modulates the AC voltage's amplitude and frequency to shape the AC grid's characteristics.

The output control strategy permits for the synchronized regulation of active and reactive power in the MMC. Figure 4.17 depicts the control method used for output power regulation. Figures 4.18–4.21 show fluctuations in output phase voltage, phase current, arm current, and circulation current when the output power changes.

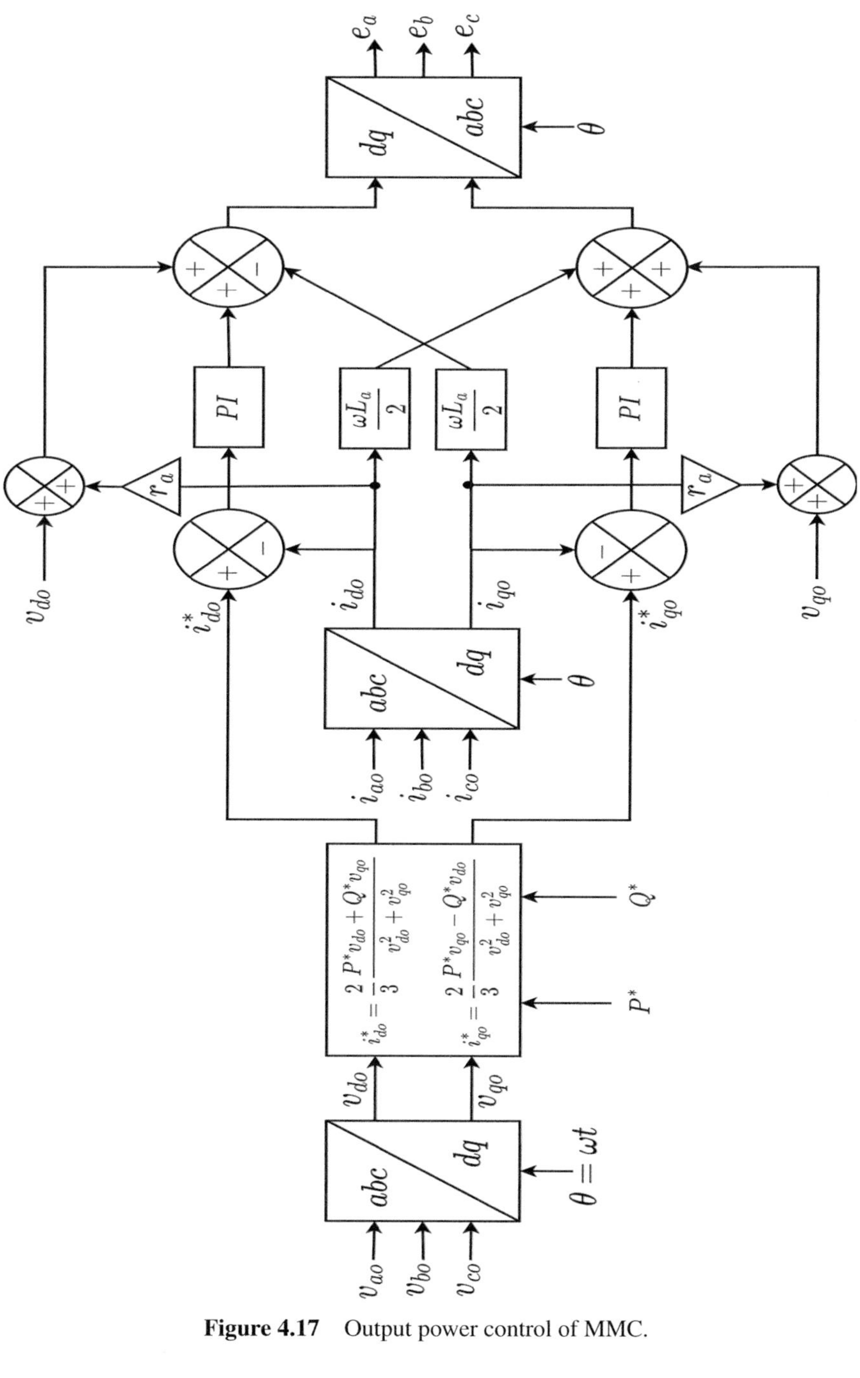

Figure 4.17 Output power control of MMC.

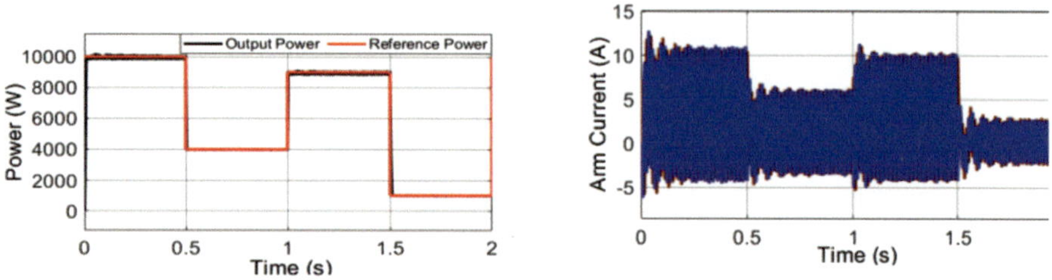

(a) (b)

Figure 4.18 (a) The reference and actual power (b) arm current of MMC.

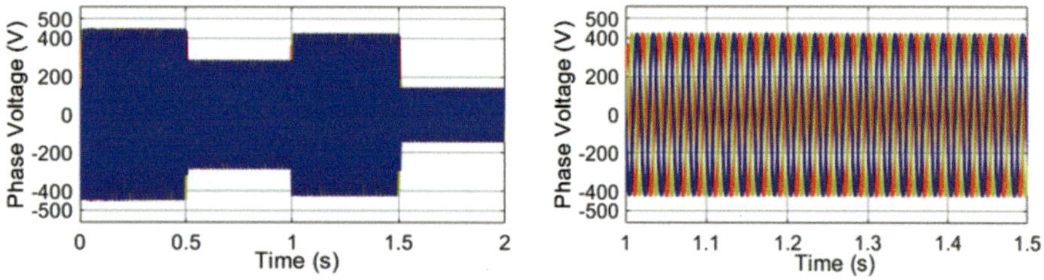

(a) (b)

Figure 4.19 The phase voltage of MMC varies with changes in output power.

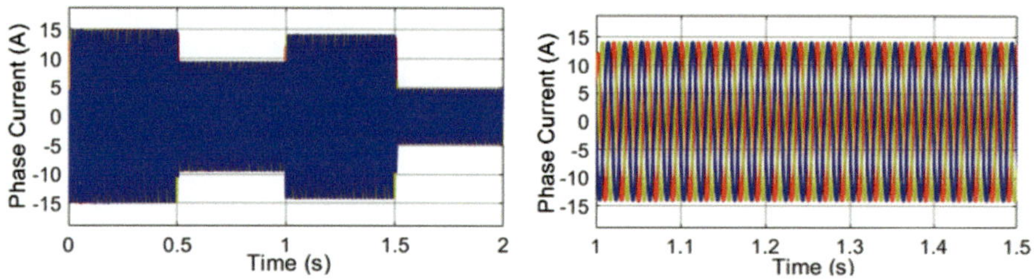

(a) (b)

Figure 4.20 The phase current of MMC with the change of output power.

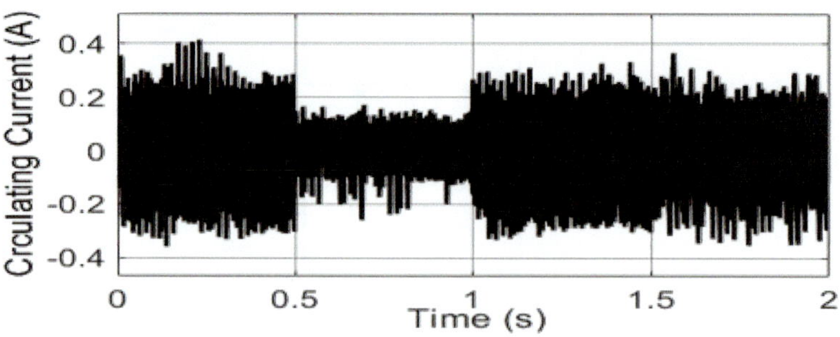

Figure 4.21 Circulating current of MMC.

4.5 Conclusion

In conclusion, this chapter has comprehensively explored the submodule structure, operation, and classical control methods applicable to modular multilevel converters (MMCs). The versatility of submodule configurations in realizing MMCs, with a significant emphasis on the half-bridge submodule for high and medium-voltage applications, has been discussed. PWM schemes, particularly the widely employed PS-PWM, are crucial in regulating MMCs.

Advanced control strategies are vital for MMC operations. These strategies leverage closed-loop controllers and sophisticated algorithms to ensure optimal management of critical aspects like submodule capacitor voltages, AC output current, and circulating currents. The management of submodule capacitor voltages was addressed by implementing a practical strategy that included voltage control in the leg and submodule capacitor voltage balancing inside the arm. This approach is adaptable to various modulation schemes.

Furthermore, a simple controller was presented for MMCs supplying power to passive loads. Leveraging reference frame theory, this controller streamlines the control process, eliminating steady-state errors. The persistent challenge of circulating current among converter legs, primarily characterized by even-order harmonic components, was addressed through a straightforward regulation approach in dq and abc frames. This approach mitigates circulating currents and enhances efficiency, ensuring the reliable performance of modular multilevel converters. Overall, these outcomes contribute beneficial perceptions to expanding MMC expertise and its practical relevance in diverse power systems.

References

[1] M. Glinka and R. Marquardt, "A new AC/AC multilevel converter family," *IEEE Transactions on Industrial Electronics*, vol. 52, no. 3, pp. 662–669, 2005, doi:10.1109/TIE.2005.843973.

[2] A. Nami, J. Liang, F. Dijkhuizen, and G. D. Demetriades, "Modular multilevel converters for HVDC applications: Review on converter cells and functionalities," *IEEE Trans Power Electron*, vol. 30, no. 1, pp. 18–36, 2015, doi:10.1109/TPEL.2014.2327641.

[3] S. Debnath, J. Qin, B. Bahrani, M. Saeedifard, and P. Barbosa, "Operation, control, and applications of the modular multilevel converter: A

review," *IEEE Trans Power Electron*, vol. 30, no. 1, pp. 37–53, 2015, doi:10.1109/TPEL.2014.2309937.

[4] J. Qin and M. Saeedifard, "Predictive control of a modular multilevel converter for a back-to-back HVDC system," *IEEE Transactions on Power Delivery*, vol. 27, no. 3, pp. 1538–1547, 2012, doi:10.1109/TPWRD.2012.2191577.

[5] A. Antonopoulos, L. Ängquist, S. Norrga, K. Ilves, L. Harnefors, and H. P. Nee, "Modular multilevel converter AC motor drives with constant torque from zero to nominal speed," *IEEE Trans Ind Appl*, vol. 50, no. 3, pp. 1982–1993, 2014, doi:10.1109/TIA.2013.2286217.

[6] Y. S. Kumar and G. Poddar, "Medium-Voltage Vector Control Induction Motor Drive at Zero Frequency Using Modular Multilevel Converter," *IEEE Transactions on Industrial Electronics*, vol. 65, no. 1, pp. 125–132, 2018, doi:10.1109/TIE.2017.2721927.

[7] Y. S. Kumar and G. Poddar, "Control of Medium-Voltage AC Motor Drive for Wide Speed Range Using Modular Multilevel Converter," *IEEE Transactions on Industrial Electronics*, vol. 64, no. 4, pp. 2742–2749, 2017, doi:10.1109/TIE.2016.2631118.

[8] A. B. Acharya, M. Ricco, D. Sera, R. Teoderscu, and L. E. Norum, "Performance analysis of medium-voltage grid integration of pv plant using modular multilevel converter," *IEEE Transactions on Energy Conversion*, vol. 34, no. 4, pp. 1731–1740, Dec. 2019, doi:10.1109/TEC.2019.2930819.

[9] A. Elsanabary, S. Mekhilef, M. Seyedmahmoudian, and A. Stojcevski, "A Novel Circuit Configuration for The Integration of Modular Multilevel Converter with Large-Scale Grid-Connected PV Systems," *IEEE Transactions on Energy Conversion*, Mar. 2023, doi:10.1109/TEC.2023.3318565.

[10] C. Gan, Q. Sun, J. Wu, W. Kong, C. Shi, and Y. Hu, "MMC-Based SRM Drives with Decentralised Battery Energy Storage System for Hybrid Electric Vehicles," *IEEE Trans Power Electron*, vol. 34, no. 3, pp. 2608–2621, Mar. 2019, doi:10.1109/TPEL.2018.2846622.

[11] D. D. C. Mendonca, A. F. Cupertino, H. A. Pereira, and R. Teodorescu, "Minimum Cell Operation Control for Power Loss Reduction in MMC-Based STATCOM," *IEEE J Emerg Sel Top Power Electron*, vol. 9, no. 2, pp. 1938–1950, Apr. 2021, doi:10.1109/JESTPE.2020.2979123.

[12] Y. C. Su, H. M. Li, P. L. Chen, and P. T. Cheng, "Integration of PV Panels and EV Chargers on the Modular Multilevel Converter Based SST," in *IEEE Transactions on Industry Applications*, Institute of Electrical and

Electronics Engineers Inc., 2022, pp. 6428–6437. doi:10.1109/TIA.2022.3189969.

[13] M. Mao *et al.*, "Multi-objective power management for EV fleet with MMC-based integration into smart grid," *IEEE Trans Smart Grid*, vol. 10, no. 2, pp. 1428–1439, Mar. 2019, doi:10.1109/TSG.2017.2766363.

[14] J. Rodríguez, J. S. Lai, and F. Z. Peng, "Multilevel inverters: A survey of topologies, controls, and applications," *IEEE Transactions on Industrial Electronics*, vol. 49, no. 4, pp. 724–738, 2002, doi:10.1109/TIE.2002.801052.

[15] D. D. C. Mendonca, A. F. Cupertino, H. A. Pereira, and R. Teodorescu, "Minimum Cell Operation Control for Power Loss Reduction in MMC-Based STATCOM," *IEEE J Emerg Sel Top Power Electron*, vol. 9, no. 2, pp. 1938–1950, Apr. 2021, doi:10.1109/JESTPE.2020.2979123.

[16] S. Haq *et al.*, "An Advanced PWM Technique for MMC Inverter Based Grid-Connected Photovoltaic Systems," *IEEE Transactions on Applied Superconductivity*, vol. 31, no. 8, pp. 8–12, 2021, doi:10.1109/TASC.2021.3094439.

[17] G. Konstantinou, G. J. Capella, J. Pou, and S. Ceballos, "Single-Carrier Phase-Disposition PWM Techniques for Multiple Interleaved Voltage-Source Converter Legs," *IEEE Transactions on Industrial Electronics*, vol. 65, no. 6, pp. 4466–4474, 2018, doi:10.1109/TIE.2017.2767541.

[18] R. Darus, J. Pou, G. Konstantinou, S. Ceballos, R. Picas, and V. G. Agelidis, "A Modified Voltage Balancing Algorithm for the Modular Multilevel Converter: Evaluation for Staircase and Phase-Disposition PWM," *IEEE Trans Power Electron*, vol. 30, no. 8, pp. 4119–4127, 2015, doi:10.1109/TPEL.2014.2359005.

[19] R. Lizana, M. A. Perez, S. Bernet, J. R. Espinoza, and J. Rodriguez, "Control of Arm Capacitor Voltages in Modular Multilevel Converters," *IEEE Trans Power Electron*, vol. 31, no. 2, pp. 1774–1784, 2016, doi: 10.1109/TPEL.2015.2426183.

[20] G. P. Adam, O. Anaya-Lara, G. M. Burt, D. Telford, B. W. Williams, and J. R. McDonald, "Modular multilevel inverter: Pulse width modulation and capacitor balancing technique," *IET Power Electronics*, vol. 3, no. 5, pp. 702–715, 2010, doi:10.1049/iet-pel.2009.0184.

[21] S. D. Joshi, M. B. Ghat, A. Shukla, and M. C. Chandorkar, "Improved Balancing and Sensing of Submodule Capacitor Voltages in Modular Multilevel Converter," *IEEE Trans Ind Appl*, vol. 57, no. 1, pp. 537–548, 2021, doi:10.1109/TIA.2020.3026664.

[22] S. Sau, S. Karmakar, and B. G. Fernandes, "Reduction of capacitor ripple voltage and current in Modular Multilevel Converter based variable speed drives," *2017 IEEE 3rd International Future Energy Electronics Conference and ECCE Asia, IFEEC - ECCE Asia 2017*, pp. 1451–1456, 2017, doi:10.1109/IFEEC.2017.7992258.

5

Mathematical Modeling and Development of a Current Estimator for Multilevel Architectures

Ritankar Kumar and Olive Ray

School of Electrical and Computer Sciences, IIT Bhubaneswar, India
E-mail: a24ee09006@iitbbs.ac.in; olive@iitbbs.ac.in

Abstract

Multilevel architectures are characterized by switch node voltages assuming different distinct voltage magnitudes within a switching interval. This chapter evaluates various types of multilevel inverters, with a particular focus on five-level inverters, detailing their operational principles. The mathematical modeling of each of the inverter topologies has been evaluated based on the switching states. The mathematical model of the switch node of the multilevel architecture has been used to formulate a voltage-based current estimation method for filter inductor currents in the topology. The novelty of the proposed approach is that it uses shunt-type measurements to obtain the instantaneous estimate of current. In this method, the status of the switch node used for current estimation is identified using a single-bit analog comparator(s) connected across the switch node, and in this chapter, different possible combinations of comparator locations have been evaluated. The dynamic performance of the proposed method has been presented using simulation-based analysis. Such systems can be used to obtain a digital twin of the multilevel architectures and used for fault evaluation.

Keywords: Analog comparator, current estimation, multilevel converter

DOI: 10.1201/9781003669845-5

5.1 Introduction

The architectures of multilevel converters enable the generation of output voltages with different distinct output voltage levels. The presence of multiple output voltage levels at the output is due to the operation of power electronic semiconductor switches having blocking voltage ratings lower than conventional single-level architectures. These topologies can be realized using lower switching frequency operated switches. With these characteristics, multilevel inverters are widely used in DC-AC power electronic conversion with typically higher output voltage ratings. These multilevel inverters are often used to interface renewable sources such as solar panels where the net output voltage is generated using a stack of solar panels connected in series. The design and operation of multilevel converters require a trade-off between the use of a higher number of devices and a lower operational switching frequency. Typically, multilevel inverters generate AC output with lower total harmonic distortion (THD) using low switching frequencies [1, 2].

The multilevel inverters (MLIs) have the following advantages over two-level voltage source inverters [3]. These inverters can be switched at fundamental frequencies to achieve output voltages within acceptable THD. Due to the presence of multiple switches in the configuration, the same output voltage magnitude can be achieved using many possible switching state configurations. Hence, this lends redundancy in converter operation which is often a beneficial characteristic to have for achieving fault-tolerant operation. Due to the operation of a converter with a lower switching frequency, the switching losses are lower compared to switched-mode inverters. These topologies provide lower electromagnetic interference and low input current distortion and hence achieve better power quality [4].

The operation of multilevel inverters is specified by the number of voltage levels at the output. In such cases, as the number of voltage levels increases, the output voltage waveform becomes more refined and closely approximates a sinusoidal waveform, resulting in lower THD in output waveshape [5]. The effect of this property is reflected in the reduction of bulky and complex filtering systems at the output interface. The reactive filter components such as the inductor and capacitors are smaller in size, leading to a more compact and cost-effective design. This is one of the key advantages of using multilevel inverters in grid-connected applications, as it reduces the overall system footprint.

Multilevel inverters can be used in compensation for reactive power by connecting in parallel (as a FACTS device) [6, 7], industrial motor drives, and active front-end converters [8, 9]. Multilevel inverters (MLIs) are being used more and more in the power systems of electric vehicles (EVs) because they can improve power quality, efficiency, and voltage control [10]. Multilevel inverters are classified into two main categories based on the type of DC sources they are interfaced with. The first category, consisting of a single DC source, includes inverter topologies like the flying capacitor type multilevel inverter and the neutral point clamped (NPC) multilevel inverter, which generate multiple voltage levels from a single DC source using capacitors or diodes to create the intermediate voltage inputs. The second category consists of separate DC sources, which include the cascaded H-bridge multilevel inverter, where multiple independent DC sources are used, each supplying an H-bridge inverter. The output voltage in this case is the sum of the voltages from each H-bridge.

One of the major challenges pertaining to the operation and control of multilevel inverters is the use of a large number of active devices and their associated gate drivers [11, 12]. Each of the devices needs to be triggered at the precise time instant with appropriate bias voltages. The use of more devices needs a highly coordinated control system which has an impact on the reliability of system operation [13]. Each of the different devices operates under different reference potentials, and hence, the operation of switches needs to have a fail-safe design.

A digital twin of any electrical hardware is used for analysis and condition monitoring of system operation [14]. These systems are also used for performance comparison in a hardware-in-the-loop environment. In order to develop a digital twin of the multilevel inverter system, it is essential to evaluate the different parameters to be sensed for the evaluation of system behavior. The mathematical model of the system is essential for system understanding. In addition, this chapter also evaluates a current estimation method for multilevel converter systems using shunt-type measurements.

5.2 Mathematical Modeling of Multilevel Inverters

In a multilevel inverter architecture, the switch node voltages can be expressed as a function of the switching states of the inverter, and its instantaneous magnitude can be associated with a voltage stiff element. This section discusses the mathematical formulation of switch node voltage expressions for three different types of five-level multilevel inverter architectures. The output voltages of each of the inverters are characterized by five distinct voltage levels. These mathematical models can be used to analyze the converter

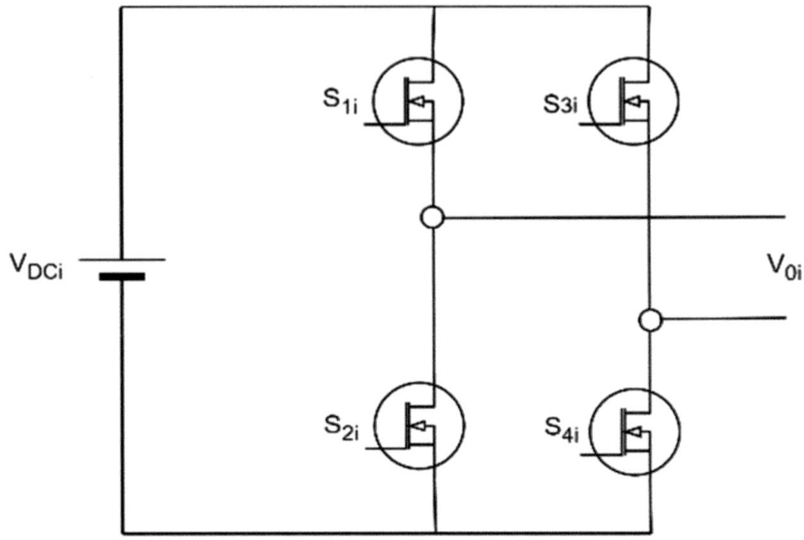

Figure 5.1 Schematic of the basic block of full-bridge cascaded H-bridge inverter.

behavior, design control strategies, and predict the output waveform under different operating conditions.

5.2.1 Five-level cascaded H-bridge inverter

The architecture of cascaded H-bridge (CHB) multilevel inverters can be obtained using single-phase full-bridge-based structures as the basic building block, as shown in Figure 5.1. For a five-level architecture, two of these basic blocks are required to obtain one phase leg of the inverter. In this topology, each of the CHB units needs to have its dedicated input DC supply, which can be either derived from a single source using isolated interfacing stages, or they can be separate sources.

The architecture of a single-phase leg of a five-level CHB inverter is shown in Figure 5.2.a. In this case, there are two units of H-bridge architecture connected in series. Each phase leg in the topology needs to be interfaced with two different DC inputs. If the two source voltages are maintained at equal voltage values (equal to V_{dc}), the output voltage of that phase assumes five distinct output voltage levels: $2V_{dc}$, V_{dc}, 0, $-V_{dc}$, and $-2V_{dc}$. Each of the different voltage levels can be related to the status of each of the switching devices. For instance, when switches S_{11}, S_{41}, S_{12}, and S_{42} (as shown in

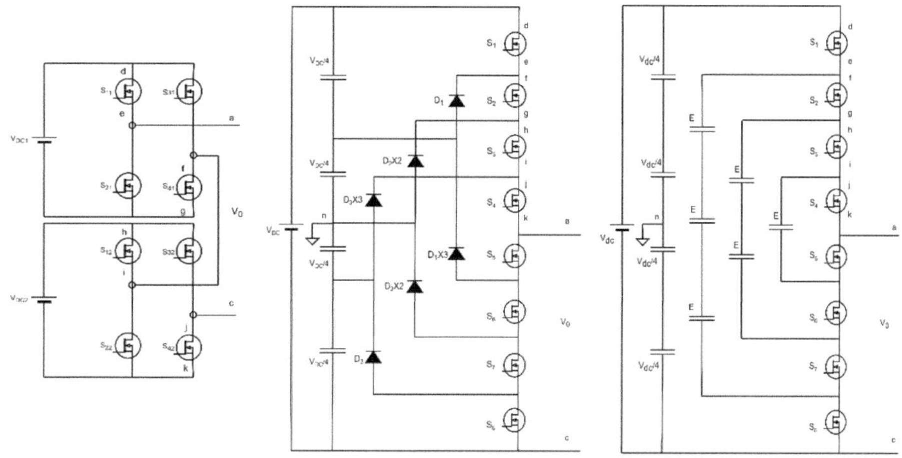

(a) (b) (c)

Figure 5.2 Schematic of different types of five-level inverters. (a) Cascaded five-level inverter, (b) diode clamped NPC five-level inverter, and (c) flying capacitor-based five-level inverter.

Figure 5.2.a) are conducting, the inverter outputs $2V_{dc}$. If switches S_{11}, S_{41}, S_{12}, and S_{32} are conducting, the output voltage is V_{dc}. The comprehensive list of all possible switching states relating output voltage to switch status is presented in Table 5.1. It can be observed that a total of 16 possible switch cases are being mapped into five distinct levels, and hence, this results in redundancy with respect to switch configuration. Some of the switch states may not be used for the optimal operation of the inverter.

By using different DC sources (V_{dc1} and V_{dc2}), CHB MLI can generate more voltage levels ($-V_{dc1}$, $-V_{dc2}$, $-V_{dc2}$, $-V_{dc1}$, 0, V_{dc1}, V_{dc2}, V_{dc1}, and + V_{dc2}) in output pattern, which is very close to a sinusoidal wave. This reduces total harmonic distortion (THD) by a large amount, which means less filtering is needed and the power quality is better.

The mathematical model of the CHB structure can be derived from the model of the basic block shown in Figure 5.1. For this topology, in *i*th two-level inverter, at a given time instant, there is one switch that is on for every leg. Hence, there are four possible combinations of switch conditions which are given in Table 5.2. The possible output voltages from this structure are zero, $+V_{dci}$, and $-V_{dci}$. The output voltage $+V_{dci}$ is obtained when the switch pair S_{1i} and S_{4i} are on. Similarly, the output value of $-V_{dci}$ is obtained when the switch pair S_{3i} and S_{2i} are on. Any other switch combination

Table 5.1 Switching states of a cascaded five-level inverter.

$\mathbf{S}_{11}$	$\mathbf{S}_{41}$	$\mathbf{S}_{12}$	$\mathbf{S}_{42}$	V_{ac}
0	0	0	0	$-2V_{dc}$
0	0	0	1	$-V_{dc}$
0	0	1	0	$-V_{dc}$
0	1	0	0	$-V_{dc}$
1	0	0	0	$-V_{dc}$
0	0	1	1	0
0	1	0	1	0
0	1	1	0	0
1	0	0	1	0
1	0	1	0	0
1	1	0	0	0
0	1	1	1	V_{dc}
1	0	1	1	V_{dc}
1	1	0	1	V_{dc}
1	1	1	0	V_{dc}
1	1	1	1	$2V_{dc}$

Table 5.2 Switching states of *i*th two-level inverter.

$\mathbf{S}_{1i}$	$\mathbf{S}_{3i}$	V_{0i}
0	1	$-V_{dci}$
0	0	0
1	1	0
1	0	V_{dci}

results in zero output voltage. Hence, the mathematical relationship between switch status and corresponding output voltage can be represented as OR-ed operation of switch combination as shown in eqn (5.1), where if the corresponding switch is on, its value is high. The coefficients in eqn (5.1), that is, A and B determine the sign of the output voltage. In this expression, V_{dci} is the supply voltage, and the output voltage V_{0i} is represented as an OR combination of all possible output states.

$$V_{0i} = \left(\mathrm{AS_{1i}S_{4i}} + \mathrm{BS_{2i}S_{3i}}\right) V_{\mathrm{dci}}. \tag{5.1}$$

The values of coefficients used in eqn (5.1) can be obtained using Table 5.1 and considering the complementary operation of switches in each leg. Hence, we can get A = 1 and B = −1, and substituting these values in eqn (5.1),

$$V_{\mathrm{oi}} = \left(\mathrm{S_{1i}\overline{S_{3i}}} - \mathrm{S_{3i}\overline{S_{1i}}}\right) V_{\mathrm{dci}}, \tag{5.2}$$

by minimizing the digital expression due to dependency only on S_{1i} and S_{3i}, the simplified relationship can be derived as eqn (5.3).

$$V_{0i} = (S_{1i} - S_{3i})\ V_{dci}. \tag{5.3}$$

The generalized expression can be extended for cascaded structure with '*n*' different voltage inputs can be shown in eqn (5.4).

$$V_0 = \sum_{i=1}^{n} V_{0i} = \sum_{i=1}^{n} (S_{1i} - S_{3i})\ V_{dci}. \tag{5.4}$$

5.2.2 Five-level neutral point diode clamped (NPC) inverter

The five-level neutral point diode clamped inverter architecture for a single-phase leg is shown in Figure 5.2.b. For this architecture, there are eight active switches, 12 diodes, and four capacitors. The total input voltage is applied to the inverter input through the series of connected capacitors, each sharing the voltage of magnitude $V_{dc}/4$. The five distinct output voltage levels for this inverter are $\frac{V_{dc}}{2}$, $\frac{V_{dc}}{4}$, 0, $-\frac{V_{dc}}{4}$, and $-\frac{V_{dc}}{2}$. The eight active switches are connected in series, with the inverter output node connected between switches S_4 and S_5. There are four complementary pairs of switches, that is, when S_1 is off, S_5 is on ($S_1 = \overline{S_5}$, $S_2 = \overline{S_6}$, $S_3 = \overline{S_7}$, and $S_4 = \overline{S_8}$). The output voltage of this inverter can be mapped to the states of the switches. For example, when switches S_1, S_2, S_3, and S_4 conduct, the inverter output voltage is $\frac{V_{dc}}{2}$. Similarly, when S_2, S_3, and S_4 conduct, the inverter output voltage is $\frac{V_{dc}}{4}$. Table 5.3 shows the inverter output voltages for all possible switch combinations.

In this architecture, the clamped diodes are subjected to different blocking voltages, and hence, depending upon the voltage rating, multiple series connected diodes with similar voltage ratings are used. As an example, when

Table 5.3 Switching states of NPC diode clamped five-level inverter.

S_1	S_2	S_3	S_4	V_{ac}
0	0	0	0	$-V_{dc}/2$
0	0	0	1	$-V_{dc}/4$
0	0	1	1	0
0	1	1	1	$V_{dc}/4$
1	1	1	1	$V_{dc}/2$

switches S_1, S_2, S_3, and S_4 conduct the diodes, D_1, D_2, and D_3 block same voltage $\frac{V_{dc}}{4}$. In order to use diodes with the same voltage blocking, two- and three-series connected diodes are required for blocking voltages of $\frac{V_{dc}}{2}$ and $\frac{3V_{dc}}{4}$.

For this type of multilevel inverter, all the controllable switches S_1, S_2, S_3, and S_4 are connected in series. Hence, to derive the mathematical model, the output voltage will be the sum of the product terms of the switching functions. Here, the status of each controllable switch is represented by a switching function which is high when the switch is on and low when it is off. Each of the complementary switch pairs can be independently controlled. The generalized expression of the output voltage can be shown in eqn (5.5), where A, B, C, D, and E are the coefficients and V_{dc} is the source voltage.

$$V_0 = \left(\mathrm{A}S_1S_2S_3S_4 + \mathrm{B}\overline{S_1}S_2S_3S_4 + \mathrm{C}\overline{S_1S_2}S_3S_4 + \mathrm{D}\overline{S_1S_2}\,\overline{S}_3S_4 + \mathrm{E}\overline{S_1S_2}\,\overline{S}_3\overline{S}_4\right) V_{dc}. \tag{5.5}$$

The five unknown values of coefficients in (5.5) can be obtained from Table 5.3 as $\mathrm{A} = \frac{1}{2}$, $\mathrm{B} = \frac{1}{4}$, $\mathrm{C} = 0$, $\mathrm{D} = -\frac{1}{4}$, and $\mathrm{E} = -\frac{1}{2}$. Hence, the expression of the output voltage can be simplified into relation (eqn (5.6))

$$V_0 = \left(S_1S_2S_3S_4 - \overline{S_1S_2}\,\overline{S}_3\overline{S}_4\right) \frac{V_{dc}}{2} + \left(\overline{S_1}S_2S_3S_4 - \overline{S_1S_2}S_3S_4\right) \frac{V_{dc}}{4}. \tag{5.6}$$

5.2.3 Five-level flying capacitor-based inverter

The architecture of the flying capacitor-type multilevel inverter is shown in Figure 5.2.c. This topology uses eight controllable switches, grouped into four pairs of complementary switches. The operation of the switch pairs can be denoted as $S_1 = \overline{S_8}$, $S_2 = \overline{S_7}$, $S_3 = \overline{S_6}$, and $S_4 = \overline{S_5}$. In this structure, in comparison to diode-clamped architecture, capacitors with different voltage ratings are used instead of clamped diodes. The five different output voltage levels of this inverter are all the same such as: $-\frac{V_{dc}}{2}$, $\frac{V_{dc}}{4}$, 0, $-\frac{V_{dc}}{4}$, and $-\frac{V_{dc}}{2}$. Each of the flying capacitors shown in Figure 5.2.c is precharged to E, which is equal to $\frac{V_{dc}}{4}$.

For inverter operation, when S_1, S_2, S_3, and S_4 conduct, the inverter output voltage is $\frac{V_{dc}}{2}$, when S_1 is off, then by conducting S_2, S_3, S_4, and S_8, the inverter output voltage is $\frac{V_{dc}}{4}$. Table 5.4 presents all possible states of the inverter leg output as a function of switch states. For inverter operation, there are 16 possible switch states which are mapped into five distinct inverter output values.

Table 5.4 Switching states of five-level flying capacitor-based inverter.

S_1	S_2	S_3	S_4	V_{ac}
0	0	0	0	$-V_{dc}/2$
0	0	0	1	$-V_{dc}/4$
0	0	1	0	$-V_{dc}/4$
0	1	0	0	$-V_{dc}/4$
1	0	0	0	$-V_{dc}/4$
0	0	1	1	0
0	1	0	1	0
0	1	1	0	0
1	0	0	1	0
1	0	1	0	0
1	1	0	0	0
0	1	1	1	$V_{dc}/4$
1	0	1	1	$V_{dc}/4$
1	1	0	1	$V_{dc}/4$
1	1	1	0	$V_{dc}/4$
1	1	1	1	$V_{dc}/2$

The expression for output voltage for a flying capacitor-based five-level inverter can be obtained by application of Kirchoff's voltage law along the path of conducting switches. When switch S_1 is on, the associated switch node voltage equal to $\frac{V_{dc}}{2}$ is part of the loop voltage expression. If the complementary switch S_8 is on, the node voltage equal to $-\frac{V_{dc}}{2}$ is involved in the loop. Similarly, if the switch S_2 is on, two nodes associated with this switch are shorted, else the capacitor voltage becomes part of the loop. The mathematical model of this inverter output can be written as eqn (5.7).

$$V_0 = (S_1 - S_8)\frac{V_{dc}}{8} + (S_2 - S_7)\frac{V_{dc}}{8} + (S_3 - S_6)\frac{V_{dc}}{8} + (S_4 - S_5)\frac{V_{dc}}{8}. \quad (5.7)$$

5.3 Control of Multilevel Inverter

The control strategy of a multilevel inverter depends upon its intended application. Typically, for a standalone inverter connected to AC loads, the control structure of the inverter consists of two cascaded loops with an output voltage and an inner current loop. The outer voltage loop compares the reference voltage to the measured output value and provides the current reference. The inner current control compares the current reference to the measured currents to generate the reference modulation indices which are used to operate the

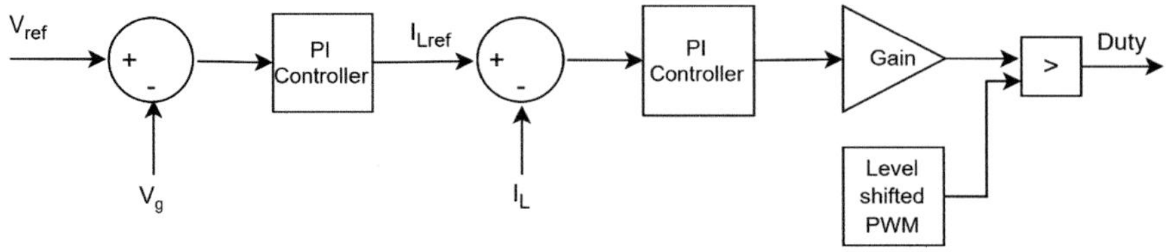

Figure 5.3 Closed-loop control of multilevel inverter.

inverter. From an implementation point of view, for closed-loop regulation, both inverter voltages and currents need to be sensed as feedback variables. The conceptual schematic diagram of the inverter control architecture is shown in Figure 5.3. For grid-connected operation, the inner loop is used for power control since the reference output voltage is available. In such cases, a phase-locked loop is used to track the phase of the source voltage. The control of the inverter can be implemented either in tracking or regulation mode. In tracking mode, the control references are time-varying, while in regulation mode, the control is implemented in a synchronous reference frame by using the transformation of the feedback signals into DC quantities.

Level-shifted pulse width modulation (PWM) can be used to generate the switching pulses for the multilevel inverters [15]. Level-shifted PWM is particularly suitable for multilevel inverters as it effectively manages the multiple voltage levels by shifting the reference waveforms for each level [16]. The reference signals are compared with a high-frequency carrier waveform, generating the appropriate switching pulses. These pulses control the switches in the inverter, ensuring that the output voltage is synthesized accurately according to the control signals from the voltage and current loops. This method enhances the quality of the output waveform and reduces harmonic distortion [17].

5.4 Current Estimation in Multilevel Inverters

The estimation of filter inductor currents within multilevel inverters can be used for the following purposes. For closed-loop control of inverters, the use of an estimator can lead to a current sensor-less design, which can be a cost-effective solution in different applications. In addition, the estimate of inductor current can be used for condition monitoring of the system and fault identification. Digital twins of power electronic converters are used to obtain a virtual model of the inverter which can be used for prediction of system performance. The inputs from the converter require dedicated sensors which monitor the voltages and currents of the system. The use of the current estimator in such cases enables a reduction in the number of interfacing elements that are required for feedback within the digital twin.

One of the methods for estimation of filter inductor currents within the multilevel inverters is through measurement of voltage stiff elements and switch node voltages present in the inverter. These voltages can be used to obtain an estimate of the current, without the need for a physical current sensor. The voltage across the filter inductor is obtained from the measured voltages through arithmetic operation, and hence, the voltage across the inductor (V_L) can be calculated [18]. Considering the inductance and resistance of the filter inductor to be L and R, respectively, the transfer function relating sensed inductor voltage to the current is given by eqn (5.8).

$$\frac{I_{\mathrm{L}}}{V_{\mathrm{L}}} = \frac{1}{R + sL}. \tag{5.8}$$

Using the relationship in eqn (5.8), inductor current can be estimated with a digital processor. The implementation of the inductor current estimate depends upon the magnitudes of voltages at both ends of the filter inductor. The output voltage of the inverter (V_{ac}) is different for different switching states and is also dependent on the type of inverter topology. Figure 5.4 shows one possible method of evaluation of filter inductor current for a grid-interfaced inverter. Measurement of both inverter output (V_a) and grid voltages (V_b or V_g), and their difference in instantaneous value results in voltage across inductor. This value can be used for current estimation.

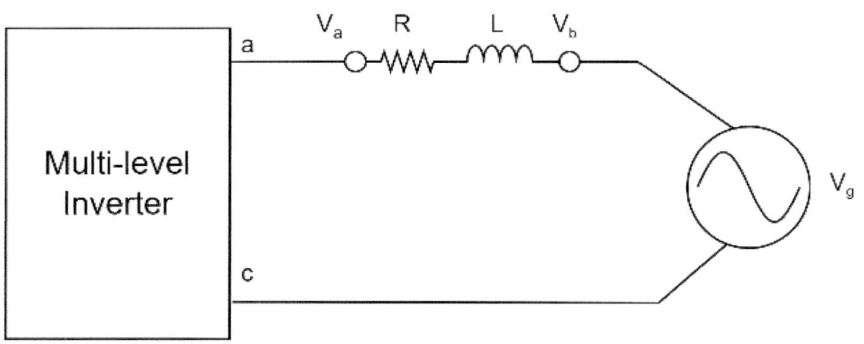

Figure 5.4 Multilevel inverter connected with the grid.

This approach effectively eliminates the need for a current sensor to measure the inductor current, which can be used for closed-loop regulation and system estimation for condition monitoring purposes. This method of inductor current estimation uses information on switch states and voltage measurement, and thus, the system avoids the complexity and cost associated with physical current sensing. The state of the inverter can be detected using comparators, which do not require digital sampling which removes the requirement for analog-to-digital (ADC) conversion stages.

In the physical implementation of a multilevel inverter, switches do not operate instantaneously upon receiving gate signals due to inherent processes and gate delays. These delays cause discrepancies between the expected and actual output voltages, leading to potential errors if the system relies solely on the gate pulses to estimate output states. To address this, an analog comparator can be used to accurately detect the actual switching sequence or voltage levels by comparing the real-time voltages with predefined reference levels, ensuring that the control system corrcctly identifies the inverter's output and compensates for any delays. This approach improves the accuracy of voltage and current estimations, leading to more reliable control performance.

For a five-level inverter, in order to detect the five levels of voltage, four comparators are required, where threshold voltage will be the intermediate value of one voltage level to another voltage level. The representative schematic of multilevel output voltage and comparator threshold values is shown in Figure 5.5. Each of the comparators detects the level transition in inverter output voltage through the output of the comparators $C_1 - C_4$. The comparator outputs C_1, C_2, C_3, and C_4 are obtained by comparison of the output voltage to the threshold reference voltages V_{th2}, V_{th1}, $-V_{th2}$, and $-V_{th3}$, respectively. If the output voltage is greater than the threshold voltage, then the comparator output will be 1, and otherwise, it will be 0. Table 5.5

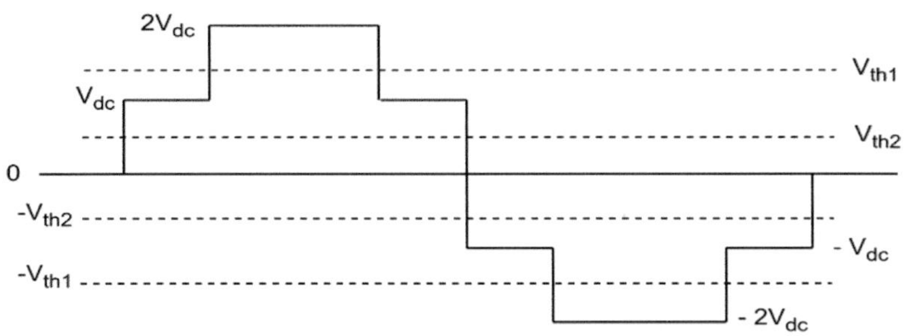

Figure 5.5 Threshold references for multilevel inverter output voltage status identification.

Table 5.5 The comparator outputs and the inductor voltage determination.

C_1	C_2	C_3	C_4	V_L
1	1	1	1	$2V_{dc} - V_g$
1	0	1	1	$V_{dc} - V_g$
0	0	1	1	$-V_g$
0	0	0	1	$-V_{dc} - V_g$
0	0	0	0	$-2V_{dc} - V_g$

maps the output of the comparators to the multilevel output voltage, and hence, the voltage across the filter inductor for a grid-tied system. In this method, two different methods can be used for the connection of comparator terminals.

5.4.1 Current estimation method 1: absolute comparison of node voltages

The status of the inverter output voltage can be obtained by detecting the transition of the inverter output node voltages by monitoring the switch node status. For a multilevel architecture, some of the control switches are independently operated, while the others operate in a complementary manner. Hence, by detecting the states of the independent switches, each of the switch node voltages can be identified. The schematic diagram for this method is shown in Figure 5.6, where the comparators are interfaced to the switch terminals for each of the different inverter topologies as shown in Figure 5.2. When a switch is on, the voltage across it will drop to 0, and when off, it will be at the peak inverse voltage (PIV). For a five-level inverter, four comparators are required, each intended to detect one voltage level transition. The reference threshold voltage can be set between 0 and PIV, allowing for precise detection of the on/off states of these switches, as shown in Figures 5.7 and 5.8. This method ensures that the correct switching sequence is identified, even in the presence of delays.

The schematic diagram of Figure 5.6 uses four comparators where comparators C_1, C_2, C_3, and C_4 compare the switch voltages, that is, the voltages across the independent switches (such as V_{de}, V_{fg}, V_{hi}, and V_{jk}) with the threshold voltage V_{th}. When the switch is on, the voltage across the switch is zero, practically it is a few volts. So, the threshold voltage is set to greater than that few volts. If the switch voltage is greater than the threshold voltage, then it means that the switch is off otherwise it is on. From the comparator outputs, the switching states can be used to identify the inverter output, and hence, inductor voltage. Tables 5.6–5.8 map the output of the comparator for the three different multilevel inverter topologies. This current estimation method requires a combination of four voltage sensors and four comparators.

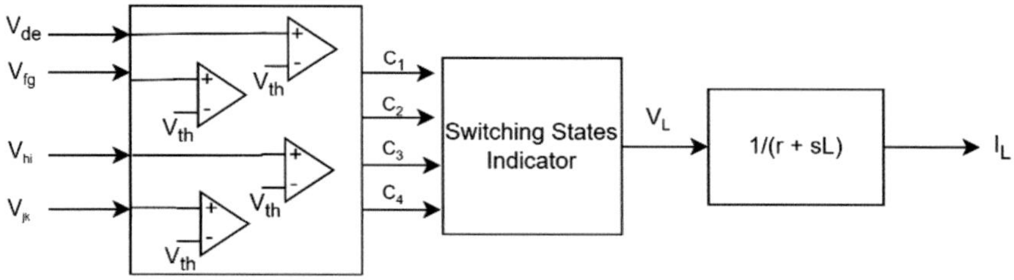

Figure 5.6 Schematic of implementation of method 1.

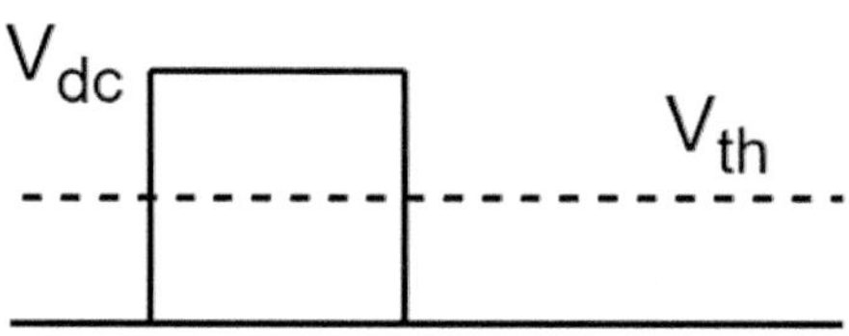

Figure 5.7 Threshold voltage selection for voltage level transition detection for cascaded multilevel inverter.

Table 5.6 Comparator outputs and the inductor voltage for cascaded inverter (method 1).

C_1	C_2	C_3	C_4	V_L
0	0	0	0	$2V_{dc} - V_g$
0	0	0	1	$V_{dc} - V_g$
0	0	1	0	$V_{dc} - V_g$
0	1	0	0	$V_{dc} - V_g$
1	0	0	0	$V_{dc} - V_g$
0	0	1	1	$-V_g$
0	1	0	1	$-V_g$
0	1	1	0	$-V_g$
1	0	0	1	$-V_g$
1	0	1	0	$-V_g$
1	1	0	0	$-V_g$
0	1	1	1	$-V_{dc} - V_g$
1	0	1	1	$-V_{dc} - V_g$
1	1	0	1	$-V_{dc} - V_g$
1	1	1	0	$-V_{dc} - V_g$
1	1	1	1	$-2V_{dc} - V_g$

Table 5.7 Comparator outputs and the inductor voltage for a five-level diode-clamped inverter (method 1).

C_1	C_2	C_3	C_4	V_L
0	0	0	0	$V_{dc}/2 - V_g$
1	0	0	0	$V_{dc}/4 - V_g$
1	1	0	0	$-V_g$
1	1	1	0	$-V_{dc}/4 - V_g$
1	1	1	1	$-V_{dc}/2 - V_g$

Table 5.8 Comparator outputs and the inductor voltage for a five-level flying capacitor-based inverter (method 1).

C_1	C_2	C_3	C_4	V_L
0	0	0	0	$2V_{dc} - V_g$
0	0	0	1	$V_{dc} - V_g$
0	0	1	0	$V_{dc} - V_g$
0	1	0	0	$V_{dc} - V_g$
1	0	0	0	$V_{dc} - V_g$
0	0	1	1	$-V_g$
0	1	0	1	$-V_g$
0	1	1	0	$-V_g$
1	0	0	1	$-V_g$
1	0	1	0	$-V_g$
1	1	0	0	$-V_g$
0	1	1	1	$-V_{dc} - V_g$
1	0	1	1	$-V_{dc} - V_g$
1	1	0	1	$-V_{dc} - V_g$
1	1	1	0	$-V_{dc} - V_g$
1	1	1	1	$-2V_{dc} - V_g$

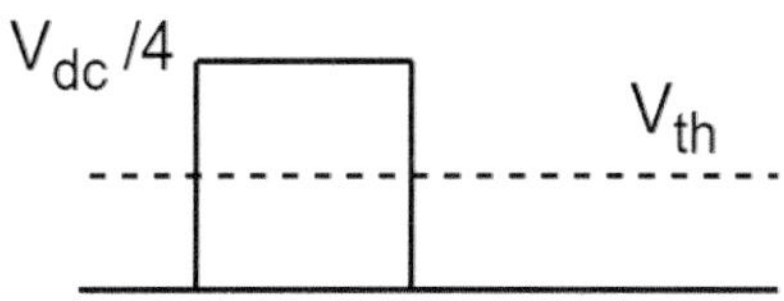

Figure 5.8 Threshold voltage selection for voltage level transition detection for NPC multi-level inverter.

5.4.2 Current estimation method 2: differential comparison of node voltages

The second method of current estimation considers the measurement of inverter output voltage instead of the switch nodes and compares the magnitude with respect to the AC side voltage. This approach requires voltage sensors but a single comparator to obtain the estimation of inductor current. Threshold voltage (V_{th}) selections of comparator for different multilevel architectures are shown in Figures 5.7 and 5.8. Figure 5.9 elaborates on the current estimation method. Here, the multilevel output (V_{ac}) is compared to load voltage or grid voltage (V_g). The output of the binary output comparator is assisted by the measurement of instantaneous grid voltage to obtain the region of operation of the inverter.

From that the switching states and the voltage across the inductor can be determined. The full schematic of this methodology is shown in Figure 5.10.

Depending upon the instantaneous values of either the load side or grid side (V_g), the output of that comparator is used to identify the different regions of inverter operation, and hence, obtain voltage across the inductor. Table 5.9 shows the three regions of operation for a five-level inverter, and relates the output of the comparator C_1 to the voltage across the inductor. In this approach, two voltage sensors are needed to measure the multilevel inverter output voltage and monitor the grid voltage or load voltage, and one comparator to compare it with the load voltage or grid voltage.

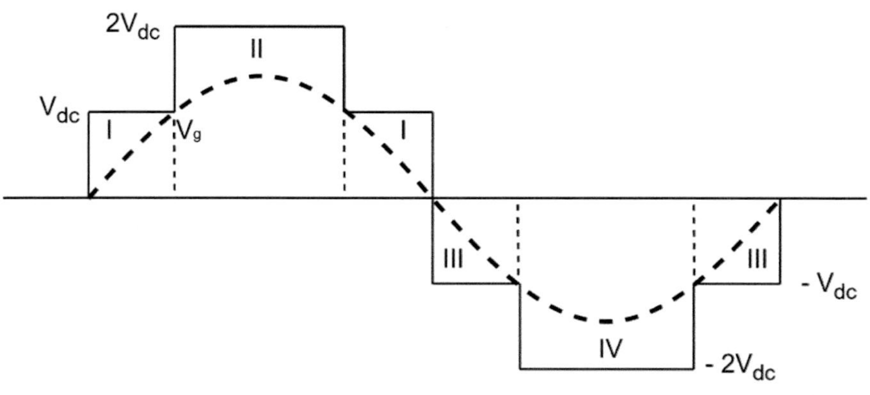

Figure 5.9 Desired load voltage or grid voltage (V_g) with inverter output voltage.

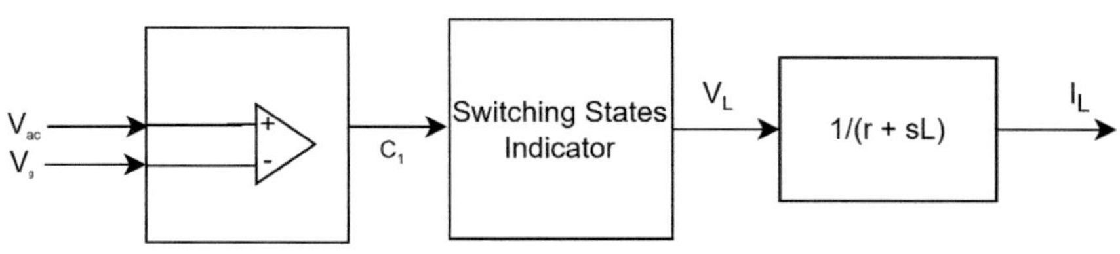

Figure 5.10 Schematic of implementation of method 2.

Table 5.9 The comparator outputs and the inductor voltage determination (method 2).

Region	Load voltage (V_0) range	C_1	V_L
I	$0 \leq V_g < V_{dc}$	0	$-V_g$
		1	$V_{dc} - V_g$
II	$V_{dc} \leq V_g$	0	$V_{dc} - V_g$
		1	$2V_{dc} - V_g$
III	$-V_{dc} \leq V_g < 0$	0	$-V_{dc} - V_g$
		1	$-V_g$
IV	$-V_{dc} \geq V_g$	0	$-2V_{dc} - V_g$
		1	$-V_{dc} - V_g$

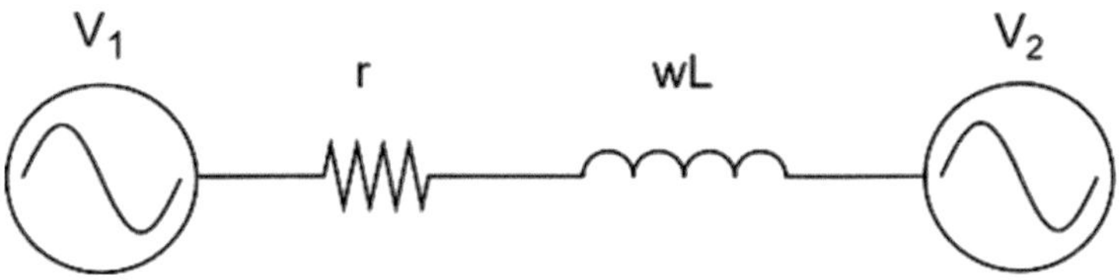

Figure 5.11 Dual AC source with line impedance.

5.4.3 Effect of parameter dependency on current estimation

The accuracy of this current estimation approach hinges on precise values for the DC resistance and inductance of the filter inductor. Otherwise, the estimated current will differ from the actual current. This section discusses those dependencies on parameter values. Since both the inverter output and load (grid) voltages across the filter inductors are both of AC in nature, the system can be explained by considering a dual AC source connected across the filter inductor as a line impedance shown in Figure 5.11.

5.4.3.1 Dependency on DC resistance of inductor (r)

If the actual resistance value is not equal to the parametric value used for current estimation, the estimate of current will have an offset depending upon the variation in parameter.

Considering V_1 to be the multilevel inverter output voltage, V_2 to be load or grid voltage (V_g), the parameters of the filter inductor to be r and L, then current through the inductor can be denoted by eqn (5.9) and increasing r, the inductor current will decrease.

$$I_{\mathrm{L}} = \frac{V_1 - V_2}{r + j\omega L}. \tag{5.9}$$

5.4.3.2 Dependency on inductance L

As shown in eqn (5.9), the magnitude of the inductor current is also inversely proportional to the inductance value. Hence, a variation in the inductor parameter used for the current estimation will result in a different value of the inductor current estimate.

Another observable dependency is the change in inductor current ripple. As we know, if V_L is the instantaneous voltage across the inductor, Δi is the current ripple and L is the inductance value, then

$$V_{\mathrm{L}} = L\frac{i}{t}. \tag{5.10}$$

So, if the actual inductance value is larger than the inductance value used in the estimation process, then the actual current ripple is less than the estimated current ripple.

5.5 Simulation Results and Discussion

The method of filter inductor current estimation for the three types of five-level inverters has been simulated using Plecs simulation software with the following system parameters as shown in Table 5.10.

The estimation of the filter inductor current of the multilevel inverter is evaluated using simulation with the first approach. For the cascaded H-bridge five-level inverter, both the DC voltages are considered equal. The simulated inverter output voltage shown in Figure 5.12 shows five distinct levels: 400, 200, 0, −200, and −400 V. The inverter is interfaced to the load terminal using an L-filter and the output voltage of the load end is shown in Figure 5.12.

Table 5.10 System specifications for five-level inverter simulation.

Parameter	Attribute
Input DC source voltage (cascaded MLI)	200 V DC (two units)
Input DC source voltage (other MLI)	800 V DC
Output voltage (RMS)	230 V DC
System frequency	50 Hz
Load resistance, inductance	50 Ω, 20 mH
Filter inductor resistance, inductance	1 mΩ, 4.42 mH
Voltage loop K_{PV}, K_{IV}	0.01, 19
Current loop K_{PI}, K_{II}	27, 7.67
Switching frequency	10 kHz

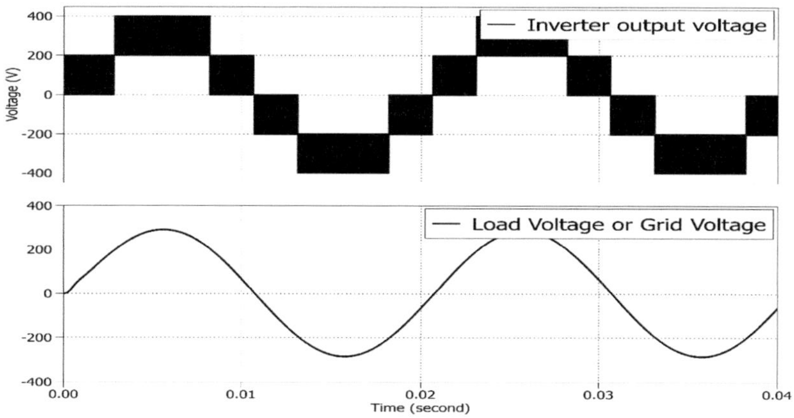

Figure 5.12 Inverter output and load terminal voltage (at a fundamental frequency of 50 Hz).

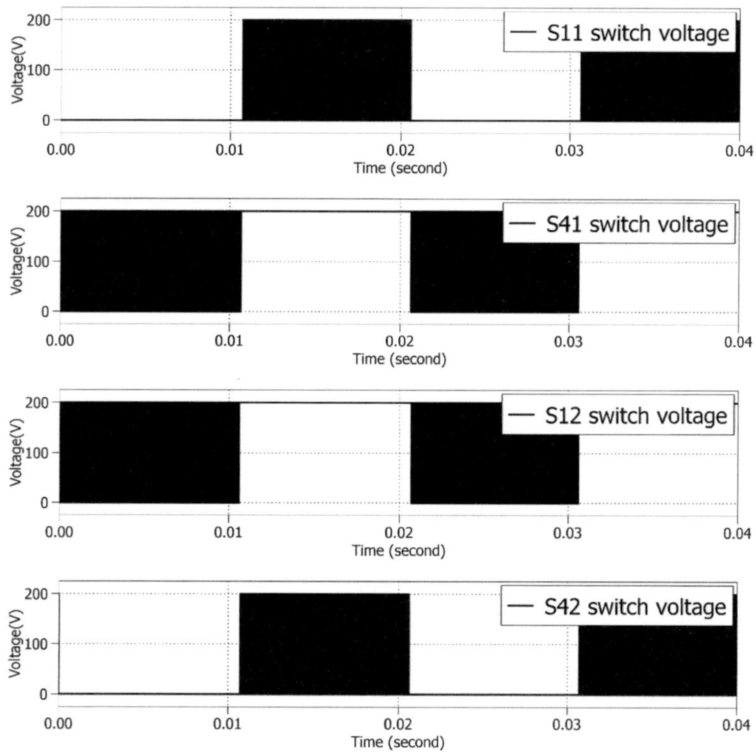

Figure 5.13 Switch voltages of cascaded five-level inverter.

Referring to Figure 5.2.a, the switch node voltages of a single leg of the inverter are monitored using sensors across the four independent switches. The switch voltages are shown in Figure 5.13, and based on a comparison of these voltages to the reference threshold, the states of the switches are determined. Here, in this figure only, the cascaded five-level inverter is mentioned, but the same type of switch voltage can be obtained in other multilevel inverters, that is, V_{de}, V_{fg}, V_{hi}, and V_{jk}, as shown in Figure 5.6. The respective comparator of that switch gives output "0" or "1" depending upon the switch voltage, which indicates the status of the switch, as shown in Figure 5.14. After that, the voltage across the filter inductor is calculated using different logics for different multilevel inverters as previously discussed. By switching state indicator, the voltage across the filter inductor is evaluated, as shown in Figure 5.15.

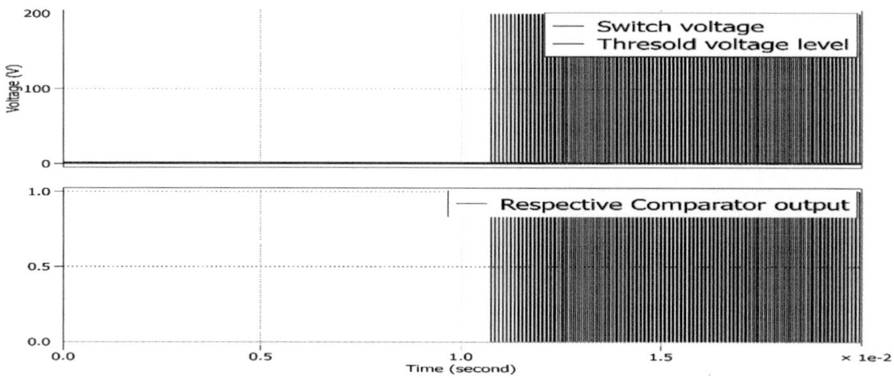

Figure 5.14 Determination of switch status using a switch voltage with its corresponding comparator output.

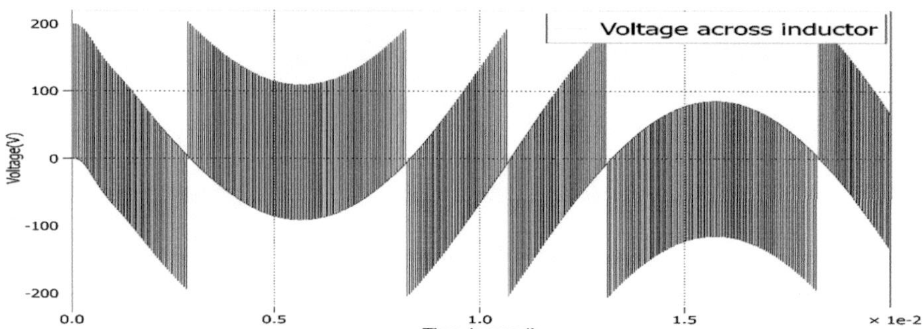

Figure 5.15 Voltage across filter inductor (for one cycle).

The computed value of inductor voltage can be used to estimate the inductor current using the system transfer function of the inductor. A comparison of the estimated and actual inductor current waveforms is shown in Figure 5.16. The switching ripples of peak magnitude of both estimated and actual filter inductor currents are shown in Figure 5.17.

If both the DC input sources of the CHB architecture are unequal, the modulation index should be adjusted to obtain zero mean at the output. The method of current estimate takes care of unequal values of DC input voltage with two voltages taken as 200 V DC and the other as 300 V DC

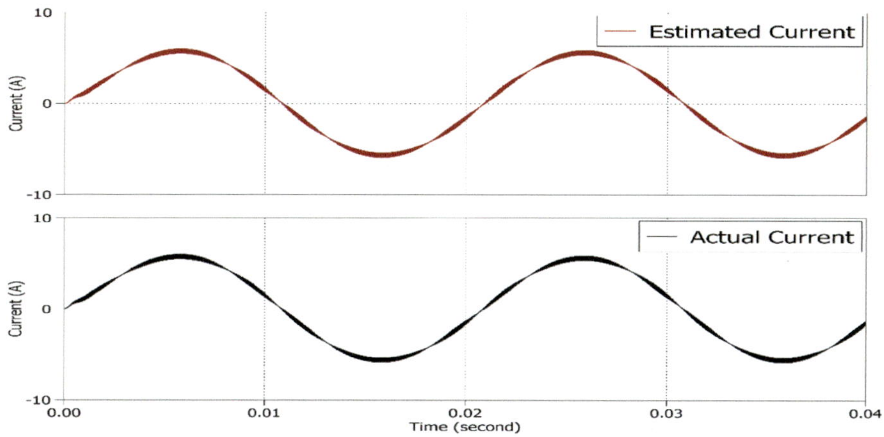

Figure 5.16 Comparison of actual load current with the estimated current (method 1).

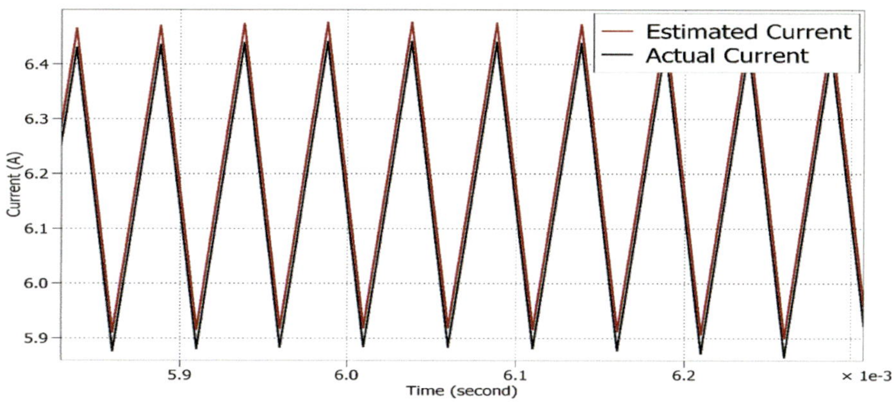

Figure 5.17 Switching ripples of actual current and estimated current (method 1).

for a cascaded five-level inverter. With unequal DC sources, the output AC voltage would be asymmetric as shown in Figure 5.18. The comparison between estimated current and actual current is shown in Figure 5.19, and peak magnitudes of actual current and estimated current are compared in Figure 5.20, a zoomed version of Figure 5.19.

The filter inductor current for the multilevel inverter is estimated in simulation using the second method discussed in this chapter. In this method, the inverter output voltage is compared to the load or grid voltage using a single comparison operation. Figure 5.21 shows the input and output behavior of the comparators. The relative position of the AC load terminal voltage is

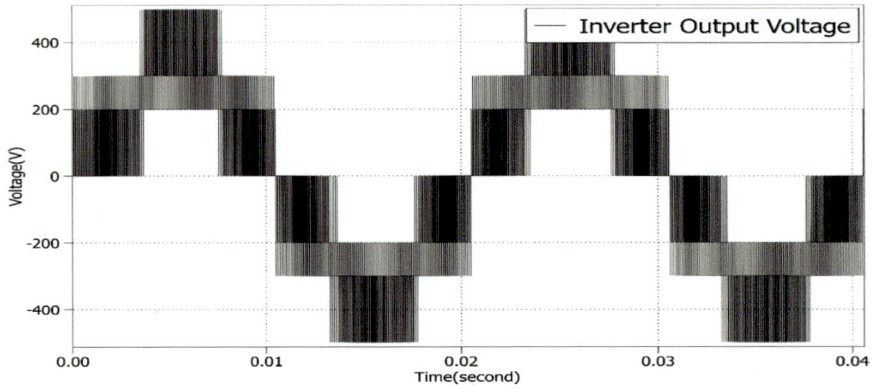

Figure 5.18 Cascaded five-level inverter output voltage for different sources.

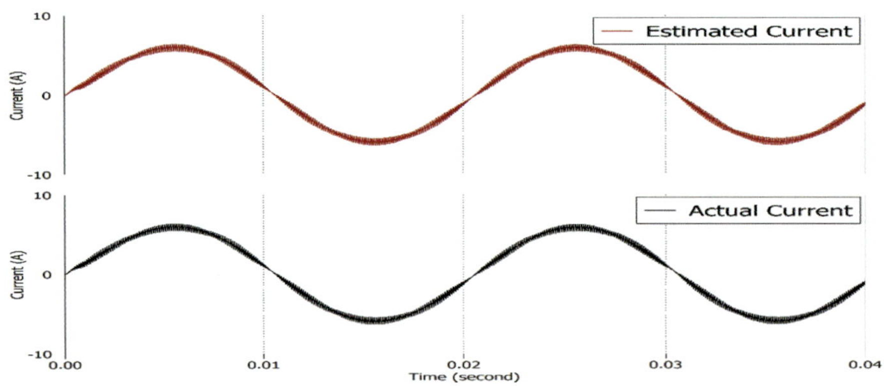

Figure 5.19 Comparison of actual load current with the estimated current for different voltage sources (method 1).

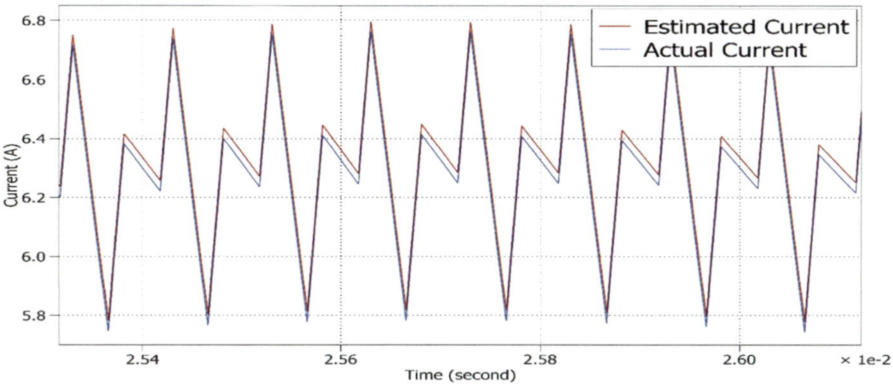

Figure 5.20 Switching ripples of actual current and estimated current for different voltage sources (method 1).

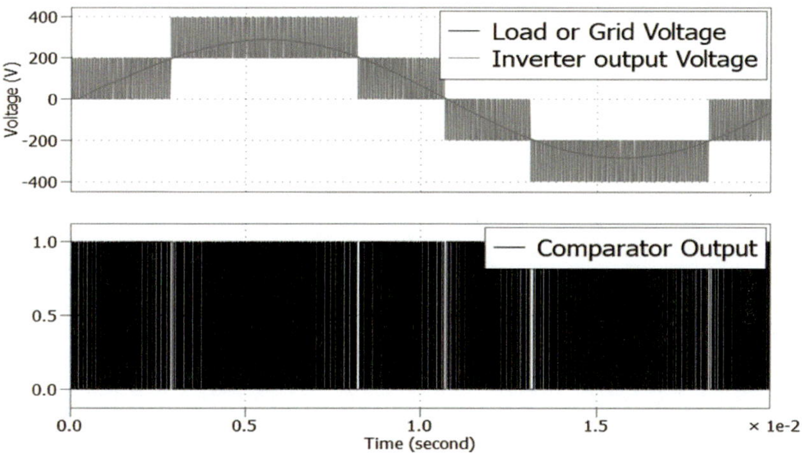

Figure 5.21 Comparator output of comparing inverter output voltage with grid voltage (for one cycle) in method 2.

coordinated with the status of the comparator to obtain the inverter output voltage. This value is used for inductor voltage and current estimation. Figure 5.22 shows the comparison of estimated and actual currents using this method. The switching ripples of the peak magnitudes of the actual current and estimated current are compared in Figure 5.23, a zoomed version of Figure 5.22.

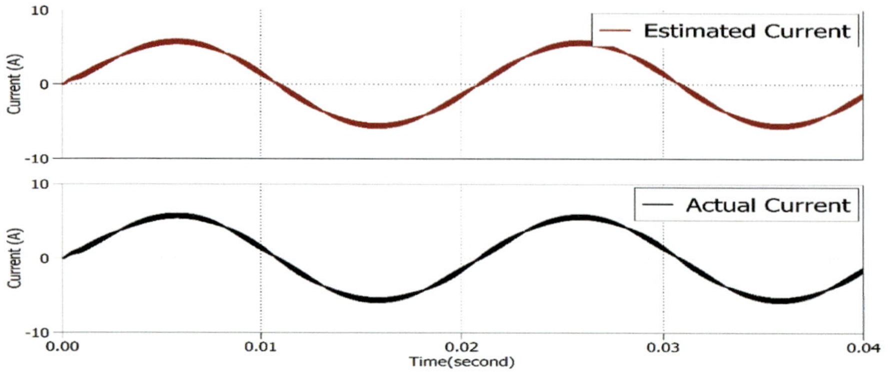

Figure 5.22 Comparison of actual current with the estimated current (method 2).

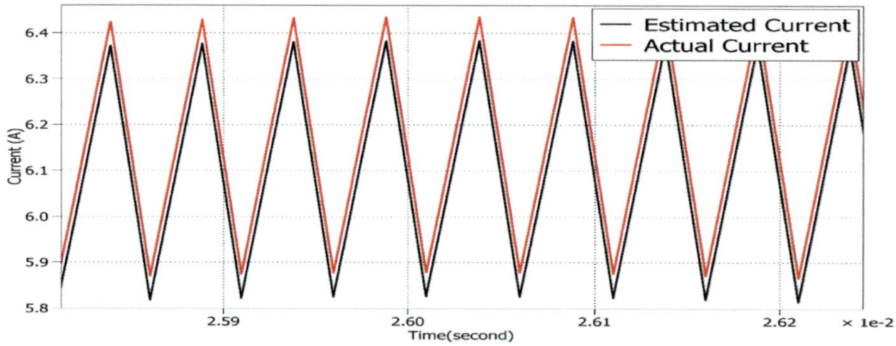

Figure 5.23 Switching ripples of actual current and the estimated current (method 2).

5.5.1 Dependency on DC resistance of the filter inductor

In the first scenario, the resistance value used in the current estimation is the same as the actual resistance of the inductor. After that, the actual resistance is increased by 400% due to aging, but the current estimation parameters are constant. That comparison is shown in Figure 5.24. After increasing the actual resistance, the actual current is decreased as discussed previously. As the resistance increases, the voltage drop increases and the output voltage will be lower. But in closed-loop control, to compensate for that voltage drop, the estimated current is higher.

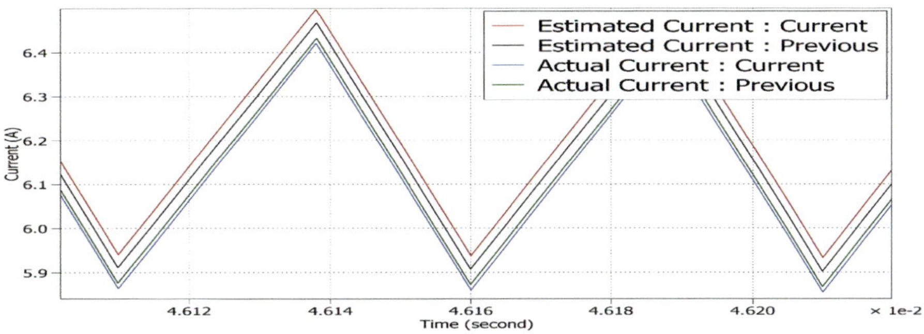

Figure 5.24 Comparison of switching ripples for resistance value mismatch.

5.5.2 Dependency on inductance of the filter inductor

In the first scenario, the inductance value used in the current estimation is the same as the actual inductance. After that, the inductance value increases by 14%, but the data used in the current estimation is not changed. As discussed previously, the actual current is decreased and the actual current ripple is also decreased in later cases, as shown in Figure 5.25. As the actual inductance increases, the voltage drop increases, and the load voltage is reduced. Due to the outer voltage loop control to regulate the output voltage at the desired value, the estimated current is increased. So, it can be said that the estimation method has a higher dependency on parameter *L*.

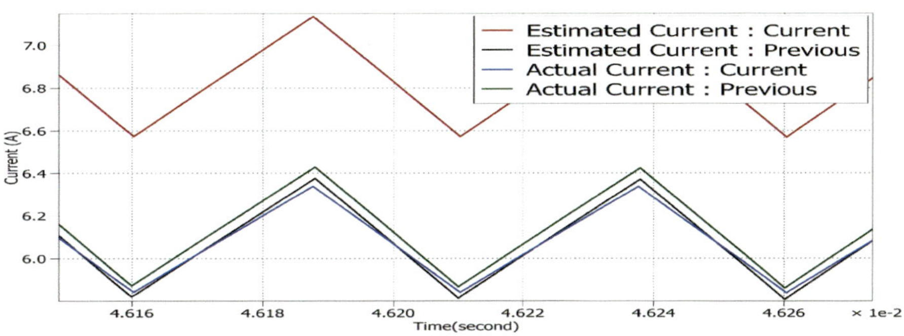

Figure 5.25 Comparison of switching ripples for inductance value mismatch.

5.6 Comparison of Two Methods

This section discusses the comparison between the two methods used in multilevel inverters, particularly focusing on component requirements and applicability to different voltage sources in cascaded inverters. Method 1 requires four voltage sensors and four comparators, making it more component-intensive but potentially offering greater monitoring and control precision. In contrast, method 2 simplifies the design by using only one voltage sensor and one comparator, reducing the overall component count, and possibly enhancing reliability. Additionally, method 1 is applicable to configurations with different voltage sources in a cascaded inverter, offering flexibility in design. On the other hand, method 2 is not suitable for such configurations, limiting its application to systems with single or uniform voltage sources.

5.7 Application

In method 1, the voltage sensors are connected across the independent switches of one leg. This voltage is actually V_{CE} of the switch, which is a few volts for the switch-on period and high voltage for the switch-off period [19]. If the IGBT develops an open-circuit fault, the switch will not conduct even when a gate signal is applied. As a result, V_{CE} remains high when the IGBT is expected to be ON, indicating a fault. In the case of a short-circuit fault, the IGBT will remain in a conducting state regardless of the gate signal. Here, V_{CE} will remain low when the IGBT is supposed to be off, indicating a fault. By continuously monitoring V_{CE}, it is possible to detect deviations from expected behavior. Thus, switch faults can be detected [20, 21]. This method is widely used due to its simplicity and effectiveness in detecting faults early, helping to improve the reliability and safety of power electronic systems.

5.8 Conclusion

This chapter has provided a brief review of many kinds of multilevel inverters, including how they are categorized and how their mathematical models work. A method has been talked about for finding switching states and the inverter output voltage, which lets the filter inductor current be estimated without the need for a current sensor. Two different methods of current estimation have been analyzed, among those, method 1 is more accurate, gives an additional layer of protection, and helps to identify the switch faults, whereas in method 2, current has been estimated with the single-bit analog comparator. So, the current estimation approach not only makes the system easier to use but also makes it more reliable by adding more voltage sensors to find the switch faults.

References

[1] Ned Mohan et al. "Power Electronics Converters Applications." (2023)

[2] D. G. Holmes and B. P. McGrath, "Opportunities for harmonic cancellation with carrier-based PWM for a two-level and multilevel cascaded inverters," *IEEE Transactions on Industry Applications*, vol. 37, no. 2, pp. 574-582, March-April 2001

[3] A. Krishna R and L. P. Suresh, "A brief review on multi level inverter topologies," 2016 International Conference on Circuit, Power and Computing Technologies (ICCPCT), Nagercoil, India, pp. 1-6. 2016

[4] M. Vijeh, M. Rezanejad, E. Samadaei and K. Bertilsson, "A General Review of Multilevel Inverters Based on Main Submodules: Structural Point of View," *IEEE Transactions on Power Electronics*, vol. 34, no. 10, pp. 9479-9502, Oct. 2019

[5] M. Trabelsi, H. Vahedi and H. Abu-Rub, "Review on Single-DC-Source Multilevel Inverters: Topologies, Challenges, Industrial Applications, and Recommendations," IEEE Open Journal of the Industrial Electronics Society, vol. 2, pp. 112-127, 2021.

[6] K. K. Monfared, Y. Neyshabouri, A. Miremad, S. Ahmadi and H. Iman-Eini, "Optimal Switching-Sequence-Based Model Predictive Control for a Hybrid Multilevel STATCOM," IEEE Transactions on Industrial Electronics, vol. 69, no. 10, pp. 9952-9960, Oct. 2022

[7] F. Simonetti, S. Mohamadian, C. Buccella and C. Cecati, "DC Current Suppression in CHB-STATCOM With Model Predictive Control Employing Current Transformers," IEEE Transactions on Industrial Electronics, vol. 71, no. 10, pp. 13135-13145, Oct. 2024

[8] H. Abu-Rub, M. Malinowski and K. Al-Haddad, "Multilevel Converter/Inverter Topologies and Applications," Power Electronics for Renewable Energy Systems, Transportation and Industrial Applications, IEEE, pp.422-462. 2014

[9] F. A. Silva, "Advanced DC/AC Inverters: Applications in Renewable Energy (Luo, F.L. and Ye, H.; 2013) [Book News]," IEEE Industrial Electronics Magazine, vol. 7, no. 4, pp. 68-69, Dec. 2013
[10] A. Poorfakhraei, M. Narimani and A. Emadi, "A Review of Multilevel Inverter Topologies in Electric Vehicles: Current Status and Future Trends," IEEE Open Journal of Power Electronics, vol. 2, pp. 155-170, 2021
[11] H. P. Vemuganti, D. Sreenivasarao, S. K. Ganjikunta, H. M. Suryawanshi and H. Abu-Rub, "A Survey on Reduced Switch Count Multilevel Inverters," IEEE Open Journal of the Industrial Electronics Society, vol. 2, pp. 80-111, 2021
[12] N. V., V. Kumar, T, GowriManohar. "A comprehensive survey on reduced switch count multilevel inverter topologies and modulation techniques", Journal of Electrical Systems and Inf Technol 10, 3, 2023.
[13] K. K. Gupta, A. Ranjan, P. Bhatnagar, L. K. Sahu and S. Jain, "Multilevel Inverter Topologies With Reduced Device Count: A Review," IEEE Transactions on Power Electronics, vol. 31, no. 1, pp. 135-151, Jan. 2016
[14] Sergio de López Diz et al. "A real-time digital twin approach on three-phase power converters applied to condition monitoring", Applied Energy, Volume 334, 120606,ISSN 0306-2619, 2023
[15] B. P. McGrath and D. G. Holmes, "Multicarrier PWM strategies for multilevel inverters," IEEE Transactions on Industrial Electronics, vol. 49, no. 4, pp. 858-867, Aug. 2002
[16] S. -M. Kim, J. -S. Lee and K. -B. Lee, "A Modified Level-Shifted PWM Strategy for Fault-Tolerant Cascaded Multilevel Inverters With Improved Power Distribution," IEEE Transactions on Industrial Electronics, vol. 63, no. 11, pp. 7264-7274, Nov. 2016
[17] B. Wu, and M. Narimani, "High-Power Converters and AC Drives". (2016).
[18] K. Biswas and O. Ray, "A Nonintrusive Digital Current Sensing Method for DC–DC Converters With Wide Load Range," in IEEE Sensors Letters, vol. 7, no. 6, pp. 1-4, Art no. 6002704, June 2023
[19] S. P. Gautam, M. Jalhotra, L. K. Sahu, M. R. Kumar and K. K. Gupta, "A Survey on Fault Tolerant and Diagnostic Techniques of Multilevel Inverter," IEEE Access, vol. 11, pp. 60866-60888, 2023

[20] B. Lu and S. Sharma, "A Literature Review of IGBT Fault Diagnostic and Protection Methods for Power Inverters," 2008 IEEE Industry Applications Society Annual Meeting, Edmonton, AB, Canada, pp. 1-8, 2008

[21] O. -S. Yu, N. -J. Park and D. -S. Hyun, "A Novel Fault Detection Scheme for Voltage Fed PWM Inverter," IECON 2006 - 32nd Annual Conference on IEEE Industrial Electronics, Paris, France, pp. 2654-2659, 2006

6

Switching Strategy Optimization in Transformer-integrated Cascaded Multilevel Inverters

Shaik Shakeera and Kunamneni Rachananjali

EEE, VFSTR, India
E-mail: shakirashaik99@gmail.com; rachananjali@gmail.com

Abstract

This chapter explores the optimization of switching strategies in transformer-integrated cascaded multilevel inverters (CMI) employing the repeating sequence method for configurations with 5, 7, 11, 15, 29, 31, 33, and 35 levels. The chapter delves into the implementation of the repeating sequence technique and compares the performance with selective harmonic elimination PWM (SHEPWM) with respect to total harmonic distortion (THD). The transformer-integrated CMI architecture aims to enhance power electronics systems by incorporating transformers, offering increased voltage levels and improved efficiency. In order to precisely modulate the output waveforms, the inverter's semiconductor components are switched on and off via the repeating sequence technique. This strategy is especially important for multi-level inverters since lowering THD is necessary to lower harmonic distortion and enhance the quality of the output voltage. To assess the significance of switching techniques for the system performance, look at CMIs with 7, 11, 15, 29, 31, 33, and 35 levels are studied. The goal is to maximize the trade-off between harmonic distortion and system complexity by putting repeating sequence PWM into practice. The results of the comparative analysis provide

DOI: 10.1201/9781003669845-6

insights into the most effective switching strategy for transformer-integrated CMIs at different voltage levels. The findings contribute to the advancement of power electronics by identifying optimal configurations that balance the complexity of the inverter system with the need for minimized harmonic distortion, paving the way for more efficient and reliable multilevel inverter designs.

Keywords: Transformer-integrated CMI, inverter configuration, power quality, PWM techniques, repeating sequence technique

6.1 Introduction

Multilevel inverters are used in electrical systems to convert power from direct current (DC) to many levels of alternating current (AC). Due to their ability to produce excellent waveforms with little to no harmonic distortion, these inverters have become more and more common in a variety of applications, including motor drives, renewable energy sources, and high-voltage power transmission. Multilevel voltage source inverters are one particular kind of multilevel inverter (VSI). Multiple voltage levels are conceivable at the output waveform of multilayer inverters, in contrast to standard two-level inverters that have a transition between two voltage levels (positive and negative). This facilitates the attainment of increased voltage and power levels while preserving a reduced harmonic content [2]. There are different configurations of multilevel inverters, but among the most common is an H-bridge inverter that cascades. In the hierarchy, several H-bridge modules are linked in a sequence, with every H-bridge capable of producing various voltage levels. By merging the outputs of these H-bridges, a staircase-like voltage waveform is generated, resulting in a stepped approximation of a sine wave.

The inverters can be categorized as follows based on the voltage levels of the output voltage waveforms:

- Two level inverters
- Multilevel inverters

A multilevel inverter is an apparatus that may generate AC output voltages at different levels.

The comparison table of two-level inverters versus multilevel inverters is shown in Table 6.1.

6.1.1 Differences between two and multilevel inverters

Table 6.1 Two-level versus multilevel inverter comparison.

Two-level inverter	**Multilevel inverter**
1. Output voltage has two levels	1. Output voltage has more number of levels
2. Higher THD	2. Lower THD
3. Easy to control and fewer components	3. Complicated and difficult to handle
4. Low cost and increased power outage	4. Cost is substantial yet power loss is minimal

6.1.2 Distinct multilevel inverter types

- Diode clamped multilevel inverter
- Flying capacitor multilevel inverter
- Cascaded H-bridge multilevel inverter

6.1.2.1 Diode clamped multilevel inverter

A diode-clamped, or neutral point-clamped (NPC) multilevel inverter, is a particular kind of versatile converter that takes advantage of clamping diodes to attain various voltage waveforms at output [7]. This topology of inverters is widely utilized in high-power applications and has become popular in various industrial systems because of its capacity to generate superior waveforms with minimal harmonic distortion in the output.

The Figure 6.1 shows the circuit diagram of a three-level diode-clamped inverter.

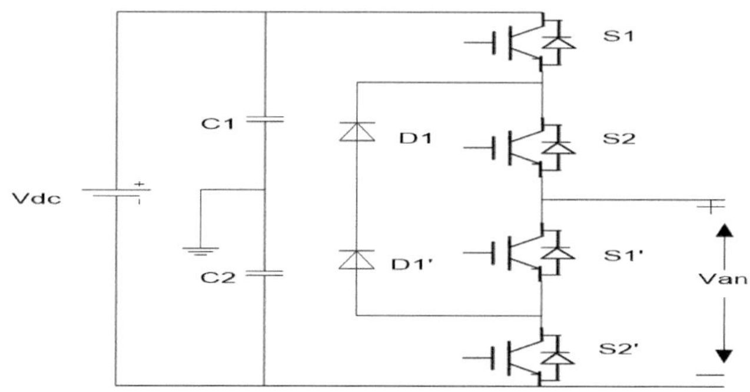

Figure 6.1 Schematic of a three-level diode-clamped inverter.

An m-level inverter with diode clamping requires

- Quantity of capacitors on DC bus $(N_{\mathrm{dc}}) \;=\; m-1.$ (6.1)
- Clamping diode count $(\mathrm{N_{cd}}) = (m-1)\,(m-2).$ (6.2)
- Quantity of semiconductor switches $(N_{\mathrm{ss}}) \;=\; 2\;(m-1)\,.$ (6.3)

6.1.2.2 Flying capacitor multilevel inverter

A flying capacitor multilevel inverter (FCMLI) is another kind of multilevel inverter that uses capacitors as energy storage elements to attain various output levels [7]. A flying capacitor in this context refers to the fact that the capacitors switch between different voltage levels to produce the intended results waveform. This type of multilevel inverter is known due to its capacity to produce excellent sinusoidal output waveforms within diminished harmonic distortion.

The Figure 6.2 shows the circuit diagram of a Three-Level Flying Capacitor Inverter.

The requirement for a flying capacitor inverter at the m-level

- Quantity of DC bus capacitors $(N_{\mathrm{dc}}) = (m-1)\,.$ (6.4)
- Total balancing capacitors $(N_{\mathrm{bc}}) = \frac{(m-1)(m-2)}{2}.$ (6.5)
- Quantity of semiconductor switches $(N_{\mathrm{ss}}) = 2\,(m-1)\,.$ (6.6)

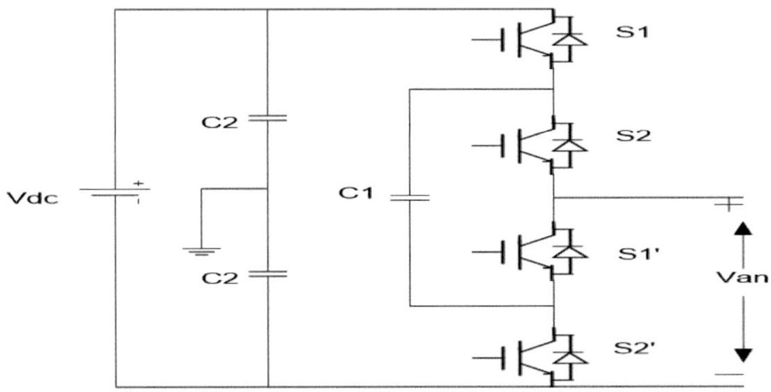

Figure 6.2 Schematic of a three-level flying capacitor inverter.

6.1.2.3 Cascaded H-bridge multilevel inverter

The cascaded H-bridge multilevel inverter was a popular, commonly employed hierarchy within the multilevel inverter family [7]. It is well-known for its capacity to generate output waveforms that feature outstanding quality and reduced harmonic distortion. The cascading H-bridge inverter is designed

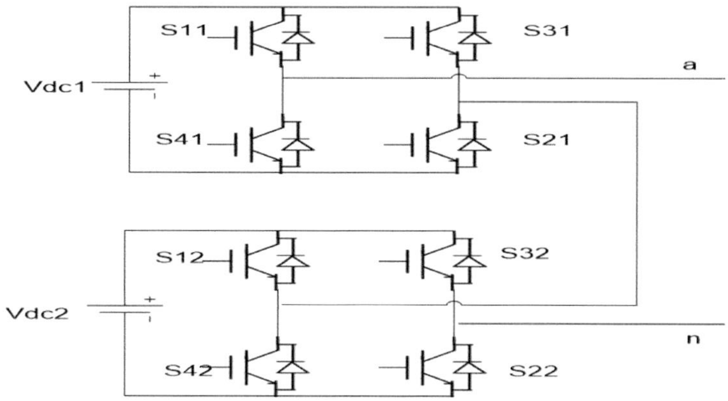

Figure 6.3 Five-level single-phase cascaded H-bridge inverter.

to produce an output that is a stepwise approximation of a sine wave by connecting multiple H-bridge modules in series.

The Figure 6.3 shows the circuit diagram of a Five-Level Single-Phase Structure Inverter.

A cascading m-level H-bridge inverter requires

- Amount of individual DC sources $(N_{\text{dc}}) = \frac{(m-1)}{2}$. (6.7)
- Quantity of semiconductor switches $(N_{\text{ss}}) = 2\,(m-1)$. (6.8)

6.2 Methodology

In cascaded topologies, transformers powered by a single DC source are classified as symmetric or asymmetric based on the transformer's turn ratio; an inverter is deemed balanced if all of its stage turn ratios are equal, and asymmetric if they are not [8–10]. The circuit diagram of a multilevel inverter with one DC source and cascading transformers is shown in Figure 6.4.

Asymmetrical MLI configurations allow for a greater number of levels determined by the inverter stage ratio, whereas symmetric configurations allow for the creation of $2N + 1$ levels of output from N stages [10].

6.2.1 Obtaining the signal for the resultant multilevel

The output voltage of the inverter for the N cascaded stage is as follows:

$$V_{\text{O}}\,(\omega t) = V_1\,(\omega t) + V_2\,(\omega t) + \cdots + V_{\text{N}}\,(\omega t) = \sum_{K=1}^{N} V_{\text{K}}\,(\omega t)\,, \qquad (6.9)$$

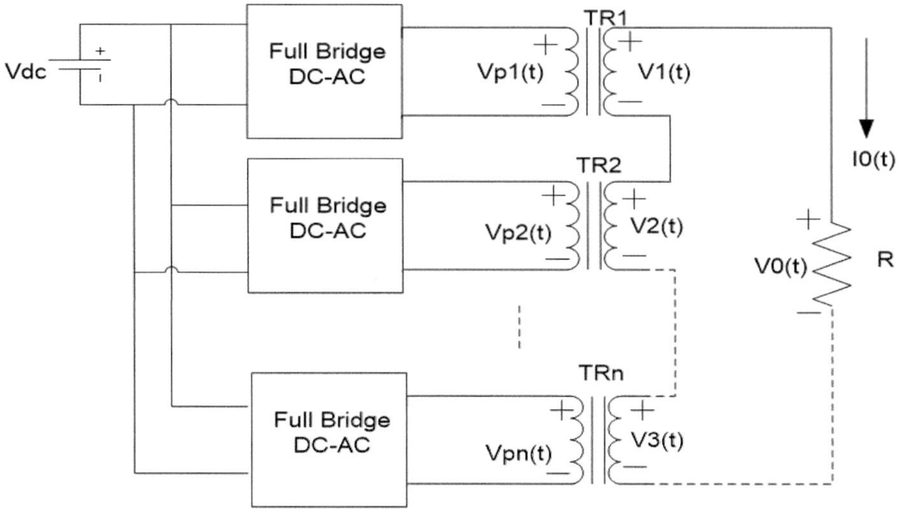

Figure 6.4 Transformer-powered multilevel cascaded inverter.

where $V_{\mathrm{K}}\,(\omega t)$ = instantaneous output voltage.

$$V_{\mathrm{K}}\,(\omega t) = V_{\mathrm{dc}}\,T_{\mathrm{R_K}}\,S_{\mathrm{F_K}}, \tag{6.10}$$

where V_{dc} = the DC input voltage's magnitude, $T_{\mathrm{R_K}}$ = transformer's turns-ratio, and $S_{\mathrm{F_K}}$ = instantaneous switching mechanism.

6.2.2 Turns ratio of transformers

$$T_{\mathrm{R_K}} = \frac{W_{\mathrm{K}} V_{\mathrm{max}}}{M V_{\mathrm{dc}}}, \tag{6.11}$$

where $T_{\mathrm{R_K}}$ = transformer turns-ratio, W_{K} = weights according to the proportion of stages $(W_1 : W_2 : W_3 : W_4)$, V_{max} = maximum output voltage, M = positive integer level, and V_{dc} = Input voltage.

6.2.3 Enhanced function of switching signal voltage levels

The study delves into the implementation of the repeating sequence technique and compares its performance with selective harmonic elimination PWM (SHEPWM) with respect to total harmonic distortion (THD) [11].

The voltage signal is defined by considering the intermediate point where two levels and a pure sine wave intersect, as demonstrated. As the *m* interval comes to an end, the beginning of the subsequent interval is determined by θ_{m} in the ωt axis, allowing the following expression of the m interval in [11] radians in lieu of $m = 1, 2, 3... M$.

$$\theta_{\mathrm{m}} = \sin^{-1}\left(\frac{2m-1}{2M}\right). \tag{6.12}$$

Every level is established regarding the range$[\theta_{\mathrm{m}}, \quad \theta_{\mathrm{m+1}}]$. The levels of voltage utilized within the semicycle that is positive are multiplied to create the semicycle that is negative by -1, and reproducing the initial inversely half of the favorable semicycle, one can obtain the latter portion of the [12] semicycle because of the sinusoidal symmetry of the signal. Table 6.2 summarizes output signal building based on θ_{m} angles.

Selective harmonic elimination pulse width modulation (SHEPWM) is a method employed to modulate the switching of multilevel inverters with the goal of selectively eliminating certain harmonics based on the voltage waveform output. Combining SHEPWM with optimization methods like particle swarm optimization (PSO) and genetic algorithms (GA) allows automated and efficient determination of the optimal switching angles, making it possible to achieve specific harmonic elimination goals [12]. PSO operates as a population-based optimization method influenced by avian social behavior

Table 6.2 Optimized switching function.

Level	Angle at interval start	Angle at interval's end
0	–	θ_1
1	θ_1	θ_2
–	–	–
$(M-1)$	$\theta_{\mathrm{M-1}}$	θ_{M}
M	θ_{M}	$\pi - \theta_{\mathrm{M}}$
$(M+1)$	$\pi - \theta_{\mathrm{M}}$	$\pi - \theta_{\mathrm{M-1}}$
–	–	–
1	$\pi - \theta_2$	$\pi - \theta_1$
0	$\pi - \theta_1$	$\pi + \theta_1$
-1	$\pi + \theta_1$	$\pi + \theta_2$
–	–	–
$-(M-1)$	$\pi + \theta_{\mathrm{M-1}}$	$\pi + \theta_{\mathrm{M}}$
$-M$	$\pi + \theta_{\mathrm{M}}$	$2\pi - \theta_{\mathrm{M}}$
$-(M-1)$	$2\pi - \theta_{\mathrm{M}}$	$2\pi - \theta_{\mathrm{M-1}}$
–	–	–
-1	$2\pi - \theta_2$	$2\pi - \theta_1$
0	$2\pi - \theta_1$	θ_1
0	θ_1	θ_2
–	–	–

and fish. It aims to determine the best response to a given issue by repeatedly modifying a population of potential fixes [13]. This integration allows for a systematic and automated approach to determining modulation strategies, reducing the need for manual tuning and providing flexibility in achieving specific harmonic performance goals [13].

$$V_{0(t)} = \frac{a_0}{2} + \sum_{i=0}^{n} a_{\mathrm{n}}\cos\left(\frac{2\pi nt}{T}\right) + b_{\mathrm{n}}\sin\left(\frac{2\pi nt}{T}\right), \tag{6.13}$$

where $a_0 = $ DC voltage, $a_{\mathrm{n}} = $ even harmonic, and $b_{\mathrm{n}} = $ odd harmonic.

Firing angles in SHEPWM method $(\theta_1,\ \theta_2,\ \theta_3,\ \ldots\theta_{\mathrm{m}})$ are calculated using the formula

$$\theta_{\mathrm{m}} = \sin^{-1}\left(\frac{2m-1}{2M}\right), \tag{6.14}$$

where $m = 0, 1\ldots M$

The maximum positive level is M.

By preserving the connection between $(0 < \theta_1 < \theta_2, \ldots, < \theta_{\mathrm{m}} < \pi/2)$

6.3 Results and Discussion

6.3.1 Three-level inverter's output voltage

Figure 6.5 shows the output waveform of a three-level inverter consisting of four switches and a single bridge integrated with the transformer.

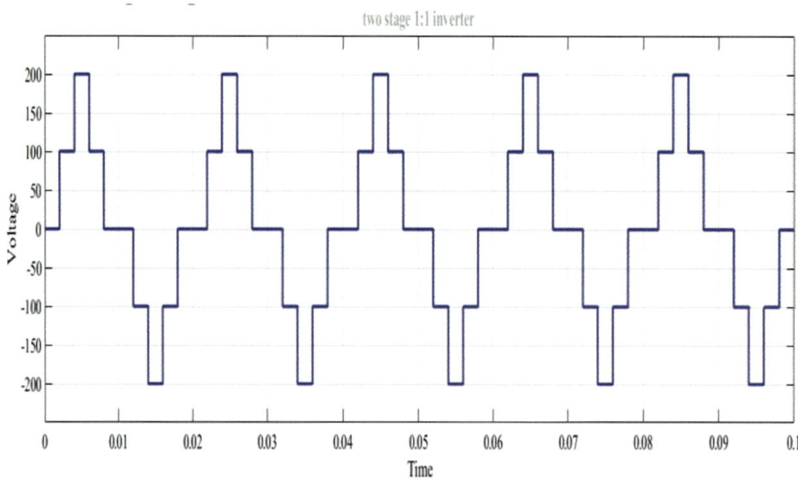

Figure 6.5 The output voltage of the three-level inverter.

6.3.2 Voltage output of five-level inverter

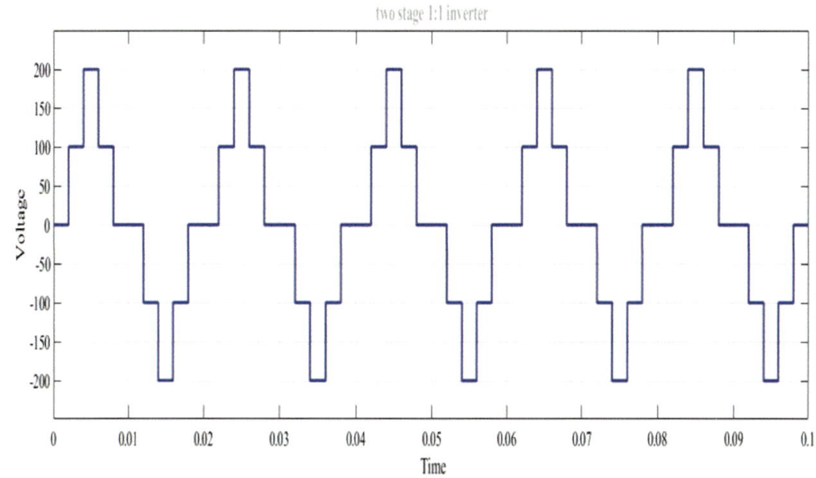

Figure 6.6 The output voltage of the five-level inverter.

Figure 6.6 shows the output waveform of the five-level inverter with two bridges integrated with two transformers, composed of eight switches with a symmetric ratio of 1:1.

6.3.3 Output voltage of seven-level inverter

Figure 6.7 shows the output voltage of a seven-level inverter for an inverter comprised of two bridges integrated with two transformers.

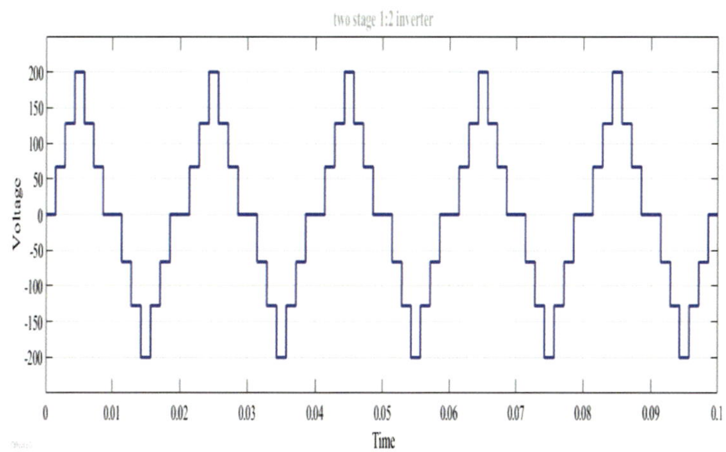

Figure 6.7 Voltage output of seven-level inverter.

Formula:

Time period = 1/(frequency) = 1/(50) = 0.02 milliseconds.

Total cycle = 360°

Number of levels = m = 14/cycle.

Pulse width = 360/m.

Time period for each level = $\frac{0.01}{m}$

6.3.4 Output voltage of the nine-level inverter

The asymmetrical nine-level inverter consists of eight IGBT switches with a single DC source that has parallel connections to H-bridges and also has two transformers that are single-phase and connected in series. Waveform as shown in Figure 6.8. Certain switches come into operation based on the phase delay given to them and pulse width is calculated. Firstly, from the DC source, current flows through the switch S_1 transformer resistor(r) to S_6 and after a certain phase delay, switch S_5 conducts through which S_2 conducts upon delay and pulse width. On account of all these, a positive half cycle is produced. Nine voltage ranges are produced 0, 1, 2, 3, 4, 3, 2, 1, and 0 V. Similarly, the same is repeated for the negative half cycle with remaining switches. Due to this, negative half cycle is produced with negative voltage

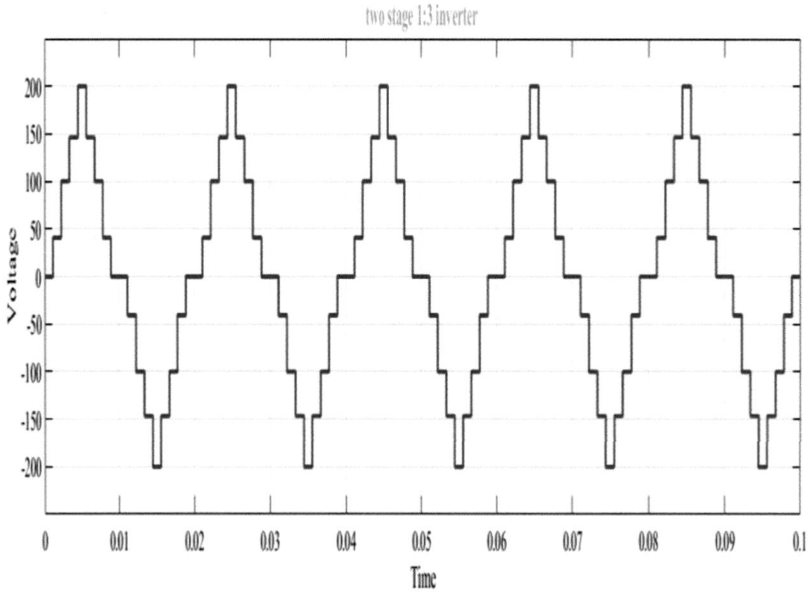

Figure 6.8 The output voltage of the nine-level inverter.

levels 0, −1, −2, −3, −4, −3, −2, −1, and 0 V. The pulse width is calculated with the help of phase delay.

6.3.5 Simulation diagram of the 11-level inverter by repeating sequence PWM method

The asymmetrical 11-level inverter consists of 12 IGBT switches with a single parallel-connected DC source to an H-bridge having a trio of series-connected single-phase transformers as shown in Figure 6.9. Certain switches come into operation based on the phase delay and pulse width.

From the DC source current flows through the switch S_1 transformer resistor(r) to S_{10} and after a certain phase delay, switch S_9 conducts through which S_6 conducts upon delay and pulse width, and switch S_5 conducts through which S_2. On account of all these, a positive half cycle is produced. 15 voltage levels are produced similarly, and the same is repeated for the negative half cycle with the remaining switches. Due to this, negative half cycle is produced with negative voltage levels.

Asymmetric ratio 1:1:3 by repeating sequence PWM method the 11-level inverter provides a 20%:20%:60% power distribution. The same PWM technique is implemented for the 19-level by changing the transformer turn ratios

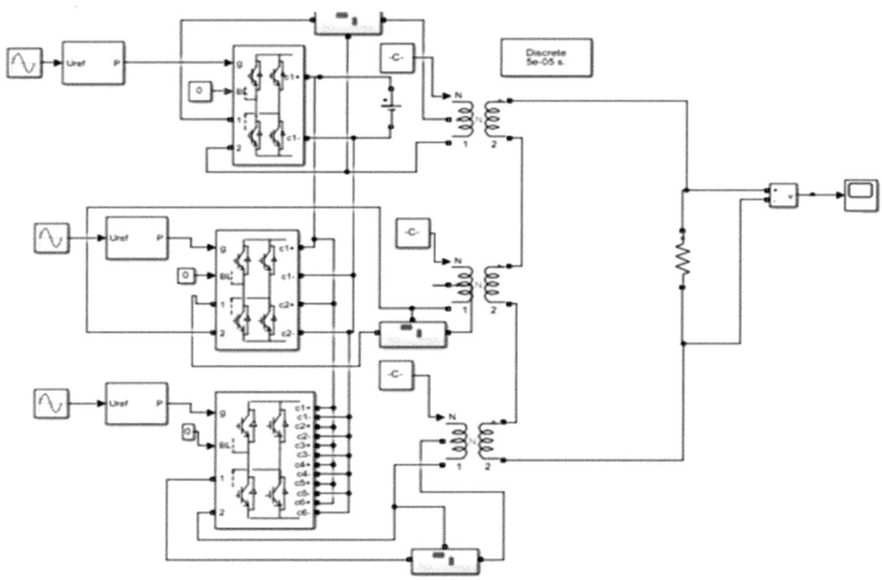

Figure 6.9 11-level inverter simulation diagram using repeating sequence method.

of 1:2:6 providing the power distribution of 11%:22%:66%. By using this method, the power balance is marginally improved at the expense of additional PWM control. The conventional inverter uses symmetric 20 switches and DC voltage sources whereas to generate the same quantity of levels, this converter utilizes a smaller quantity of switches.

6.3.6 Simulation diagram of selective harmonic elimination pulse width modulation (SHEPWM) 11-level inverter

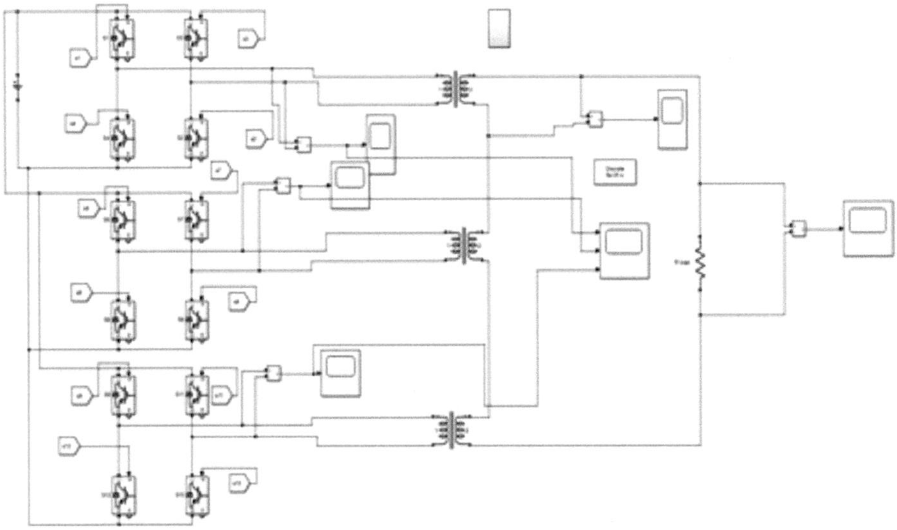

Figure 6.10 11-level inverter simulation diagram using SHEPWM approach.

$$\theta_m = \sin^{-1}\left(\frac{2m-1}{2M}\right) \qquad \theta_3 = \sin^{-1}\left(\frac{2\times3-1}{2\times5}\right) = 30°$$

$$\theta_1 = \sin^{-1}\left(\frac{2\times1-1}{2\times5}\right) = 5.379° \qquad \theta_4 = \sin^{-1}\left(\frac{2\times4-1}{2\times5}\right) = 44.42°$$

$$\theta_2 = \sin^{-1}\left(\frac{2\times2-1}{2\times5}\right) = 17.457° \qquad \theta_5 = \sin^{-1}\left(\frac{2\times5-1}{2\times5}\right) = 64.158°$$

6.3.7 Output voltage waveform of 11-level inverter

Figure 6.10 shows the 11-level inverter simulation diagram by using SHEPWM approach consists of 12 IGBT switches with a single parallel-connected DC source to h-bridges having a trio of series connected single-phase transformers. Certain switches come into operation based upon the phase delay given to them and pulse width is calculated. Figure 6.11 shows the output voltage waveform of 11-level inverter by SHEPWM method.

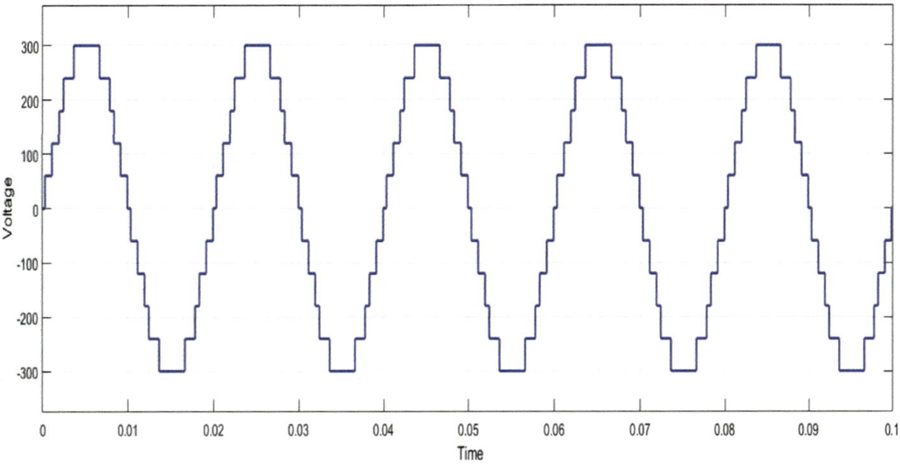

Figure 6.11 Eleven-level inverter's output voltage.

6.3.8 Output voltage of 15-level inverter

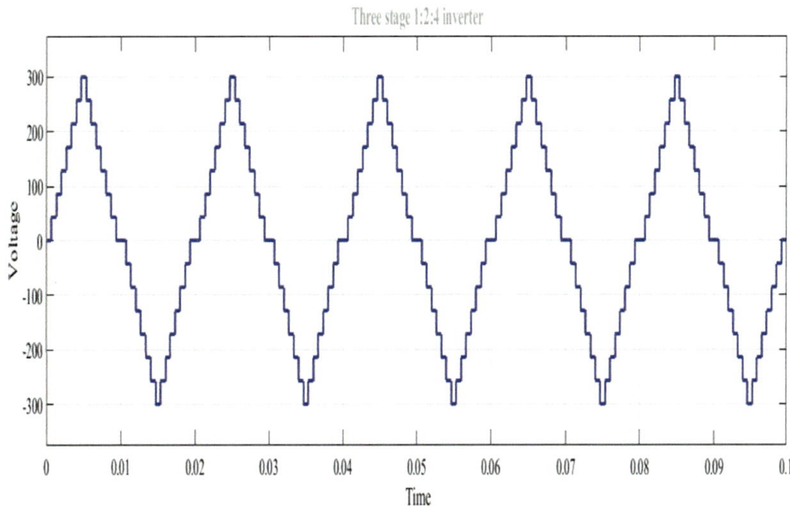

Figure 6.12 Output voltage of 15-level inverter.

Here, the repeating sequence is implemented through the delay repetition, repeated sequence via delay pattern. Asymmetric binary ratio 1:2:4 enables the acquisition of a 15-level signal and produces an imbalanced 15.01%:36.03%:48.96% power distribution. Figure 6.12 shows the output voltage waveform of 15-level inverter by using repeating sequence method.

The same repeating sequence technique is implemented for the 27-level by changing the transformer turn ratio of 1:3:9 and yields an inadequate power distribution of 15.14%:81.39%: 3.47%.

6.3.9 Switching pattern synthesis

The suggested asymmetrical relation in this work creates an arithmetic sequence that poses an enhancement challenge within the distinct field and makes use of $W_K \in N$. [14] The choice of W_K denotes the discovery of a relationship between the stages' levels that permits the extraction of both the associated redundancy, which offers an optimization degree, and the complete signal levels via their algebraic sum.

6.3.9.1 Acquiring signal levels through redundancy

Since the inverter's single DC source powers each stage, all semiconductors experience the same voltage stress. Consecutive integer weights in semiconductor devices are used to ensure a similar current stress rating. To acquire a partial cycle of the *M*-level signal, solve $w_k = w,\ w+1,\ w+2,\ w+3$; varying *w*, multiple solutions are found in which the voltage range from 0 to *M* interval [14]. The desired levels are achieved by using redundancy. Redundancy in the standards is set as $\{w,\ w \in N : 3 \le w \le 7\}$ for $M = 15$.

For weight $w = (4{:}5{:}6{:}7)$ and a shorter time period, a high degree of lower levels of redundancy is to be used. All levels exhibit distributed redundancy. There is no redundancy at the 13th level for $w = 7$ (7:8:9:10), so it does not reach the highest level [14]. There are more possible switching patterns for lower values of *w*. Insufficient redundancy at the highest level makes it less likely that stages will have equal power distribution.

It is possible to generate a 35-level signal by using the four-stage cascaded multilevel inverter's asymmetric ratio, which is 6:7:8:9, results in a highly balanced power distribution through switching pattern optimization [15].

The parameter table for four stage inverter (6:7:8:9) is shown in Table 6.3.

6.3.9.2 Output voltage of asymmetric (6:7:8:9) 29-level inverter

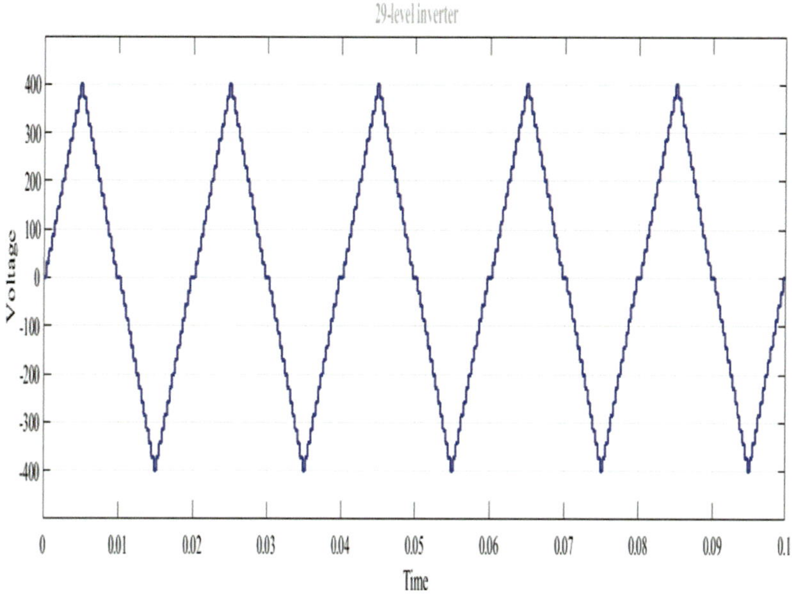

Figure 6.13 Output voltage of the inverter with 29 levels.

6.3.9.3 Switching pattern of asymmetric (6:7:8:9) 29-level inverter

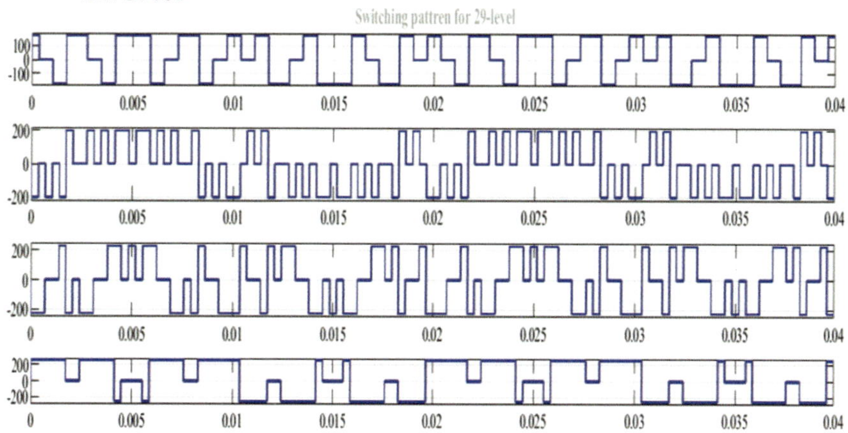

Figure 6.14 Switching pattern of 29-level inverter.

Figure 6.13 shows the output waveform of the 29-level inverter, inverter topology consists of eight switches and four bridges combined with the four transformers with an uneven ratio of 6:7:8:9. Figure 6.14 shows the switching pattern of the 29-level inverter for each bridge stage.

6.3.9.4 Output voltage of asymmetric (6:7:8:9) 31-level inverter

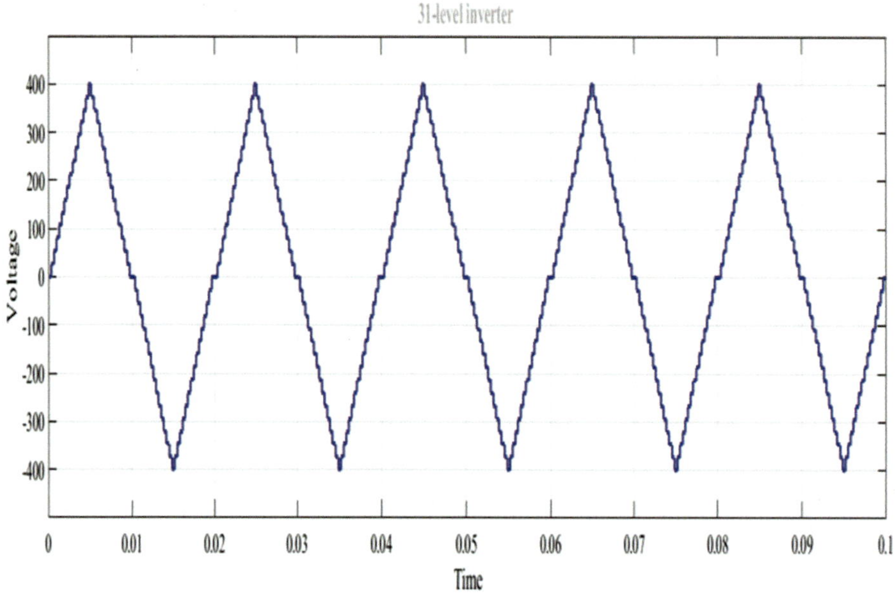

Figure 6.15 Output voltage of the inverter with 31 levels.

6.3.9.5 Switching pattern of asymmetric (6:7:8:9) 31-level inverter

Figure 6.15 shows the output waveform of the 31-level inverter. Inverter topology consists of eight switches and four bridges combined with the four transformers with an uneven ratio of 6:7:8:9. Figure 6.16 shows the switching behavior of the 31-level inverter for each bridge stage.

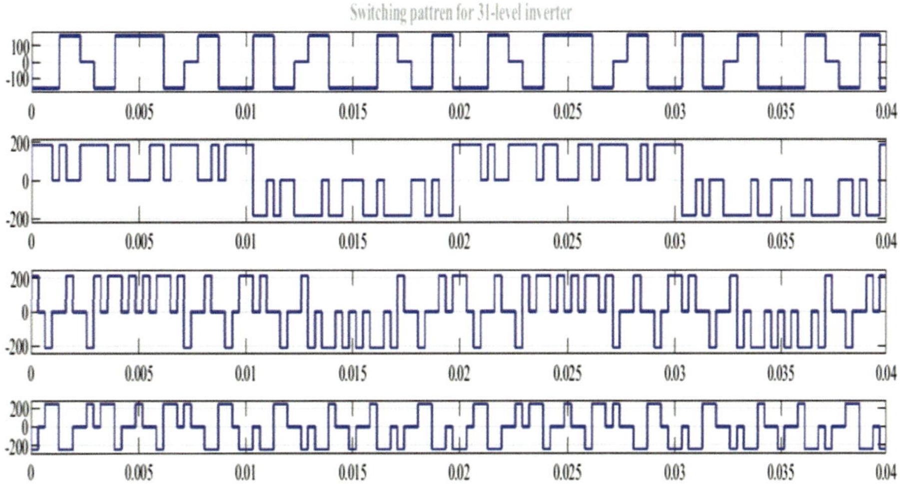

Figure 6.16 31-level inverter switching pattern.

6.3.9.6 Output voltage of asymmetric (6:7:8:9) 33-level inverter

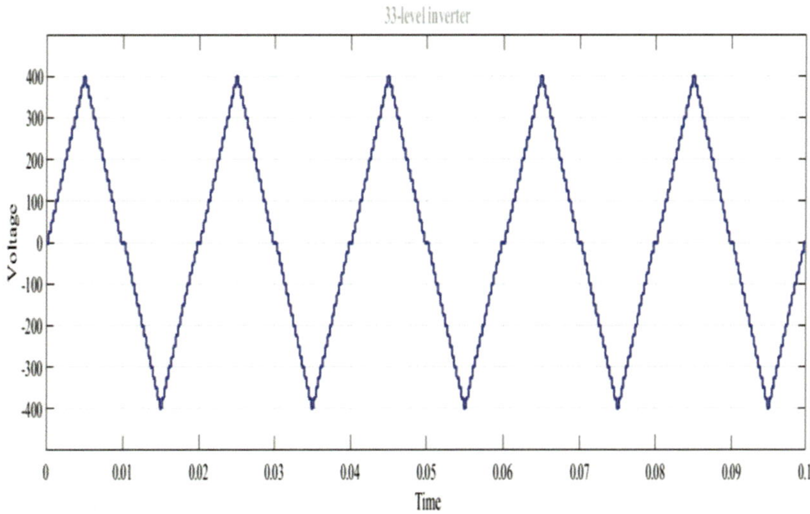

Figure 6.17 Output voltage of the inverter with 33 levels.

6.3.9.7 Switching pattern of asymmetric (6:7:8:9) 33-level inverter

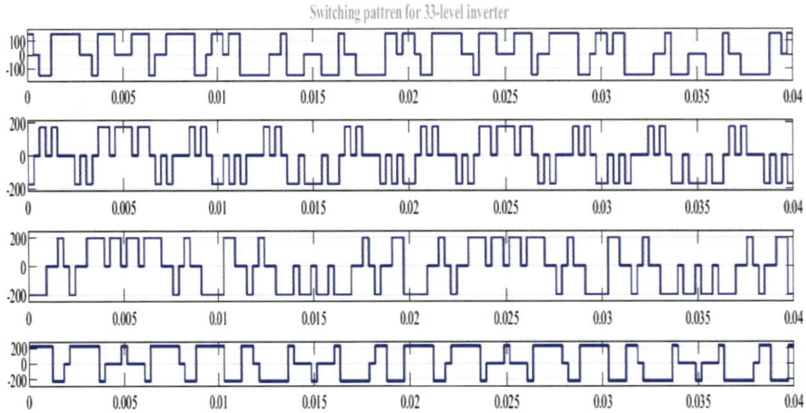

Figure 6.18 Sequential switching of 33-level inverter.

Figure 6.17 shows the output waveform of the 33-level inverter. Inverter topology consists of eight switches and four bridges combined with the four transformers with asymmetrical 6:7:8:9 ratios. Figure 6.18 shows the switching behavior of the 33-level inverter for each bridge stage.

6.3.9.8 Output voltage of asymmetric (6:7:8:9) 35-level inverter by SHEPWM

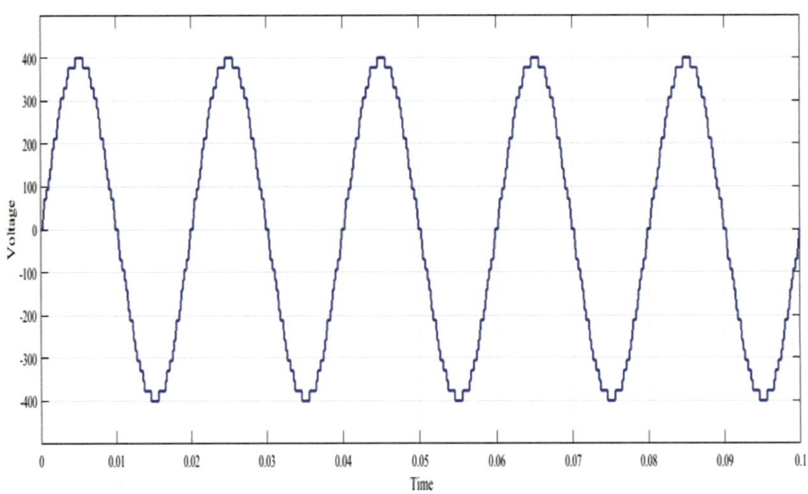

Figure 6.19 Output voltage of the inverter with 35 levels.

$$\theta_{\mathrm{m}} = \sin^{-1}\left(\frac{2m-1}{2M}\right)$$
$$\theta_1 = \sin^{-1}\left(\frac{2\times 1-1}{2\times 17}\right)$$
$$\theta_1 = 1.685^\circ$$
$$\theta_2 = \sin^{-1}\left(\frac{2\times 2-1}{2\times 17}\right)$$
$$\theta_2 = 5.062^\circ$$
$$\theta_3 = \sin^{-1}\left(\frac{2\times 3-1}{2\times 17}\right)$$
$$\theta_3 = 8.456^\circ$$
$$\theta_4 = \sin^{-1}\left(\frac{2\times 4-1}{2\times 17}\right)$$
$$\theta_4 = 11.881^\circ$$
$$\theta_5 = \sin^{-1}\left(\frac{2\times 5-1}{2\times 17}\right)$$
$$\theta_5 = 15.349^\circ$$
$$\theta_6 = \sin^{-1}\left(\frac{2\times 6-1}{2\times 17}\right)$$
$$\theta_6 = 18.876^\circ$$
$$\theta_7 = \sin^{-1}\left(\frac{2\times 7-1}{2\times 17}\right)$$
$$\theta_7 = 22.479^\circ$$
$$\theta_8 = \sin^{-1}\left(\frac{2\times 8-1}{2\times 17}\right)$$
$$\theta_8 = 26.178^\circ$$
$$\theta_9 = \sin^{-1}\left(\frac{2\times 9-1}{2\times 17}\right)$$
$$\theta_9 = 30^\circ$$
$$\theta_{10} = \sin^{-1}\left(\frac{2\times 10-1}{2\times 17}\right)$$
$$\theta_{10} = 33.974^\circ$$
$$\theta_{11} = \sin^{-1}\left(\frac{2\times 11-1}{2\times 17}\right)$$
$$\theta_{11} = 38.144^\circ$$
$$\theta_{12} = \sin^{-1}\left(\frac{2\times 12-1}{2\times 17}\right)$$
$$\theta_{12} = 42.568^\circ$$
$$\theta_{13} = \sin^{-1}\left(\frac{2\times 13-1}{2\times 17}\right)$$
$$\theta_{13} = 47.332^\circ$$
$$\theta_{14} = \sin^{-1}\left(\frac{2\times 14-1}{2\times 17}\right)$$
$$\theta_{14} = 52.571^\circ$$
$$\theta_{15} = \sin^{-1}\left(\frac{2\times 15-1}{2\times 17}\right)$$
$$\theta_{15} = 58.533^\circ$$
$$\theta_{16} = \sin^{-1}\left(\frac{2\times 16-1}{2\times 17}\right)$$
$$\theta_{16} = 65.750^\circ$$
$$\theta_{17} = \sin^{-1}\left(\frac{2\times 17-1}{2\times 17}\right)$$
$$\theta_{17} = 76.06^\circ$$

6.3.9.9 Switching pattern of asymmetric (6:7:8:9) 35-level inverter by SHEPWM

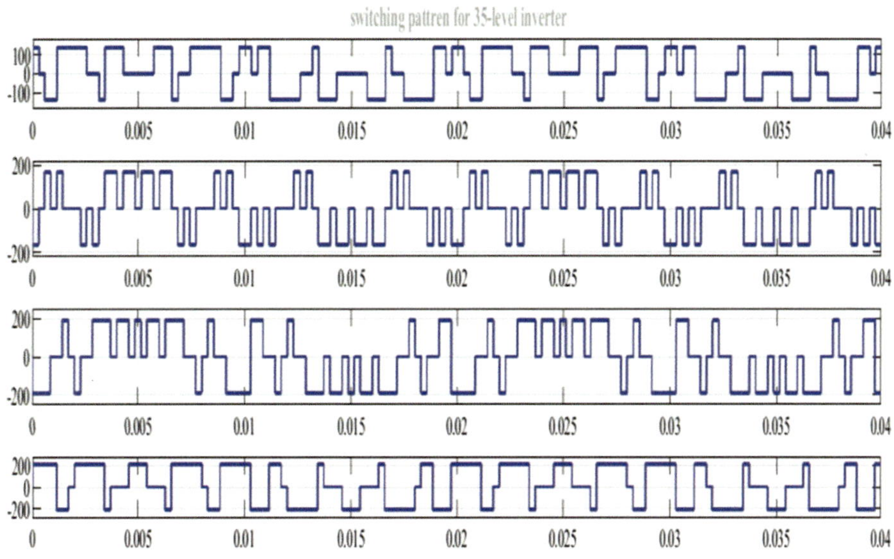

Figure 6.20 35-level inverter switching pattern.

The 35-level inverter's output waveform with four bridges integrated with the four transformers is presented in Figure 6.19. The topology consists of eight switches, and has an asymmetrical of 6:7:8:9 ratios. Figure 6.20 shows the 35-level inverter switching pattern for each bridge stage.

6.3.9.10 Parameter table

Table 6.3 Parameters of four-stage inverter

Parameter	Value
Voltage source	100 V
Fundamental frequency	50 Hz
Load resistor	100 Ω
Load inductance	200 mH
Switching frequency	3 KHz

6.4 Conclusion

This chapter presents a multilevel inverter cascade approach on transformers to enforce symmetrical power distribution. The method is based on the selection of an optimal switching pattern for each inverter. The inverter can be changed for a variety of extra tiers (27, 29, 31, 33, and 35) because the suggested asymmetry ratio of 6:7:8:9 provides for optimal switching patterns. Revolutionary multilevel inverter architecture uses fewer switches than the traditional topology, producing more levels with the same switches in the customary topology. To reduce the harmonic content, a new repeating

Table 6.4 THD comparison table.

Inverter configuration	THD level (%)
Three-level cascaded H-bridge multilevel inverter (CHBMLI)	78.65
Five-level cascaded H-bridge multilevel inverter	42.92
Seven-level cascaded H-bridge multilevel inverter	31.13
Nine-level cascaded H-bridge multilevel inverter	27.65
29-level cascaded H-bridge multilevel inverter	15.05
31-level cascaded H-bridge multilevel inverter	14.87
33-level cascaded H-bridge multilevel inverter	14.47
35-level cascaded H-bridge multilevel inverter	14.13
11-level CHBMLI with repeating sequence method	5.55
11-level CHBMLI with SHEPWM method	4.04
15-level CHBMLI with repeating sequence method	3.16
35-level CHBMLI with SHEPWM method	1.98

sequence mechanism and selective harmonic elimination (SHEPWM) are applied, which basically changes each level's step angle to make it more like a sine waveform. The multilevel inverter with switching sequence uses MATLAB to convert an asymmetrical turn's ratio into a 35-level waveform. A comparison of the obtained data shows that the proposed topology reduces the harmonic content.

As more levels are added, the output waveform approaches a sinusoidal shape, and also, THD decreases. Table 6.4 indicates that the SHEPWM technique reduces THD when compared to repeating sequence and sine PWM.

6.5 Future Scope

The technology proposed can have a big impact on smart grid applications, electric car charging, and the integration of renewable energy sources because of the growing need for small, energy-efficient power conversion devices. The transformer-integrated MLI lowers costs, improves reliability, and streamlines the design by eliminating the requirement for numerous DC sources or intricate control systems. Subsequent investigations may concentrate on enhancing the power density and efficiency of these inverters, as well as refining the transformer design to minimize losses and improve thermal regulation. The dynamic performance and fault tolerance of these inverters could be further improved by the integration of sophisticated control algorithms, such as machine learning-based optimization and model predictive control, making them appropriate for a variety of utility-scale, residential, and industrial applications.

Acknowledgment

Any task that is successfully done must be acknowledged by the people whose leadership and unwavering support enabled our efforts to be effective.

We use this chance to convey my sincere gratitude and deep regards to *Dr. Lavu Rathaiah*, Chairman and Chancellor, Vignan's Foundation for Science, Technology, and Research University, Vadlamudi, for giving this project the facilities it needs to be finished.

We would want to express my gratitude to *Col. Prof. P. Naga Bhushan*, Vice-Chancellor, Vignan's Foundation for Science, Technology, and Research University, Vadlamudi, for giving this project the facilities it needs to be finished.

We express my sincere thanks to *Dr. P. V. S. Sobhan*, Associate Professor, Department of Electrical and Electronics Engineering of Vignan's Foundation for Science, Technology, and Research (Deemed to be University) for his precious guidance, regular encouragement, and prop up during the course of my work.

We articulate my gratitude and sincere thanks to *Dr. K. Rachananjali*, Associate Professor, Department of Electrical and Electronics Engineering of Vignan's Foundation for Science, Technology, and Research (Deemed to be University) for her valuable guidance, constant encouragement, and support during the duration of our work. From the start of this thesis to the finish, I sincerely appreciate her excellent advice and support. His expertise and help in a time of need will always be remembered.

We extend my thanks to all the *staff and nonteaching members* of the Electrical and Electronics Engineering Department, *Vignan's Foundation for Science, Technology, and Research, Vadlamudi, Guntur* for the constant support and encouragement throughout project work.

References

[1] Jagabar Sathik Mohamed Ali, Vijayakumar Krishnasamy, "Compact Switched Capacitor Multilevel Inverter (CSCMLI) With Self-Voltage Balancing and Boosting Ability," *IEEE Transactions on Power Electronics*," vol. 34, no. 5, pp. 4009-4013, 2019, DOI:10.1109/TPEL.2018.2871378

[2] Concettina Buccella, Maria Gabriella Cimoroni, Mario Tinari, Carlo Cecati, "A New Pulse Active Width Modulation for Multilevel Converters," *IEEE Transactions on Power Electronic,* vol. 34, no. 8, pp. 7221-7229, 2019, DOI:10.1109/TPEL.2018.2878967

[3] Junfeng Liu , Xiangkai Zhu, and Jun Zeng, "A Seven-Level Inverter With Self-Balancing and Low-Voltage Stress," *IEEE Journal Of Emerging And Selected Topics In Power Electronics*, vol. 8, no. 1, pp. 685-696, 2020, DOI:10.1109/JESTPE.2018.2879890

[4] Marif Daula Siddique, Atif Iqbal, Mudasir Ahmed Memon, Saad Mekhilef, "A New Configurable Topology for Multilevel Inverter with Reduced Switching Components," *IEEE Access,* vol. 8, pp. 188726 – 188741, 2020, DOI:10.1109/ACCESS.2020.3030951

[5] Marif Daula Siddique, Saad Mekhilef, Sanjeevikumar Padmanaban, Mudasir Ahmed Memon , Chandan Kumar, "Single-Phase

Step-Up Switched-Capacitor-Based Multilevel Inverter Topology With SHEPWM," *IEEE Transactions On Industry Applications,* vol. 57, no. 3, pp. 3107-3119, 2020, DOI:10.1109/TIA.2020.3002182

[6] Anilkumar Chappa , Shubhrata Gupta, Lalit Kumar Sahu, Shivam Prakash Gautam , Krishna Kumar Gupta, "Symmetrical and Asymmetrical Reduced Device Multilevel Inverter Topology," *IEEE Journal Of Emerging And Selected Topics In Power Electronics* , vol. 9, no. 1, pp. 885-896, 2021, DOI:~10.1109/JESTPE.2019.2955279

[7] Peeyush Kala, Sudha Arora, "Implementation of Hybrid GSA SHE Technique in Hybrid Nine-Level Inverter Topology," *IEEE Journal Of Emerging And Selected Topics In Power Electronics,* vol. 9, no. 1, pp. 1064-1074, 2021, DOI:10.1109/JESTPE.2019.2963239

[8] Yuanmao Ye, Guoxin Zhang, Xiaolin Wang , Yong Yi , and K. W. E. Cheng, "Self-Balanced Switched-Capacitor Thirteen-Level Inverters With Reduced Capacitors Count," *IEEE Transactions On Industrial Electronics,* vol. 69, no. 1, pp. 1070-1076, 2021, DOI:10.1109/TIE.2021.3050378

[9] Pallavee Bhatnagar, Ankit Kumar Singh, Krishna Kumar Gupta, Yam P.Siwakoti, "A Switched-Capacitors-Based 13-Level Inverter," *IEEE Transaction on Power Electronics,* vol. 37, no.1, pp. 644-658, 2021, DOI:10.1109/TPEL.2021.3098827

[10] Amirhosein Akbari, Javad Ebrahimi, Yousefreza Jafarian, Alireza Bakhshai, "A Multilevel Inverter Topology with an Improved Reliability and a Reduced Number of Components," *IEEE Journal of Emerging and Selected Topics in Power Electronics,* vol. 10, no. 1, pp. 553-563, 2022, DOI:10.1109/JESTPE.2021.3089867

[11] S. Shakeera and K. Rachananjali, "Advancing Power Conversion: A Comprehensive Survey on Reduced Multilevel Inverters, Switching Techniques, And Controllers," 2024 International Conference on Advancements in Power, Communication and Intelligent Systems (APCI), KANNUR, India, 2024, pp. 1-6, DOI:10.1109/APCI61480.2024.10617107

[12] Ailton Do Egito Dutra, Montie Alves Vitorino, Mauricio Beltrao De Rossiter Correa, "A Survey on Multilevel Rectifiers With Reduced Switch Count," *IEEE Access,* vol. 11, pp. 56098-56141, 2023, DOI: 10.1109/ACCESS.2023.3283215

[13] S. Shakeera and K. Rachananjali, "Study of Multilevel Inverter with SHE-PWM Using PSO Algorithm," 2024 International Conference on

Intelligent and Innovative Technologies in Computing, Electrical and Electronics (IITCEE), Bangalore, India, 2024, pp. 1-5, DOI:10.1109/IITCEE59897.2024.10467266

[14] Ashraf Ali Khan, Usman Ali Khan, Shahnawaz Khan, Yun W. Lu, Shehab Ahmed, "Cascaded Inverters Increasing the Number of Levels and Effective Switching Frequency in Output Using Coupled Inductors," *IEEE Journal Of Emerging And Selected Topics In Power Electronics*, vol. 11, no. 4, pp. 4045-4056, 2023, DOI:10.1109/JESTPE.2022.3178702

[15] Shakeera, S, Rachananjali, K, "An innovative 11-level multilevel inverter topology with rotating trapezoidal SPWM for industrial and renewable applications", *Sci Rep* 14, 22359 (2024), DOI:10.1038/s41598-024-73791-0

7

Z-Source-based Multilevel Inverter Topologies for Photovoltaic Applications

Jammy Ramesh Rahul[1], Kuncham Sateesh Kumar[2], T. Abhilash[3], A. Kirubakaran[4], and B. Hemanth Kumar[5]

[1]Department of Electrical and Electronics Engineering, Gayatri Vidya Parishad College of Engineering (A), India
[2]Department of Electrical and Electronics Engineering, National Institute of Technology Trichy, India
[3]Caterpillar, India
[4]Department of Electrical Engineering, National Institute of Technology Warangal, India
[5]Department of Electrical Engineering, Mohan Babu University, India
E-mail: rahuljammy1925@gvpce.ac.in; sateeshkuncham@gmail.com; abhilash.tirupathi@cat.com; kiruba81@nitw.ac.in; hemub09@gmail.com

Abstract

Emerging renewable-based energy sources such as photovoltaic (PV), wind energy, and fuel cells (FC) are gaining much attention due to dwindling fossil fuels and increasing environmental concerns. However, photovoltaic energy has witnessed tremendous growth over recent years due to reduced installation costs and advancements in semiconductor technology. The power conditioners play a crucial role in injecting PV power into the stand-alone/grid. In this context, the objective of this chapter is to introduce the state-of-the-art single-stage power conversion of Z-source (impedance-source)-based power converter topologies for PV applications in both three-level and multilevel generation. Integration of Z-source with inverter can achieve benefits of voltage boosting and bucking capability with power conversion in one stage, enhanced reliability provided by short circuit protection, and better input voltage regulation. In this chapter, primarily, the derivation, classification, and comparison of various Z-source networks will be presented. Later, a

DOI: 10.1201/9781003669845-7

detailed review of Z-source-based three-level and multilevel topologies will be presented, compared, and evaluated based on the number of components, input current pattern, pros, and cons. At last, some of the recent proposals are presented. Hence, the reader would be able to make an estimated choice of a Z-source inverter for PV applications.

Keywords: Impedance source networks, isolated power converters, multilevel inverters, nonisolated power converters, pulse width modulation scheme, photovoltaic power, shoot-through control, three-level inverters, Z-source-based power converter topologies

7.1 Introduction

The classical two-level voltage source inverter (VSI) is well-known and popular for its simple structure and control. It is utilized in various applications like distribution power generation systems, electric vehicle discharging, motor drives, and others. However, the output of VSI and current source inverter (CSI) are basically buck in nature; they require an intermediate DC-DC converter in order to increase the lower input DC voltage to the desired voltage at the DC link or a 50 Hz transformer at the load end for obtaining desired load voltage. This multistage conversion results in increased losses and reduced efficiency due to additional power conversion stages, whereas the use of a transformer results in bulkiness of the overall system and increased cost. In some inverters, a dead band is provided among the switches of a phase so as to avoid shoot-through, which leads to waveform distortion. At this juncture, F.Z. Peng introduced an impedance source inverter [1] which overcomes the shortcomings of conventional inverters. It offers a boost-buck feature with one-stage DC-AC power conversion, increased reliability, and improved input voltage regulation. It can be employed for all forms of power conversion for both three-phase and single-phase, however, mostly used in DC to AC and DC to DC converters. The shoot-through (ST) is achieved by the simultaneous conduction of the switches of any of the phases or all three phases of the inverter. The output voltage is controlled by varying the ST duty cycle D_o.

Several shoot-through control strategies are reported for varying the output voltage gain of the impedance source inverter [2–4] such as constant boosting, simple boosting, and maximum boosting control schemes. The shoot-through states play a crucial role in the increase of output voltage gain and it will be included in the null states, by keeping the working states intact and without changing the inverter output voltage. However, it requires additional circuitry to measure the pulsating DC link voltage. However, various control strategies are reported [5] so as to preserve the peak DC-link voltage as constant through proper control of capacitor voltages of the impedance network. Despite the various advantages, the ZSI operates with an input source current of discontinuous nature and it is a limitation to use it for PV applications. Another improvised structure, q-ZSI (quasi Z-source inverter) [6] is introduced, which inherits the benefits of Z-source inverters and also provides input current of continuous nature, compactness, minimized voltage across electrolytic capacitors, and common ground point among the output and input terminals which makes the system to be compliable for distributed power generation systems.

At the same time, multilevel inverters (MLI) have witnessed their usage even for low power ratings and medium voltage applications [7] in recent days. The traditional configurations such as neutral-point clamped (NPC) inverter, flying capacitor (FC), and cascading H-bridge (CHB) are utilized in various applications. Besides this, various other topologies are also developed in terms of reduced control complexity, component count, number of isolated DC sources, and higher levels-based output voltage. They provide a better quality of output power due to a higher number of levels, lower voltage and current stress across the switches, better electromagnetic compatibility, operation at lower switching frequency with reduced switching losses, and small filter size. Despite the various advantages, MLIs are associated with limited output voltage gain and input voltage regulation ratio of less than two. This requires an appropriate power converter to be integrated with MLI to overcome the abovementioned limitations.

In this context, the insertion of impedance source networks with MLI topologies provides reliable solutions with the following advantages:

- Increased voltage gain with good regulation of output voltage
- Better quality waveforms due to a higher number of voltage levels across the output
- Improved reliability with shoot-through exemption capability
- Improved input voltage regulation ratio
- Suitability for renewable energy sources

7.2 Literature Review on Impedance Source-based MLI

Several impedance source-based MLI topologies are reported, which are basically derived from the upgradation of different impedance source (IS) networks and popular MLIs in order to attain the above-mentioned advantages.

The various IS networks utilized for integration as shown in Figure 7.1 include qZs, embedded Z-source, switched boost, trans Z-source, LCCT Z-source, *G*-Z-source, and switched inductor. Similarly, the MLI configurations include CHB, NPC, T-type, multilevel DC-link (MLDCL) inverter, DC link cascading, and other hybrid inverters.

The improvised impedance source networks as depicted in Figure 7.2 are derived using inductors in conventional form (Figure 7.2.a–f), switched type (Figure 7.2.d) and coupled magnetically type (Figure 7.2.g–j), and transformer-based (Figure 7.2.k,l) in order to obtain higher output gain. Several configurations have been developed from the basic IS network through suitable combinations. Various other topologies are also reported, however, cannot be integrated with MLIs.

Table 7.1 shows the comparison among the popular Z-source networks based on the input current pattern, boosting factor. The pros and cons are mostly based on the component count, blocking voltage of the switches, presence of common grounding between the source and load, number of DC sources, and adoptability for renewable power applications. Moreover, the leakage inductance present in the ZSI network leads to the production of higher voltage ripples across the output.

At first, a Z-source-based NPC inverter is proposed for a three-phase three-level operation [8, 9]. The input voltage is boosted to the desired level across the DC-link in comparison with conventional NPC. A LCCT ZSI with three-level NPC topology is reported in [10]. Significant improvement in the voltage gain can be achieved by using a high-frequency transformer (HFT)

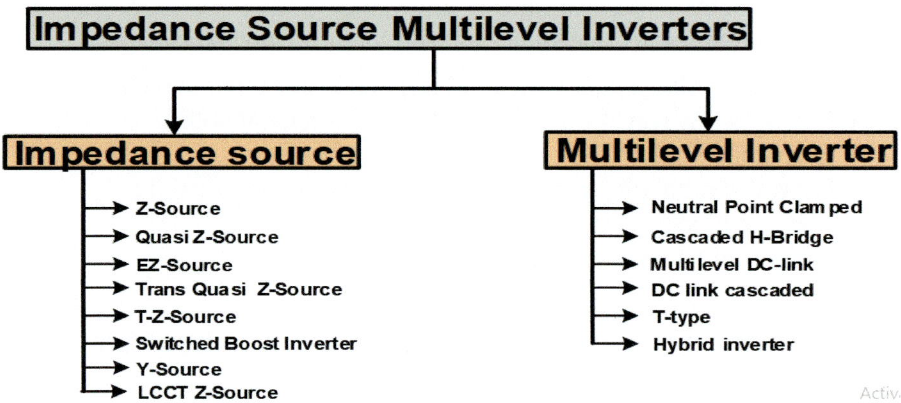

Figure 7.1 Classification of different IS networks and MLI topologies compatible with Z-source.

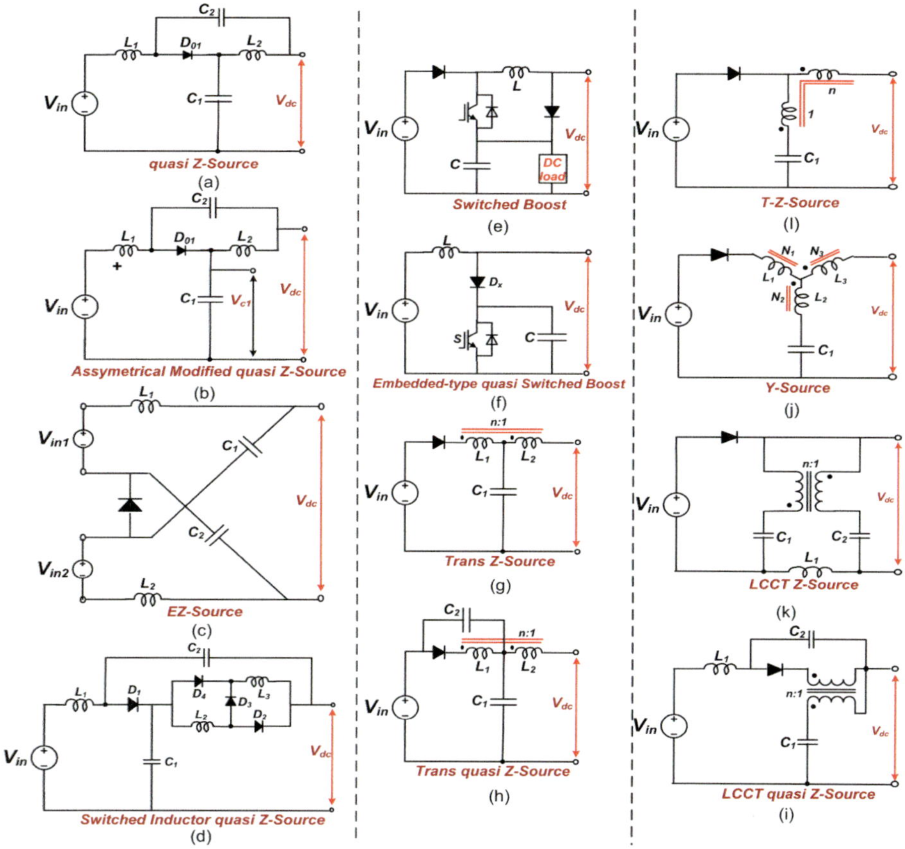

Figure 7.2 Different types of impedance source networks.

with a higher turns ratio, which also gets better usage of DC link voltage and minimizes the size of passive components when compared with basic ZSI. Further improvements are embedded in the three-level NPC topology with *G*-Z-source and trans-Z-source networks [11] as depicted in Figure 7.3. Magnetically coupled inductors are used in this topology to provide higher boosting and increased utilization of DC link voltage. Further, it smoothens the input current waveform and reduces the size of the inverter. However, the voltage spikes are generated at the DC link due to the presence of leakage inductance in the coupled inductors. The diode-clamped inverter with a Z-source network is presented for four-level and five-level operation in [12], which utilizes more number of isolated DC sources, Z-source networks, and component count.

Table 7.1 Comparison of different impedance source networks.

Type of Z-source network	Gain factor	Input current shape	Merits	Demerits
qZSI	$\frac{1}{1-2D}$	CIC	• Issues of leakage current are absent since the load and source are connected to common ground • Voltage stress across the capacitor is less	–
Asymmetrical MqZSI	$\frac{1}{1-2D}$	CIC	• Issues of leakage current are absent since the load and source are connected to common ground • Voltage stress across the capacitor is less • Improved output quality	• Increased current stress across shoot-through circuit
EZ-source	$\frac{1}{1-2D}$	DIC	• Smaller capacitor size • Smaller voltage stress	• Requirement of isolated sources • Input current discontinuity
SL-qZSI	$\frac{1+D}{1-2D+D^2}$	CIC	• Higher voltage gain	• Increased power losses due to excessive passive components
Switched boost	$\frac{1-D}{1-2D}$	CIC	• Reduced passive components	• Reduced voltage gain • Applicable only for low-power applications
Embedded type qSB	$\frac{1}{1-2D}$	CIC	• Improved efficiency • Lower voltage stress switch and diodes High power density and low cost	• Suitable for low-power applications
Trans Z-source	$\frac{1}{1-(n+1)D}$	DIC	• Variable voltage gain can be obtained by changing turns ratio of the coupled inductor	• Leakage inductance may raise the spikes in DC-link voltage
Trans quasi Z-source		CIC		

Table 7.1 Continued.

Type of Z-source network	Gain factor	Input current shape	Merits	Demerits
G-Z-source	$\frac{1}{1-\frac{n}{n-1}D}$	DIC	• Higher voltage gain	• Leakage inductance may raise the voltage spikes in DC-link
Y-source	$\frac{1}{1-\left(\frac{N_3+N_1}{N_3-N_2}\right)D}$	CIC	• Higher voltage gain at lower duty cycle • Better degree of control flexibility	• Leakage inductance may raise the spikes in DC link voltage
LCCT-based Z-source and quasi Z-source	$\frac{1}{1-(n+1)D}$	DIC	• Better voltage gain can be achieved by adjusting the turns ratio of HFT • Low voltage-rated switching devices	• Discontinuous input current • Transformer leakage inductance

A three-level qZSI NPC inverter is reported in [14] which utilizes a double quasi-Z-source network. By utilizing a fixed boost modulation scheme, the voltages across the capacitors C_2 and C_3 are balanced. The remaining capacitors are balanced using the ST duty cycle. This configuration provides continuous input current, reduced passive component size when compared to a three-level NPC-based Z-source network, and requires only a single DC source. Moreover, this topology can also be implemented in to three-phase supply system [15] and is also suitable for PV applications. The above topology is slightly modified as reported in [16] in order to obtain lower capacitor voltages and lower inrush current. A three-level NPC configuration with a modified qZs network is reported in [17]. It is derived by the use of a nonstandard inverter structure in which the capacitor is connected parallel to the inverter leg. Moreover, shoot-through operation is obtained with only one leg and the other leg is utilized to produce output-level voltage. In addition, reduced output voltage THD and reduced switching stress are obtained in comparison with three-level qZSI NPC. However, the increased current stress in the ST branch leads to an increase in the conduction losses during the boost mode. An embedded-type qSB network-based three-level NPC topology is reported in [17], which uses reduced passive components and is feasible for applications of lower voltage and medium power, which demand reduction in size and weight. A three-phase three-level configuration with a Z-source network and nine-switch inverter is shown in [18]. The configurations with

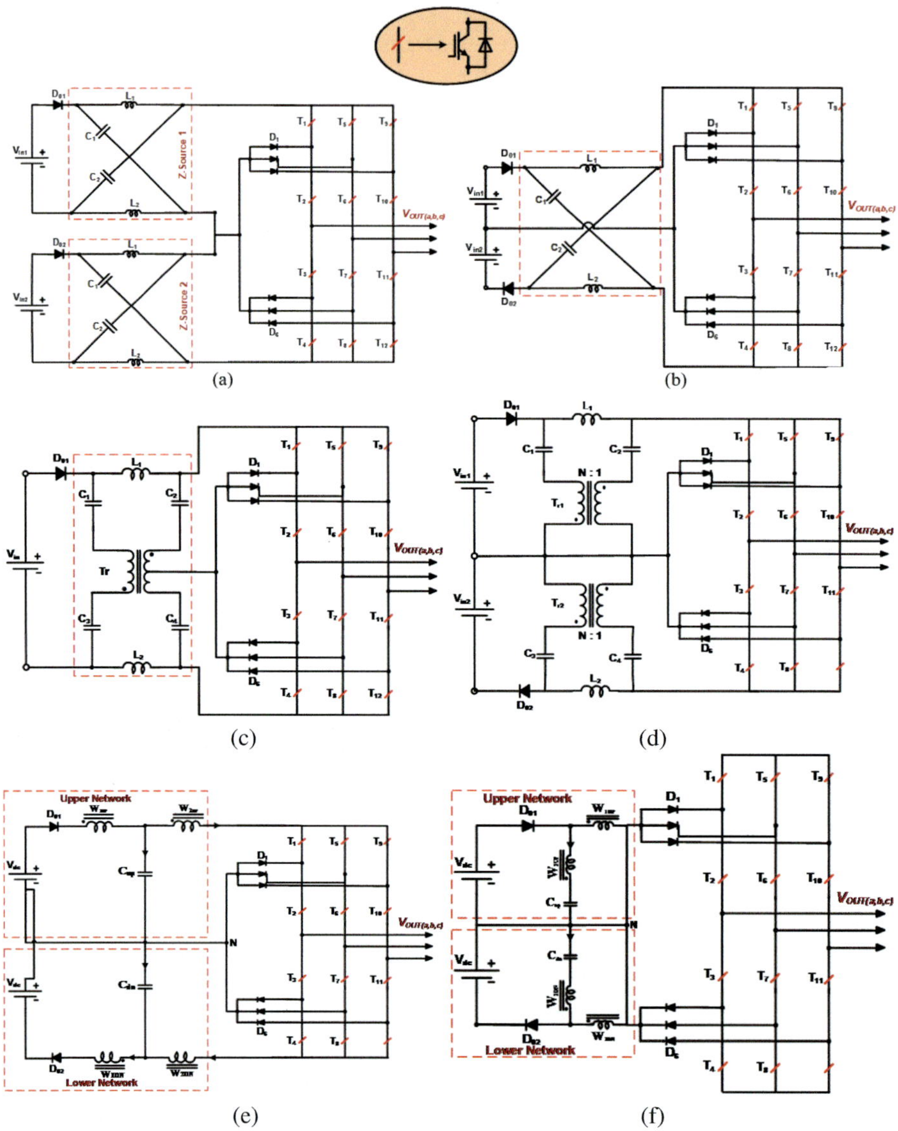

Figure 7.3 Topologies based on NPC structure (a) with two isolated DC sources and Z-source networks [8], (b) with two isolated DC sources and a Z-source network [9], (c) with HF transformer and a DC source [10], (d) with HFT and two nonisolated DC sources [10], (e) with trans Z-source network [11], and (f) with *G*-Z-source network [11].

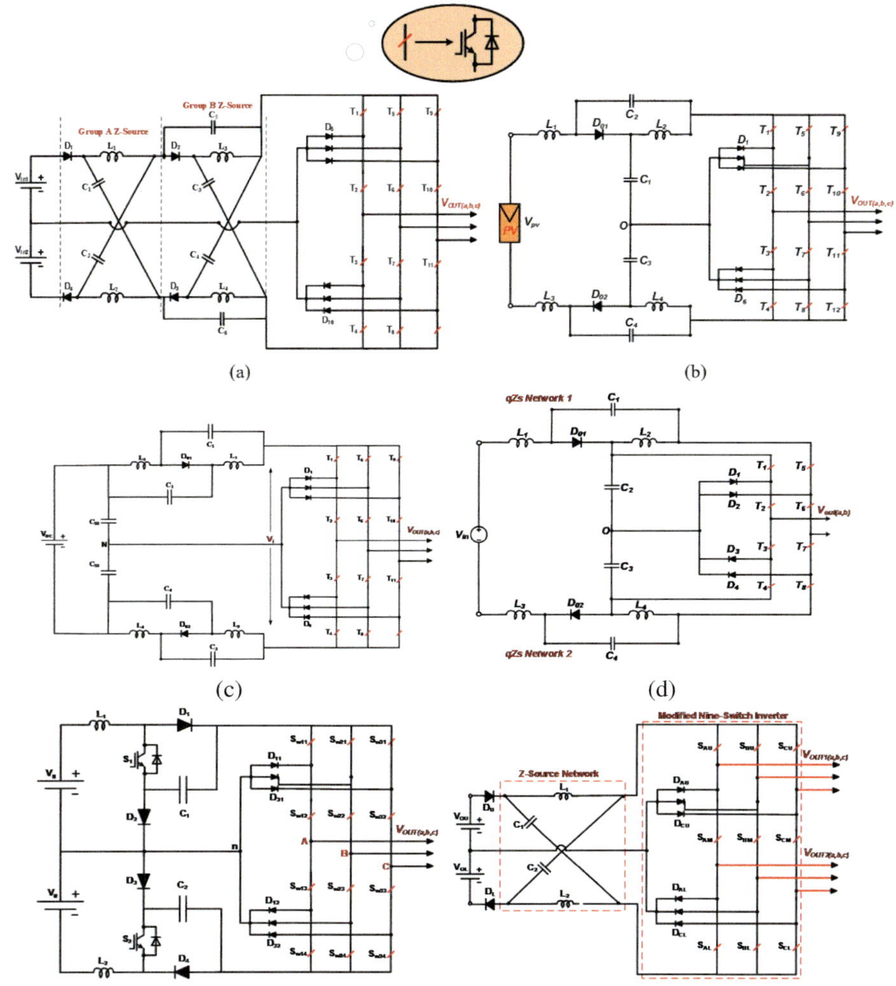

Figure 7.4 Three-level NPC inverter configurations with (a) a cascaded ZSI [12], (b) two qZs networks [14], (c) derived two qZs network [15], (d) modified qZs network [16], (e) an embedded based qSB [17], and (f) a multi-input-multioutput ZSI [18].

respect to three-level NPC inverter are depicted in Figure 7.4. This inverter is obtained by integrating two VSI structures along with three sets of commonly switches. Also, it is able to drive two different independent AC loads. This configuration is suitable for multiple port applications at both the input and output sides.

Figure 7.5.a illustrates the four-level diode clamped-based Z-source network, which requires several DC sources, Z-source networks, and switches. However, it is convenient to use for distributed energy sources because of its independent DC-link voltage control characteristic. The extended version of the previous topology is presented in [13], and illustrated in Figure 7.5.b.

A three-level-DCLC based on a Z-source inverter is reported in [21]. It also needs two Z-source branches and the same number of switches similar to three-level NPC structures. However, it requires switches with unequal voltage ratings, and the clamping diodes are not required, which further reduces the cost when compared to three-level-NPC topologies. ZSI-based dual inverter configuration is introduced in [21] as depicted in Figure 7.6. The passive component size is the same for both cases. The usage of an HF

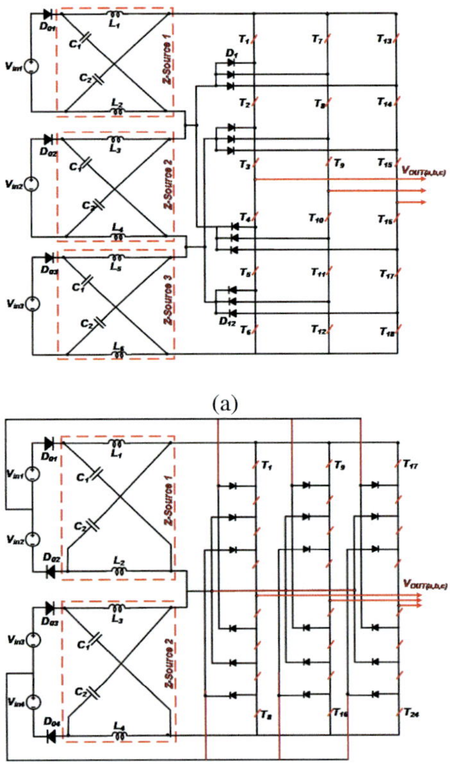

(a)

(b)

Figure 7.5 Diode clamped-based structure with (a) four-level ZSI [13] and (b) five-level ZSI [13].

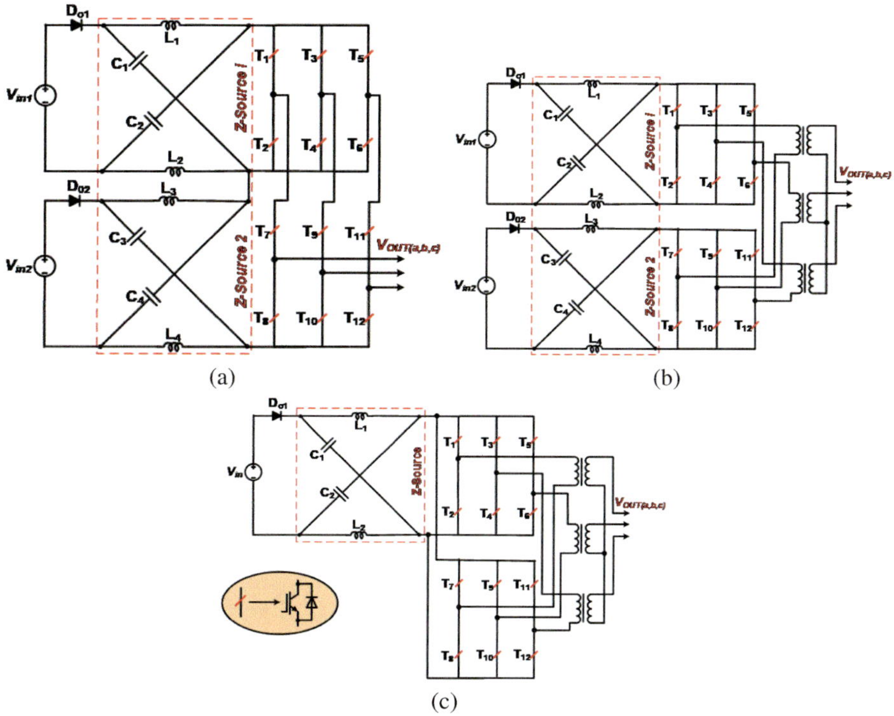

Figure 7.6 Z-source-based (a) three-level DCLC inverter [21], (b) three-level dual inverter [21], and (c) cascaded three-level dual inverter [21].

transformer maximizes the overall size, as well as cost, but it is an essential requirement in certain applications like PV systems.

A Z-source-based multilevel DC-link (MLDCL) inverter is reported in [22]. In order to realize the seven-level output voltage, three basic units are required. Each unit comprises a CHB in combination with a Z-source network. This H-bridge is used for the reversal of the polarity of the peak DC voltage. The configuration requires a reduced switch count and possesses an extendibility feature, which can maximize a higher number of levels across output with the addition of each basic unit. The same topology is extended with switched inductor and coupled inductor arrangement, in order to obtain an improved boost factor, and reported in [23] as depicted in Figure 7.7.

A seven-level qZS-based CHB inverter is reported in [24] with three different qZS networks integrated with an H-bridge to produce a seven-level output. This configuration has a modularity feature and can also extended for more number of output voltage levels. It also incorporates distributed

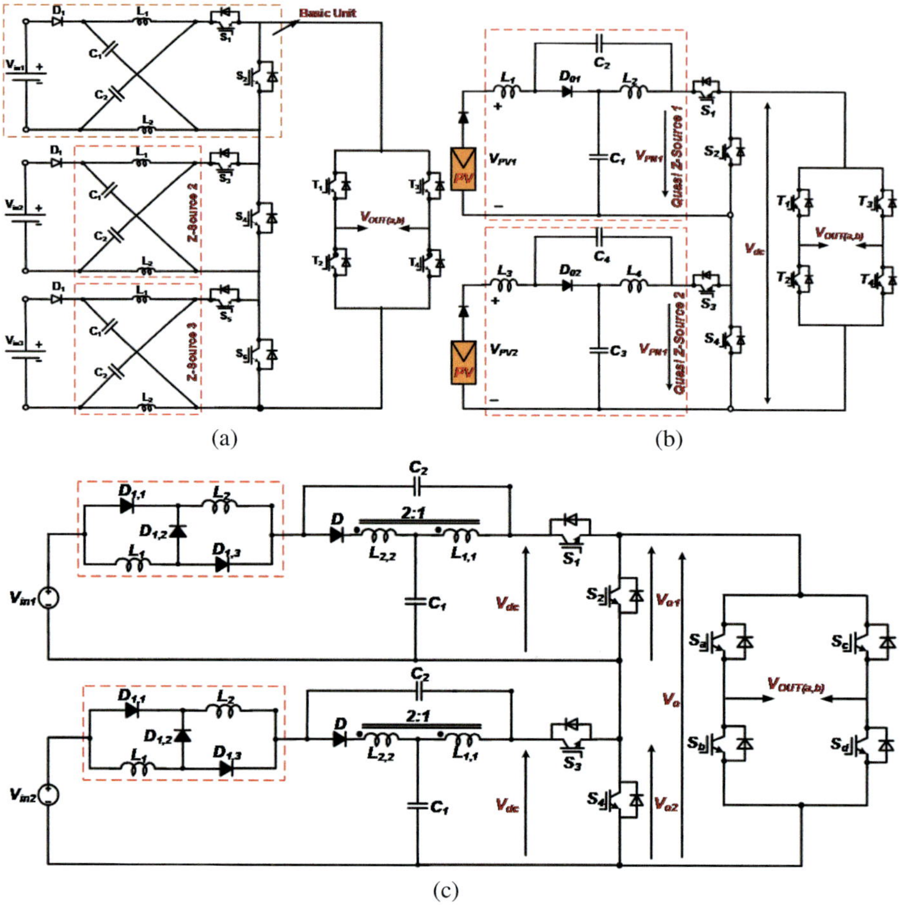

Figure 7.7 (a) ZSI-based seven-level inverter [22], (b) qZSI-based five-level inverter [23], and (c) qZSI-based cascaded five-level inverter [23].

MPPT [25] with independent DC link voltage balancing, which is compliable for PV-based applications. A three-level modified hybrid qZs inverter is reported in [26] with a single DC source, reduced number of inductors, and double the boosting ratio in comparison with three-level NPC qZSI topologies. These basic inverter structures and the H-bridge are series-cascaded to form a five-level modified qZs cascaded hybrid inverter to produce a nine-level output. A quasi-switched boost inverter (qSBI)-based five-level CHB is reported in [27, 28]. All the above configurations with respect to cascading are depicted in Figure 7.8. The qSBI utilizes a lower component count in comparison to the qZs network and is more compatible with low-power

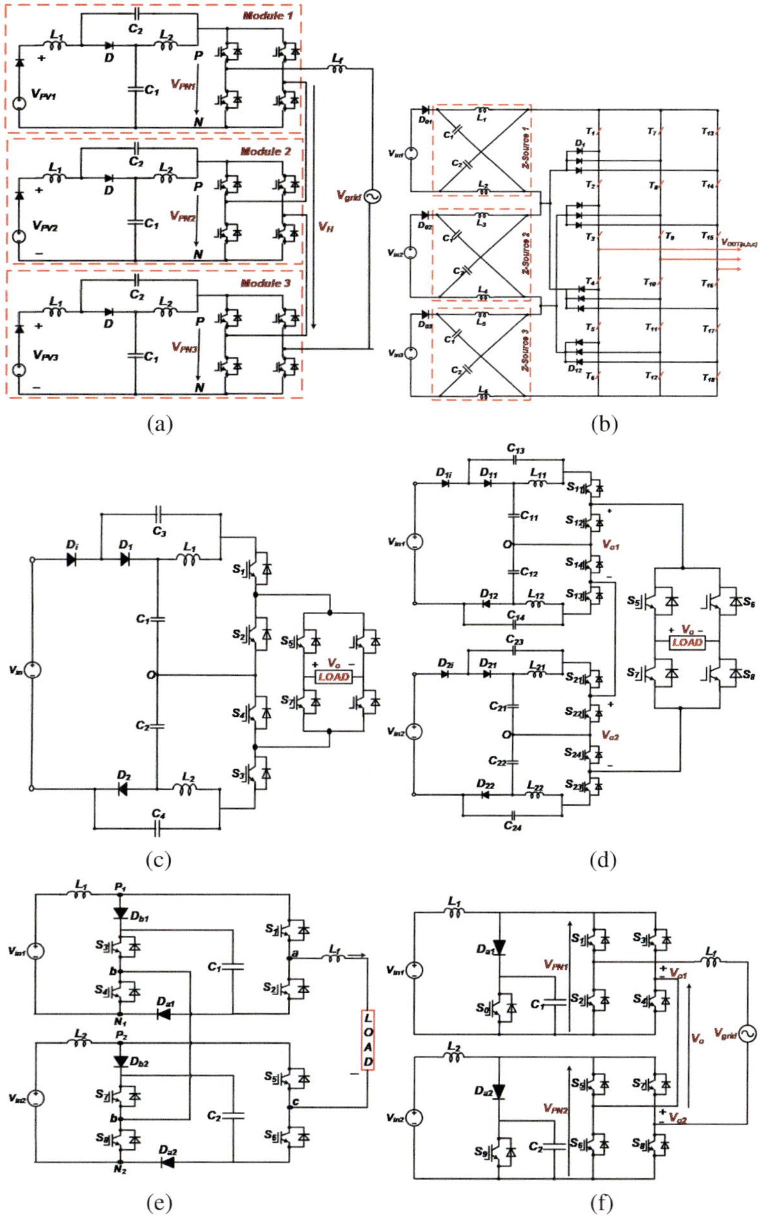

Figure 7.8 (a) Seven-level qZs-based H-bridge inverter [24], (b) cascaded four-level ZSI [25], (c) modified qZs hybrid three-level inverter [26], (d) modified qZs hybrid five-level inverter [26], (e) quasi H-bridge five-level boost inverter [27], and (f) five-level qSBI [27].

applications. High-quality output voltage can be obtained due to the simple control complexity of DC link voltage.

A three-level T-type inverter with ZSI-based is reported in [29] with the benefits of both a three-level NPC converter and a Z-source inverter. It employs SVM-based pulse generation to reduce the imbalance in voltage (at a neutral point) and provides very good quality voltage with improved gain. Similarly, a five-level T-type inverter with ZSI is reported in [30]. The output voltage gain of the above topology and the traditional Z-source network are the same. It needs a lower component count in relationship with traditional ZSI-based NPC and CHB topologies. A three-phase seven-level T-type inverter with qZSI is reported in [31]. A three-phase seven-level qSBI-based T-type inverter is reported in [32]. All the configurations with respect to T-type structure are depicted in Figure 7.9. It provides increased output gain, lower current ripple of the inductor, and enhanced reliability.

A five-level coupled inductors-based ZSI is reported in [33]. The three-phase VSI is integrated with coupled inductors in order to produce a five-level output voltage with improved quality and a similar boosting factor as that of traditional ZSI. It employs a single DC source and space vector modulation for the control of output voltage. A five-level ZSI based on the cascaded transformer is reported in [34]. It possesses a modularity feature. However, the cost and size of the inverter increase due to the use of more HF transformers. A seven-level ZSI is constructed using the switched capacitors concept and reported in [35]. The configurations reported in [33–35] are depicted in Figure 7.10. However, this configuration is not feasible, as it is associated with balancing issues in the capacitor voltage and the use of more semiconductor devices.

7.3 Comparative Analysis of ZSI-based MLI

The Z-MLI topologies can be compared based on the component count which includes passive components such as inductors, coupled inductors, transformers, and capacitors, and active components such as switches and diodes. The nature of input current whether continuous or discontinuous is also vital, which determines the suitability for distributed generation applications. Some of the configurations are reported on the magnetically coupled inductors, in order to obtain a high output gain ratio. However, these configurations are associated with leakage current issues and spikes in DC-link voltage. Some of the configurations such as CHB-based and MLDCL, which are modular in nature and also suitable for multistring applications are available. Generally,

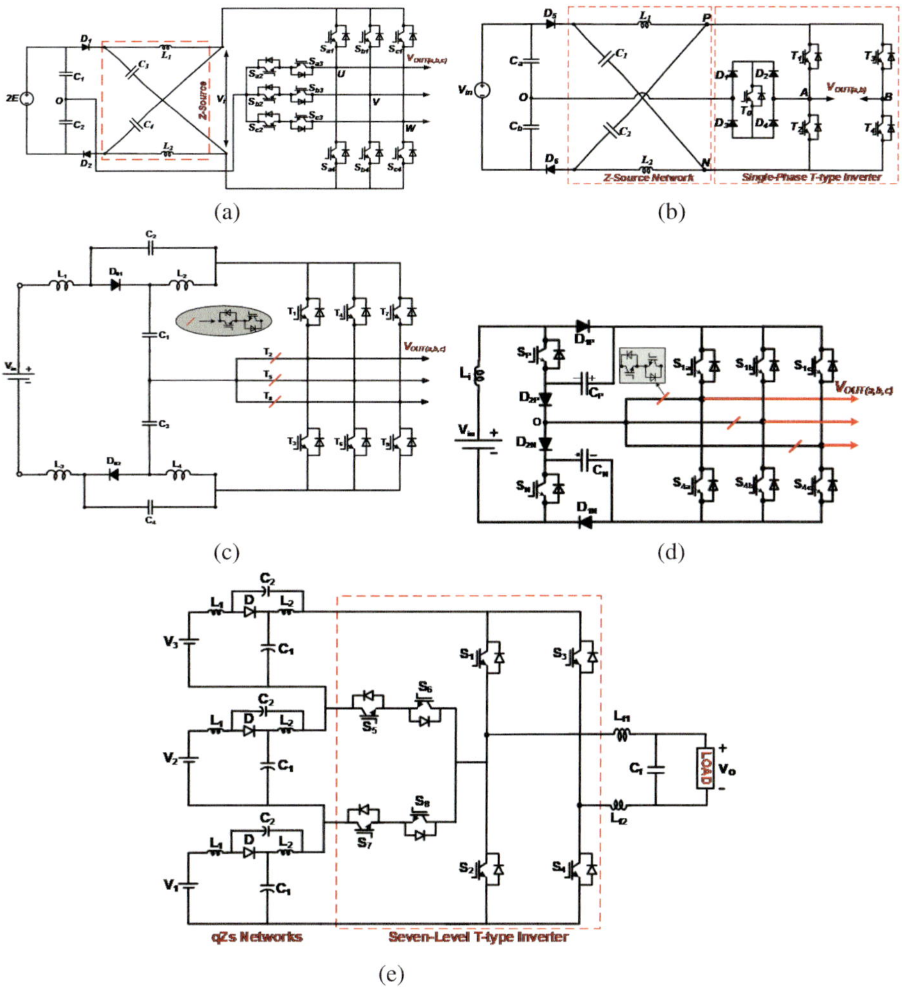

Figure 7.9 T-type structured (a) three-level ZSI [29], (b) five-level ZSI [30], (c) seven-level qZSI [31], and (d) three-level qSBI [32].

the reliability of an inverter depends on the number of components, and configurations with lower components possess better reliability. The voltage stresses of the switches and capacitors, and current stresses of the inductors determine the overall cost and performance of the inverter. Moreover, the impedance network size is dependent on the switching frequency and power rating. For high-power applications, the impedance network size is huge which makes the inverter bulky.

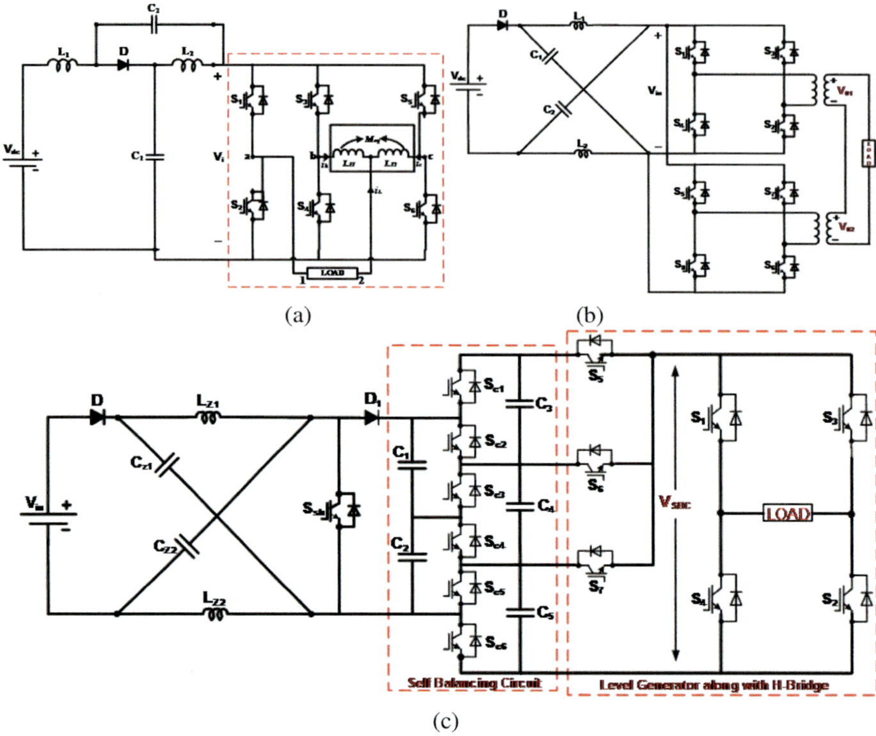

Figure 7.10 (a) Coupled inductor-based five-level qZSI [33]. (b) Transformer-based five-level ZSI [34]. (c) Switched capacitor-based seven-level Z-source inverter [35].

Therefore, end users can select appropriate and feasible configurations from the set of Z-MLI topologies based on their application such as low component count, highly reliable, compact size, lower cost, and compliable for a wide range of input voltages.

7.4 Methodology and System Design

From the literature review on various topologies of impedance-source-based MLI, the following observations are made:

- Topologies based on traditional configurations such as NPC and CHB demand a higher number of switch counts.
- Some configurations provide discontinuous input current which makes the configuration to be unsuitable for renewable applications.

- Some configurations require more passive component count to realize a higher number of voltage levels which increases the overall size and weight of the system.
- Most of the configurations do not have isolation between the photovoltaic (PV) panel and grid, which is mandatory in some applications.
- Although few configurations are reported based on the multistring applications, their DC-link voltage control is complex.

Hence, there is adequate scope for further research in the area of ZSI-based MLI configurations. The proposed research work focuses on establishing new impedance source topologies based on packed U-cell configuration and T-type inverters with enhanced reliability suitable for PV applications. It aims to provide high efficiency, high boost output voltage gain, improved input voltage regulation, reduced voltage and current stress of the switches, reduced switch count, suitability for all types of loads, continuous input current, and low THD. Improved modulation techniques are also investigated in order to reduce the switching commutations, thus reducing the switching losses.

7.4.1 System design

The size of the inductors and capacitors are selected to reduce the current and voltage ripples, respectively, either at the converter's input or output side. The input side inductor in ZSI is arranged to suppress the low-frequency ripple produced during the ST state $D_{\mathrm{o}}T$ and the voltage is positive as per eqn (7.1) (where, D_{o} is shoot-through duty ratio).

$$V_{\mathrm{L}} = L\,\frac{\Delta i_{\mathrm{L}}}{D_{\mathrm{o}}T}. \tag{7.1}$$

Let the input current ripple is assumed to be $x_{\mathrm{L}}\,\%$

$$x_{\mathrm{L}}\,\% = \frac{\Delta i_{\mathrm{L}}}{I_{\mathrm{L}}} \times 100. \tag{7.2}$$

The equivalent inductance can be evaluated as

$$L_1 = L_2 = \frac{(V_{\mathrm{in}} + V_{\mathrm{c2}})D_{\mathrm{o}}T}{x_{\mathrm{L}}\,I_{\mathrm{L}}}. \tag{7.3}$$

Example: $x_{\mathrm{L}} = 20\%$, $D_{\mathrm{o}} = 0.25$, $I_{\mathrm{L}} = 10$ A, $V_{\mathrm{in}} = 60$ V, $f_{\mathrm{s}} = 10$ kHz, the value of inductance is

$$L_1 = L_2 = \frac{V_{\text{in}}(1 - D_{\text{o}}).\, D_{\text{o}}}{x_{\text{L}}\,.I_{\text{L}}\,(1 - 2D_{\text{o}})\,f_{\text{s}}} \approx 1.125\ \text{mH}. \tag{7.4}$$

Similarly, the capacitance is selected to preserve constant DC-link voltage and to reduce the voltage ripple. The capacitor current under the nonshoot-through period can be written as eqn (7.5) (where, $(1 - D_{\text{o}})$ is the non-ST duty ratio).

$$i_{\text{c}} = C\frac{\Delta V_{\text{c}}}{(1 - D_{\text{o}})T}. \tag{7.5}$$

The value of capacitance can be evaluated as

$$C_1 = C_2 = \frac{2 \times \frac{I_{\text{L}} \times D_0}{1 - D_0} \times (1 - D_0)T_{\text{s}}}{\Delta(V_{\text{c}1} + V_{\text{c}2})}. \tag{7.6}$$

Let the voltage ripple be $x_{\text{C}}\ \%$, and expressed as

$$x_{\text{c}}\ \% = \frac{\Delta V_{\text{c}}}{V_{\text{c}}} \times 100. \tag{7.7}$$

From eqn (7.5) to eqn (7.7), assuming the boost factor $x_{\text{c}} = 0.01$, $B = 2$, $I_{\text{L}} = 10$ A, $D_{\text{o}} = 0.25$, $V_{\text{in}} = 30$ V, and $f_{\text{s}} = 10$ kHz, and calculated as

$$C_1 = C_2 = \frac{2 \times I_{\text{L}} \times D_0}{B \times x_{\text{c}} \times V_{\text{in}} f_{\text{s}}} \approx 830\ \mu F. \tag{7.8}$$

7.5 Findings With Discussions

Contemporarily, power generation through distributed energy sources is getting more attention to satisfy the increased energy demands. The integration of distributed energy sources with the conventional power system is a great challenge in the present trends due to lower output gain and requirement of more components and lower efficiency as clearly reported in section 7.2. It has been identified that the use of impedance sources and MLI together can inherit the pros of both. Therefore, in this section, two different packed U-cells based isolated and nonisolated topologies are presented, respectively. The proposed configuration is an upgradation of quasi ZSI with an asymmetrical seven-level inverter to obtain significant advantages such as higher boost gain, lesser number of devices, and increased efficiency. Further, the frontend quasi-Z-source part offers wide input voltage control for the PV standalone/grid-tied applications.

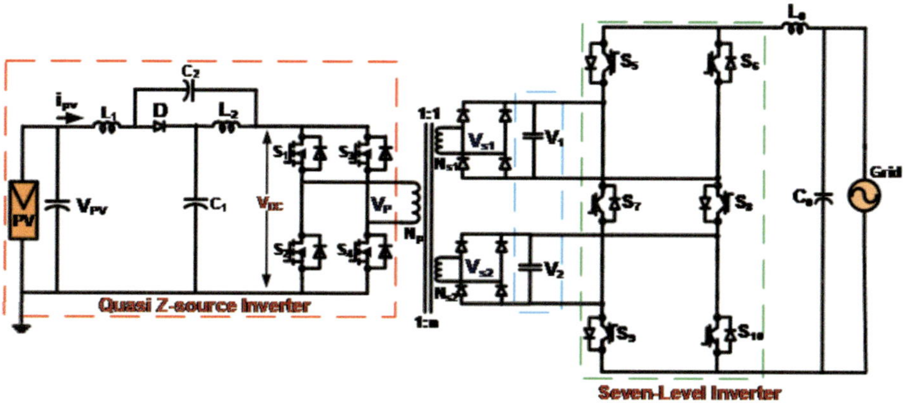

Figure 7.11 qZs-based seven-level inverter with PUC structure and HF transformer for grid-tied PV system [35].

7.5.1 A qZSI-based seven-level isolated inverter

Figure 7.11 illustrates the qZs-based seven-level inverter with a packed U-cell (PUC) structure [36] in which a three-winding HF transformer is employed between the source and load/grid for isolation purposes. Power flow in the circuit started from the source, qZs network, H-bridge inverter, HF transformer, diode bridge rectifiers, PUC, passive filters, and grid. Regulated voltage boosting can be obtained by changing the ST duty ratio and transformation ratio of the HF transformer. In this topology, a 1:2 HF transformer has been selected to multiply the output voltage obtained from the qZs network. Further, the boosted voltage is rectified and applied to PUC to generate seven-level output. The entire setting is established to construct a PV-based power conditioner for a 110 V, 50 Hz, single-phase grid-connected system.

Major pros of the proposed inverter are:

- Quasi Z-source network enables continuous input current and high voltage gain.
- HF transformer enhances the voltage gain further and provides an isolated supply to the PUC.
- The configuration is well suited for PV power extraction due to its wide voltage gain adjustment by qZs and HF transformer.
- Higher efficiency, lower size, and lower cost are possible due to reduced switch count in the seven-level inverter as compared to NPC and CHB.
- Issues associated with leakage current can be eliminated permanently with the isolation provided by the HF transformer.

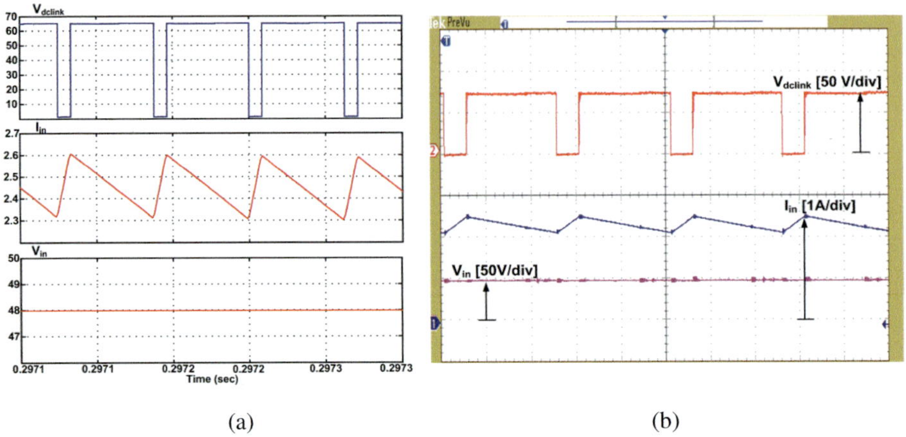

Figure 7.12 Input voltage, input current, and DC-link voltage (a) simulation and (b) experimentation.

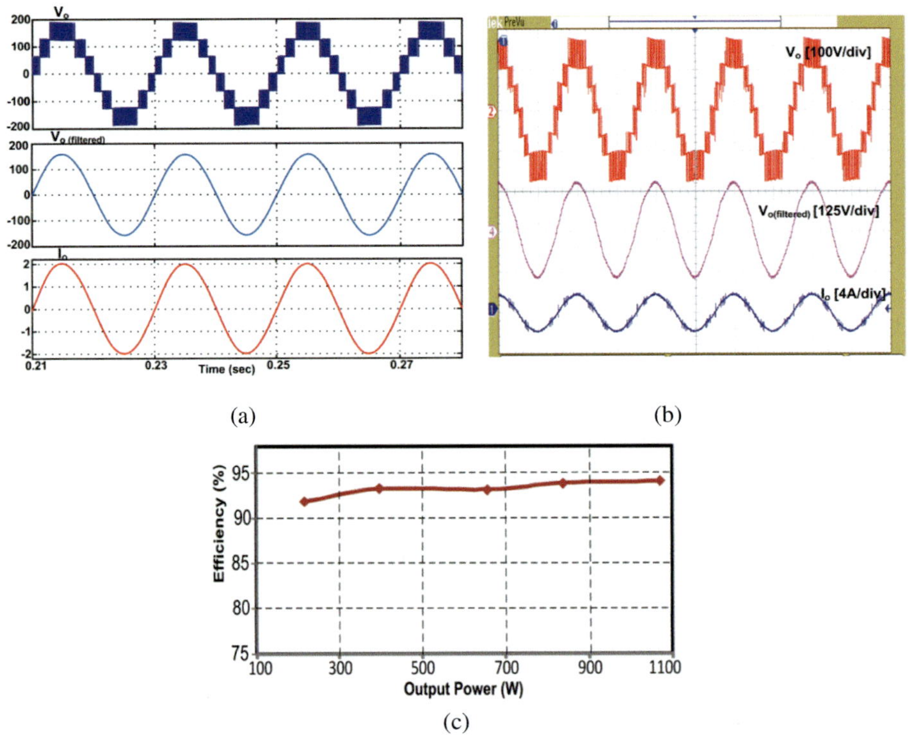

Figure 7.13 Seven-level voltage, grid voltage, and current (a) simulation, (b) experimentation, and (c) efficiency curve.

An improved control modulation technique is utilized for generating control pulses in qZSI. Multicarrier level shifted SPWM scheme is employed for control pulse generation of the inverter and to generate shoot-through duty pulse. Moreover, each switch is designed to operate at a switching frequency to normalize the power losses. The distribution of shoot-through states on both sides of the active states enables in reduction of the input current ripple. The performance of the power conditioner is tested under steady state in both simulation studies and experimental test setup under the same operating conditions as illustrated in Figures 7.12 and 7.13, respectively. The selection of inductors and capacitors and the development of a simple PI controller are studied.

7.5.2 A qZSI-based seven-level cascaded inverter

The qZSI-based seven-level cascaded inverter topology for PV-based multi-string applications [37, 38] is proposed as depicted in Figure 7.14. It contains a qZSI as a front-end converter with a seven-level PUC inverter. Two different PV strings along with the qZs network supply input to the inverter ($V_2 = 2 \times V_1$). The voltage gain can be regulated by changing the ST duty ratio D_o of the switches S_1 and S_2. Moreover, the proposed configuration can also be extended to more output levels by the addition of a front-end qZs

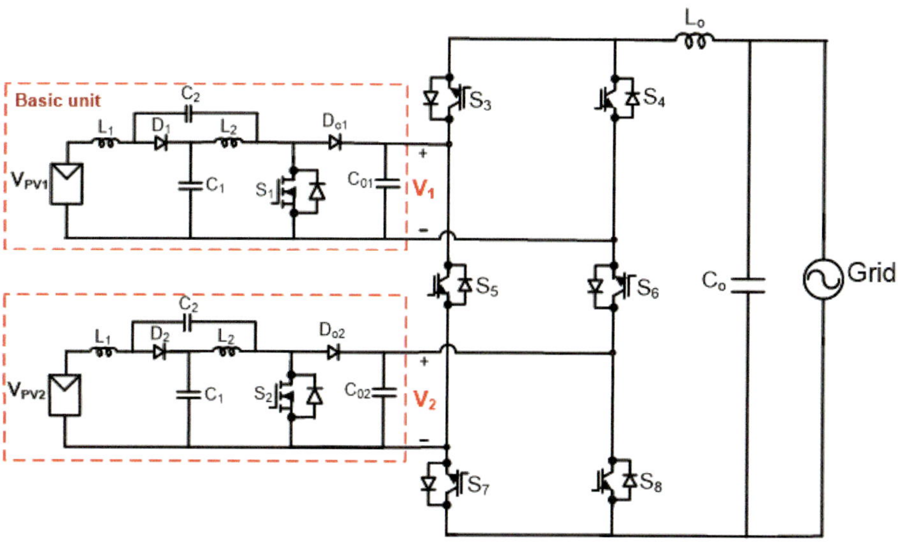

Figure 7.14 Quasi Z-source-based seven-level cascaded MLI.

network along with the necessary modifications in PUC. The entire setting is established to construct a PV-based power conditioner for a 110 V, 50 Hz, single-phase system. Further, the boundary condition between the input current in discontinuous or continuous mode is derived. The performance of the power conditioner is tested for both the modes in simulation and experimentation under the same conditions. The simulation studies and experimentally validated results are similar and are shown in Figures 7.15 and 7.16. The dynamic performance is also tested by implementing a simple PI controller for changes in input voltage and load.

The major features of the above configuration are described as follows:

- Reduction in switch count, isolated DC sources when compared to seven-level qZSI CHB inverter.
- Facilitate control of DC link voltage independently for each qZs unit.

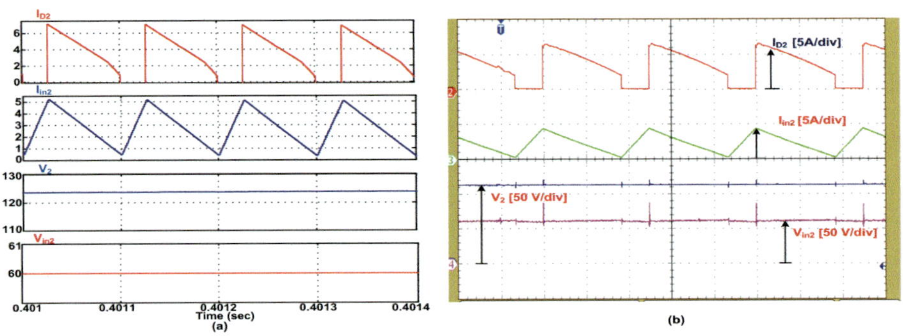

Figure 7.15 Input voltage, DC bus voltage, input current, and diode current (a) simulation and (b) experimentation.

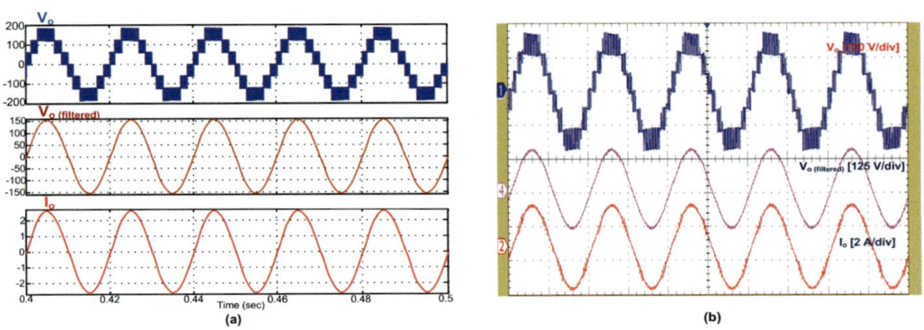

Figure 7.16 Seven-level output voltage, voltage (filtered), and output current (a) simulation and (b) experimentation.

- Provides better efficiency and compliable for PV-based multistrings.
- Both DC and AC loads can be supplied simultaneously.

7.5.3 qZS-based seven-level T-type inverter

The qZs-based seven-level T-type inverter (7L qZST^{2}I) is introduced [39] as shown in Figure 7.17. It is basically derived from the integration of qZS networks with a T-type inverter. It basically contains eight switches. The switches of the H-bridge inverter (S_3 and S_4) are operated at a 50 Hz frequency, whereas the switches S_1 and S_2 are employed for ST insertion in the impedance network and are basically operated at higher frequencies. The bidirectional switches (S_5 and S_6) and (S_7 and S_8) are utilized for the generation of output voltage levels and also act as an intermediate connection between the H-bridge inverter and qZs networks. The DC-link voltage for all the qZs networks is regulated and maintained constant for different input voltages in order to achieve a seven-level output voltage.

Two different switching schemes based on full ST and hybrid ST methods are studied. The different switching states for level generation and ST are discussed. The performance of the model under steady state is tested using simulation and experimental validation for both methods. It is noticed that

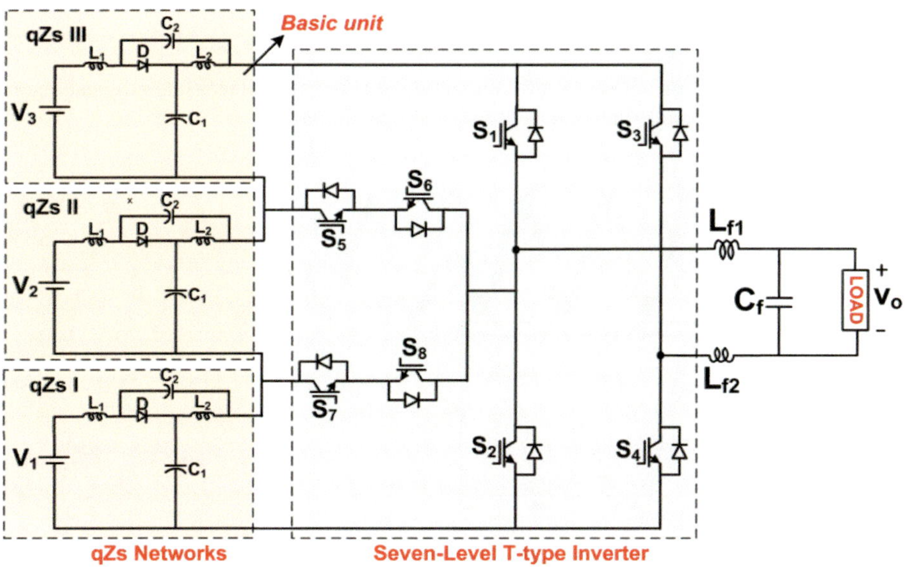

Figure 7.17 qZS-based seven-level T-type inverter.

the output voltage THD in the full ST method is more due to the insertion of the FST state for each switching cycle.

In the HST method, the THD of output voltage is decreased because of the inclusion of upper ST and lower ST when there is a transition of output voltage between upper and middle voltage levels. The simulation waveforms are in validation with the experimentation results as depicted in Figures 7.18 and Figure 7.19. The significant features of the proposed topology are:

- Reduction in the number of switches in comparison with equivalent 7L qZSI-based MLI-based configurations.
- Facilitates control of DC-link voltage independently for each qZs unit.
- Provides better efficiency and is feasible for PV-based multistring applications.
- Two separate switching strategies are investigated.

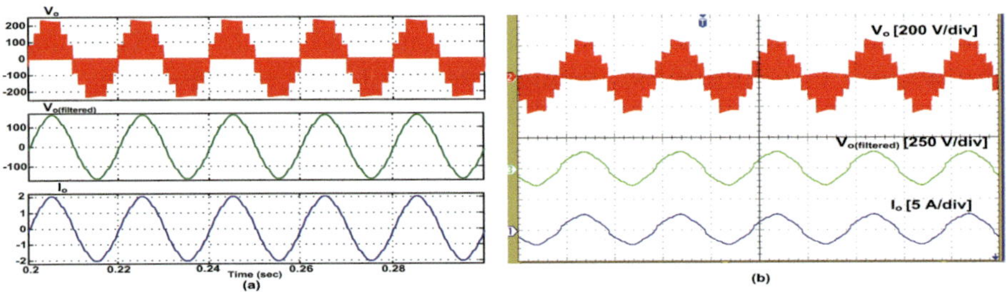

Figure 7.18 Waveforms of seven-level output voltage, output voltage (filtered), and output current of FST method (a) simulation and (b) experimentation.

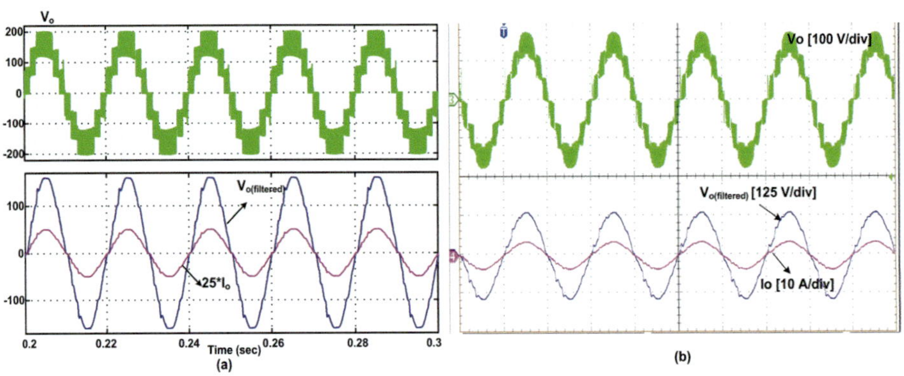

Figure 7.19 Waveforms of seven-level output voltage, output voltage (filtered), and output current of HST method (a) simulation and (b) experimentation.

Table 7.2 Comparison of the proposed inverter with existing topologies.

Item	Proposed inverter 1 [36]	Proposed inverter 2 [38]	Proposed inverter 3 [39]	CHB-qZSI [24]	Cascaded qZSI [22]	Transformer-qZSI [34]
Input DC sources	1	2	3	3	3	1
Inductors	2	4	6	6	6	2
Capacitors	4	6	6	6	6	2
Diodes	9	4	3	3	3	1
Switches	10	8	8	12	10	12
HF transformer	1	0	0	0	0	3
Total blocking voltage	16 V DC	15 V DC	18 V DC	12 V DC	18 V DC	12 V DC
Power balancing capability	Yes	Yes	Yes	Yes	No	No

7.5.4 Comparison of the proposed configurations with existing seven-level qZSI-based MLI

Table 7.2 shows the comparison of the proposed configurations with existing 7L CHB-qZSI, 7L Cascaded qZSI, and 7L transformer-based qZS in terms of number of active and passive components. It can be observed that the proposed configurations require fewer DC sources, components, and switching components in comparison with other existing inverters [40]. The total blocking voltage of some of the configurations is more due to reduced switch count. The presence of an HF transformer provides galvanic isolation and power balancing capability exists in all proposed inverters.

7.6 Conclusion

This chapter has covered the requirements and importance of the ZSI for PV applications. The next part of the chapter will be dedicated to the development of various ZSI networks based on their configuration, boosting factor, merits, and demerits. Moreover, a literature review of quasi-Z-source-based three-level and MLI topologies suitable for PV applications is reported in this chapter. Further, recently derived topologies with results are presented. It has been observed that the proposed configurations have been reported to possess features such as reduced switch count, better input voltage regulation, better reliability, and efficiency, compatible with PV applications. Finally, a comparative assessment of all the topologies on the basis of their component count, output voltage gain, type of input current pattern, and suitability of application, merits, and demerits, along with the detailed analysis has been discussed for the educated choice of Z-source inverter.

7.7 Future Directions

From the literature review on various topologies of impedance-source-based MLI, the following observations are made:

- Most of the configurations are based on traditional topologies, CHB and NPC topologies, which require more semiconductor switches for an increased number of output voltage levels.
- Some configurations provide a discontinuous input current, making it unsuitable for renewable applications.
- Some configurations require more passive components at higher output voltage levels that make the system bulky.
- Most of the configurations do not provide isolation between PV and grid, which is mandatory in some applications.
- Although few configurations are reported based on the multistring applications, their individual DC-link voltage control is complex.

Hence, there is adequate scope for further research in the area of ZSI-based MLI topologies with reduced leakage current and common-mode voltage (CMV) in grid-connected PV systems.

References

[1] F. Z. Peng, "Z-Source Inverter," *IEEE Transactions on Industrial Applications.*, vol. 39, no. 2, pp. 504-510, March 2003.

[2] P. C. Loh, F. Gao, F. Blaabjerg, S. Y. C. Feng and N. J. Soon, "Pulse-width-Modulated Z-Source Neutral Point Clamped Inverter," vol. 43, no. 5, pp. 1295-1308, Sept 2007.

[3] F. Z. Peng, M. Shen and Z. Qian, "Maximum boost control of the Z-source inverter", *IEEE Transactions on Industrial Applications.*, vol. 20, no. 4, pp. 833-838, 2005.

[4] M. Shen, J. Wang, A. Joseph, F. Z. Peng, L. M. Tolbert, D. J. Adams, "Constant boost control of the Z-source inverter to minimize current ripple and voltage stress",*IEEE Transactions on Industrial Applications.*, vol. 42, no. 3, pp. 770-778, 2006.

[5] Y. P. Siwakoti, F. Z. Peng, F. Blaabjerg, P. C. Loh, G. E. Town, "Impedance Source Networks for Electric Power Conversion Part-II: Review of Control Method and Modulation Techniques," *IEEE Transactions on Power Electronics.,* vol. 30, no. 4, pp. 1887-1906, April 2015.

[6] J. Anderson, F. Z. Peng, "Four quasi Z-Source inverters," *IEEE Power Electronics Specialists Conference, PESC*, pp. 2743-2749, 2008.

[7] J. Rodriguez, J. S. Lai, F. Z. Peng, "Multilevel Inverters: A Survey of topologies, controls and applications", *IEEE Transactions on Industrial Electronics.,* vol. 49, no. 4, pp. 724-738, August 2002.

[8] P. C. Loh, S. W. Lim, F. Gao and F. Blaabjerg, "3-level Z-Source Inverters Using a Single LC Impedance Network," *IEEE Transactions on Power Electronics.,* vol. 22, no. 2, pp. 706-711, March 2007.

[9] R. Strzelecki, M. Adomowicz, D. Wojciehowski " Buck-Boost Inverters with Symmetrical Passive Four-terminal Networks", *Proceedings Compatibility in Power Electronics, CPE* pp. 1-9, May 2007.

[10] R. Strezelecki, W. Bury, M. Adomowicz and N. Strzelecka, " New alternative passive networks to improve the range of output voltage regulation of the PWM inverters", *Proceedings of IEEE APEC*, pp. 857-863, 2009

[11] W. Mo, P. C. Loh, F. Blaabjerg, P. Wang, "Trans Z-source and *G*-Z-source neutral point clamped inverters", *IET Power Electronics,* vol. 8, no. 4, pp. 371-377, 2015.

[12] Yuyao H, Hailong L, Wei F, Jie L. Novel Cascaded Z-Source Neutral Point Clamped Inverter, *Chineese Journal of Electronics*, 2016;25:965-973.

[13] F. Gao, P. C. Loh, F. Blaabjerg, R. Teodorescu, D. M. Vilathgamuwa, "Five-level Z-source diode-clamped inverter", *IET Power Electronics,* vol. 3, no. 4, pp. 500-510, 2010.

[14] O. Husev, C. Roncero-Clemente, E. Romero-Cadaval, D. Vinnikov, S. Stepenko," Single phase 3-level neutral point clamped quasi Z-source inverter", *IET Power Electronics,* vol. 8, no. 1, pp. 1-10, 2015.

[15] O. Husev, C. Roncero-Clemente, E. Romero-Cadaval, D. Vinnikov, T. Jalakas, " 3-level three-phase quasi Z-source neutral point clamped

inverter with novel modulation technique for photovoltaic application", *Electric Power Systems Research*, vol. 130, no. 1, pp. 10-21, 2016.

[16] D. Yu, Q. Cheng, J Gao, F Tan, Y Zhang, " 3-level neutral-point clamped quasi Z-source inverter with reduced Z-source capacitor voltage", *IET Electronic Letters*, vol. 53, pp. 185-187, 2017.

[17] O. Husev, R Strzelecki, F Blaabjerg, V. Chopyk, D. Vinnikov, "Novel Family of Single-Phase Modified Impedance-Source Buck-Boost Multilevel Inverters With Reduced Switch Count," *IEEE Transactions on Power Electronics.,* vol. 31, no. 2, pp. 7580-7591, 2016.

[18] S. Manoranjan, K. Sivakumar, "A Three Level LC-Switching Based Voltage Boost NPC Inverter," *IEEE Transactions on Power Electronics.,* vol. 64, no. 2, pp. 2876-2883, 2017.

[19] S. M. Dehghan, M. Mohamadian, A. Yazdian, F. Ashrafzadeh, "A Dual-Input-Dual-Output Z-Source Inverter", *IEEE Transactions on Power Electronics.,* vol. 25, pp. 360-368, 2010.

[20] P. C. Loh, F. Gao, F. Blaabjerg, "Topological and Modulation Design of 3-level Z-source Inverters," *IEEE Transactions on Power Electronics.,* vol. 23, no. 5 pp. 2268-2277, September 2008.

[21] F. Gao, P. C. Loh, F. Blaabjerg and D. M. Vilathgamuwa, "Dual Z-Source Inverter With 3-level Reduced Common Mode Switching", *IEEE Transactions on Industrial Applications*., vol. 43, no. 6, pp. 1597-1608, Dec 2007.

[22] M. R. Banaei, A. R. Dehghanzadeh, E. Salary, H. Khounjahan and R. Alizadeh, "Z-source-based multilevel inverter with reduction of switches. *IET Power Electronics,* vol. 5, no. 3, pp. 385-392, 2012.

[23] E. Babaei, M. S. Zarbil, M. Sabahi, "A New Structure of Quasi-Z-Source-Based Cascaded Multilevel Inverter", *Journals of Circuits, Systems and Computers*, vol. 26, pp. 1-26, 2017.

[24] D. Sun, B. Ge, X. Yan, D. Bi, H. Zhang, Y. Liu, H. Abu-Rub, L. Ben-Brahim, F. Z. Peng, "Modeling, Impedance Design and Efficiency Analysis of Quasi-Z Source Module in Cascaded Phtotovoltaic Power System"., *IEEE Transactions on Industrial Electronics*., vol. 61, no. 11, pp. 6108-6117, 2014.

[25] Y. Liu, B. Ge, H. Abu-Rub, F. Z. Peng, "An Effective Control Method for Quasi-Z-Source Cascade Multilevel Inverter-Based Grid-Tie Single-Phase Photovoltaic Power System," *IEEE Transactions on Industrial Informatics*., vol. 10, pp. 399-407, 2014.

[26] A. V. Ho, T. W. Chun, "Single-Phase Modified Quasi-Z-Source Cascaded Hybrid Five-level Inverter", *IEEE Transactions on Industrial Electronics.,* vol. 65, no. 4, pp. 5125-5134, 2018.

[27] M. K. Nguyen, T. T. Tran, "Quasi Cascaded H-Bridge Five-Level Boost Inverter", *IEEE Transactions on Industrial Electronics.,* vol. 64, no. 4, pp. 8525-8533, 2017.

[28] T. T. Tran, M. K. Nguyen, "Cascaded five-level quasi switched-boost inverter for single-phase grid-connected system," *IET Power Electronics*, vol. 10, pp. 1896-1903, 2017.

[29] X. Xing, C. Zhang, A. Chen, J. He, W. Wang, C. Du, "Space-Vector-Modulated Method for Boosting and Neutral Voltage Balancing in Z-Source 3-level T-Type Inverter" *IEEE Transactions on Industrial Applications.*, vol. 52, pp. 1621-1631, 2016.

[30] H. T. Luong, M. K. Nguyen and T. T. Tran, "Single-Phase five level Z-source T-type inverter", *IET Power Electronics.*, vol. 11, pp. 2367-2376, 2018.

[31] V. F. Pires, A. Cordeiro, D. Foito, J. F. Martins, "Quasi-Z-Source Inverter with a T-type Converter in Normal and Failure Mode", *IEEE Transactions on Power Electronics*, vol. 11, pp. 7462-7470, 2016.

[32] D. T. Do, M. K. Nguyen "3-level Quasi-Switched Boost T-Type Inverter: Analysis , PWM Control and Verification", *IEEE Transactions on Industrial Electronics.*, vol. 65, pp. 8320-8329, 2018.

[33] M. R. Banaei, A. B. Oskouei, A. R. Dehghanzadeh, "Extended Switching algorithms based space vector control for five-level quasi-Z-source inverter with coupled inductors", *IET Power Electronics*, vol. 7, pp. 1509-1518, 2014.

[34] M. R. Banaei, A. R. Dehghanzadeh, A. Fazel, A. B. Oskouei, "Switching algorithm for Single Z-source boost multilevel inverter with ability of voltage control", *IET Power Electronics*, vol. 6, no. 7 pp. 1350-1359, 2013.

[35] Matam MB, Venkata AKD, Mallupu VK. Analysis and implementation of impedance source based Switched Capacitor Multi-Level Inverter, An International Journal of Engineering, Science and Technology, 2018;21.:869-885.

[36] J. R. Rahul, A. Kirubakaran, "A New Configuration of 7-level quasi-Z-source based isolated inverter for renewable applications", *International Transactions on Electrical Energy Systems*, vol. 29, pp. 2833, 2019.

[37] J. R. Rahul, A. Kirubakaran, "Multistring 7-level Quasi Z-Source Based Asymmetrical Inverter," *Indonesian Journal of Electrical Engineering and Computer Science,* vol. 15, pp. 88-94, 2019.
[38] J. R. Rahul, A. Kirubakaran, "FPGA Based Implementation of Single-Phase Seven-Level Quasi Z-Source Inverter", *International Journal of Circuit Theory and Applications,* vol. 47, pp. 1970-89, 2019.
[39] J. R. Rahul, A. Kirubakaran, Chinmay Kumar Das, "Operation, Control and Verification of 7-level quasi-Z-source based T-type Inverter," *Journal of Circuits, Systems and Computers,* vol. 29, pp. 2050023, 2020.
[40] J. R. Rahul, Chinmay Kumar Das, A. Kirubakaran, V. T. Somasekhar, "Impedance Source-based Multilevel Inverter: A State-of- the-Art Review," *Journal of Circuits, Systems and Computers,* vol. 29, pp. 2030011, 2020.

8

A Comparative Analysis of Multicarrier PWM Techniques in a Seven-level Asymmetrical Multilevel Inverter

Sangita Das Biswas[1], Bikram Das[2], and Champa Nandi[1]

[1]Department of Electrical Engineering, Tripura University, India
[2]Department of Electrical Engineering, NIT Agartala, India
Email: sdasbiswas@tripurauniv.in; powerelectronicsbd81@gmail.com; cnandi@tripurauniv.in

Abstract

This chapter provides a comparative analysis of multicarrier pulse width modulation (PWM) techniques utilized in a seven-level asymmetrical multilevel inverter (MLI). Multilevel inverters have garnered considerable attention owing to their capacity for harmonic reduction and achieving higher voltage levels compared to traditional two-level inverters. Yet, the selection of the PWM technique significantly influences the performance of the inverter system in rapports of harmonic distortion, switching losses, and overall efficiency. Multicarrier PWM techniques, such as phase disposition (PD), phase opposition disposition (POD), alternate phase opposition (APOD), and phase shifted (PS), have been used for seven-level asymmetrical cascaded MLI and the simulation results were obtained using Matlab/Simulink. Performance metrics, including total harmonic distortion (THD) and voltage waveform quality, are evaluated for each PWM technique. The comparative analysis aims to provide insights into the advantages and limitations of each PWM technique to assist researchers and practitioners in choosing the most appropriate PWM strategy for seven-level asymmetrical MLI applications. Simulation results demonstrate the effectiveness of each PWM technique in achieving desired voltage output characteristics and minimizing harmonic content, thereby contributing to the advancement of multilevel

DOI: 10.1201/9781003669845-8

inverter technology. APOD minimizes total harmonic distortions (THDs) of output voltages and currents at higher frequencies keeping the modulation index as unity.

Keywords: Asymmetrical cascaded multilevel inverter, multicarrier PWM techniques, total harmonic distortions, MLI

8.1 Introduction

Presently, MLIs have garnered significant attention in the power electronics sector, for their capability to satisfy the growing need for high-power applications and high-voltage conversion with improved efficiency and reduced harmonic distortion. The emergence of multilevel inverters over the last few decades has been driven by the increasing demand for high- and medium-power applications. These inverters not only enable high power ratings but also facilitate the integration of renewable energy sources with the multilevel inverter (MLI). There are several advantages of MLI which include synthesizing of nearly sinusoidal output voltage waveform with a reduction in the total harmonic distortions (THDs), minimization of switching losses, and improved electromagnetic compatibility. The prevalent topologies of MLI include the diode-clamped, flying capacitor, and cascaded MLI. The CMLI is the most commonly preferred MLI because of its modularity feature, which connects a single-phase full-wave inverter, also known as an H-bridge inverter, in series so as to obtain the multilevel output. Each H-bridge uses an isolated DC input. These separate DC sources can be renewable energy sources or batteries.

Among various multilevel inverter topologies, the seven-level asymmetrical MLI has gathered significant attention for its ability to achieve higher voltage levels while employing a reduced number of semiconductors compared to its symmetrical counterparts. This topology is particularly advantageous in applications requiring high-voltage output and compact designs, such as machine drives and grid-connected power converters. The performance of a multilevel inverter is inherently tied to the modulation strategy for switching signals for semiconductor devices. Pulse width modulation (PWM) techniques serve as the foundation of such modulation strategies, allowing precise control over output by adjusting the widths of switching pulses. The choice of PWM technique significantly influences crucial aspects of inverter operation, including harmonic distortion, voltage regulation, switching losses, and overall efficiency.

In this context, multicarrier PWM techniques have emerged as effective solutions for achieving low harmonic distortion and improving the quality of the output voltage waveform in multilevel inverters. These techniques utilize multiple carrier signals to synthesize the desired output voltage waveform by selectively controlling the switching instants of the semiconductor devices. In the realm of multicarrier PWM techniques, carrier-based sinusoidal PWM (SPWM), selective harmonic elimination PWM (SHEPWM), and space vector PWM (SVPWM) stand out and are widely recognized for their effectiveness and versatility in addressing diverse application requirements.

This chapter focuses on conducting a comprehensive comparative analysis of these multicarrier PWM techniques within the framework of a seven-level asymmetrical MLI. The selected PWM techniques are evaluated based on various performance metrics, including total harmonic distortion (THD), voltage waveform quality, and switching losses, under different operating conditions. By systematically comparing these PWM techniques, this study aims to provide valuable insights into their respective strengths and limitations, enabling researchers and assisting practitioners in making informed decisions when selecting the most appropriate PWM strategy tailored to specific applications of seven-level asymmetrical MLIs. Furthermore, the findings of this comparative analysis are expected to contribute to the ongoing research and development efforts aimed at advancing multilevel inverter technology. By identifying the performance characteristics and trade-offs associated with different PWM techniques, this study lays the groundwork for future optimization and refinement of modulation strategies, thereby facilitating the continued evolution and widespread adoption of multilevel inverters in various industrial and commercial applications.

8.2 Asymmetrical Cascaded Multilevel Inverter (ACMLI)

CMLI are categorized as symmetrical and asymmetrical, depending on the DC sources they employ. Symmetrical CMLI utilizes equal DC sources, whereas asymmetrical CMLI utilizes unequal DC sources. The ACMLI, a type of power electronics device, is utilized in electrical systems. The ACMLI is based on a cascaded configuration, where multiple power electronic switches are arranged in a series to generate multiple voltage levels. It typically consists of several H-bridge inverters linked in series. In symmetric multilevel inverters, each phase has the same number of voltage levels; ACMLIs have different numbers of voltage levels for each phase. Figures 8.1 and 8.2 show the structure of symmetrical and asymmetrical MLI,

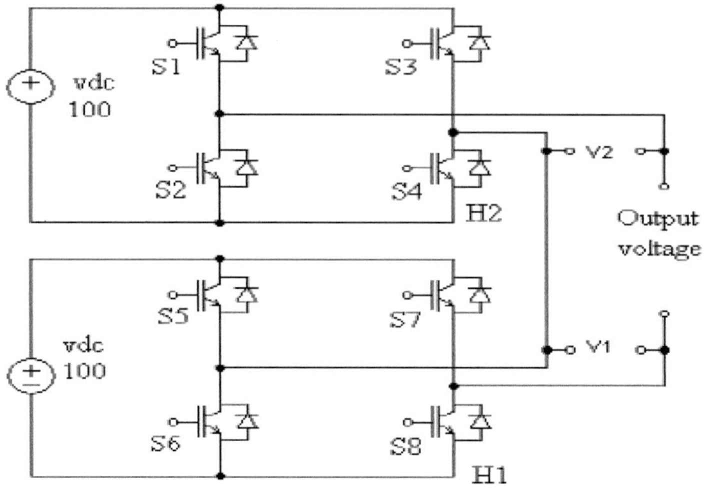

Figure 8.1 Symmetrical CMLI.

respectively, where in SCMLI, input DC sources of 100 V each have been used and in ACMLI of Figure 8.2 uses 100 and 200 V for two individual H-Bridges in binary configuration.

The ACMLI topology is further classified as binary and trinary structure. If the proportion of DC sources in asymmetric topologies doubles from the preceding one, it is termed a binary structure; if it triples, it is termed a trinary structure. The output voltage levels are calculated using eqn (1) for binary (N_{binary}) and eqn (2) for trinary ($N_{trinary}$) structures.

$$N_{\text{binary}} = 2^{n+1} - 1. \tag{8.1}$$

$$N_{\text{trinary}} = 3^{n}, \tag{8.2}$$

where, n = number of H-bridges allied in series. The same number of DC sources in an asymmetrical cascaded multilevel inverter not only produces more output voltage levels but also obtains a higher level than the symmetrical cascaded multilevel inverter. Here, a seven-level ASCMLI topology is accomplished with a binary structure. This asymmetrical operation allows for improved efficiency and reduced THD in output. The various modulation techniques are used to control the switching of switches in the ACMLI to achieve the desired output voltage waveform. Popular modulation techniques include PWM and SHE.

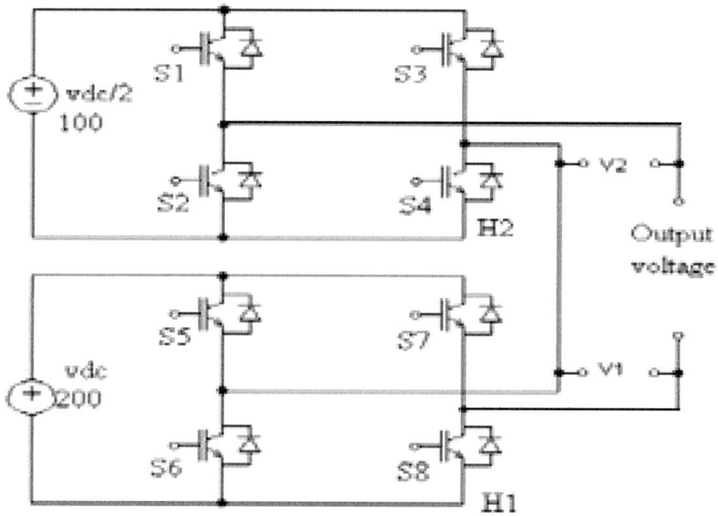

Figure 8.2 Asymmetrical CMLI.

ACMLI can generate multiple voltage levels by combining the voltage outputs of individual H-bridge cells. This capability enables it to produce high-quality AC output voltage with reduced harmonic content compared to traditional inverters. The ACMLIs offer several advantages over traditional inverters, including higher efficiency, improved output waveform quality, reduced harmonic distortion, and better voltage regulation. These features make them suitable for applications where high-quality AC power is required. Asymmetrical CMLIs (ACMLIs) have diverse applications across various industries, including renewable energy systems (such as solar and wind power), adjustable speed drives, UPS, and grid-connected power systems. The ACMLI is a sophisticated power electronics device that offers improved performance and efficiency compared to traditional inverters, making it a preferred choice for various applications in the field of electrical engineering.

8.3 Multicarrier PWM Techniques (MC-PWM)

Numerous such techniques are employed in multilevel inverter (MLI) topologies, categorized as illustrated in Figure 8.3. These techniques fall into two fundamental categories: low switching frequency (fundamental switching frequency) and high switching frequency. Among these, space vector PWM

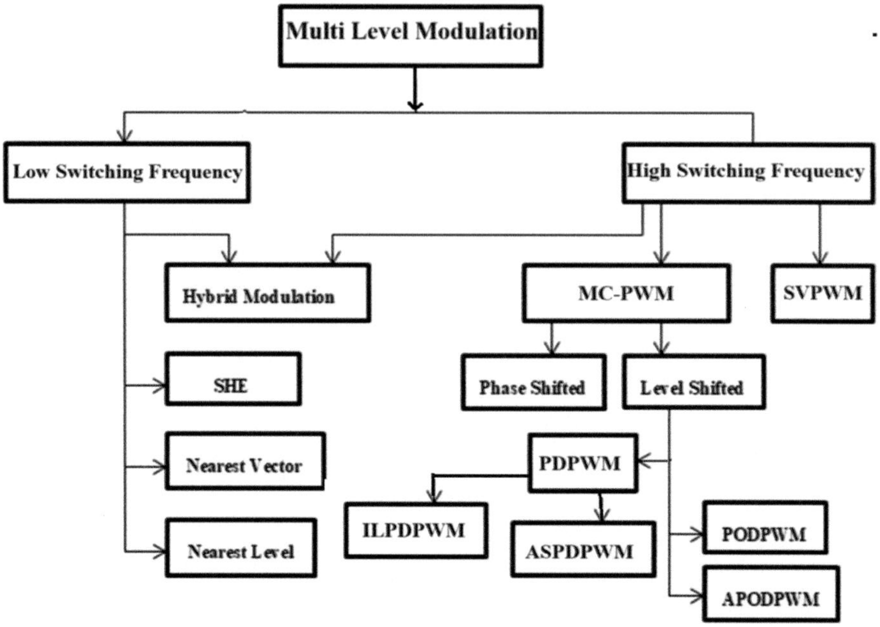

Figure 8.3 Different modulation techniques.

(SVPWM), selective harmonic PWM (SHE-PWM), and MC-PWM are the most commonly utilized in MLI topologies. The multicarrier PWM technique is a high-frequency switching method, where $(n - 1)$ carriers are juxtaposed with a single modulating signal within an n-level inverter. It employs six triangular carriers and one sinusoidal modulating signal. The MC-PWM techniques encompass various types.

8.3.1 Phase disposition (PD)

In this PWM, all carriers above and below the zero-reference point are in phase as given in Figure 8.4. It is a type of PWM method used for controlling the switching operation in inverters and motor drives. In PD modulation, the switching patterns of power devices are determined based on the relative positions of reference and multiple carrier signals in the time domain. Based on the comparison results, the duty cycles of the carrier signals are modulated

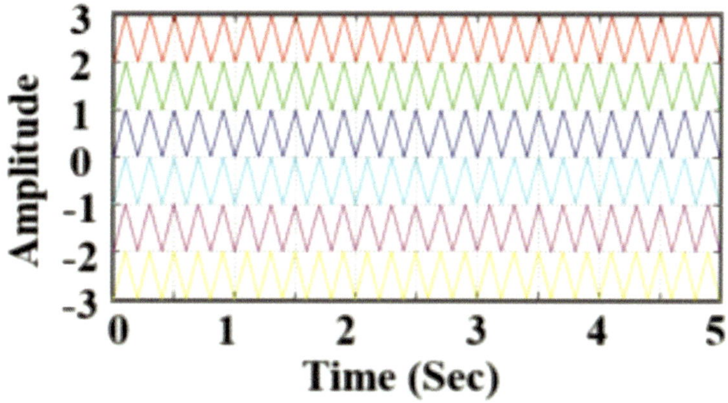

Figure 8.4 MC-PWM comparison signals for PD.

to synthesize the desired output waveform. The modulation process involves turning the power devices on or off at specific times to accomplish the chosen output voltage or current.

In phase disposition modulation, the carrier signals are typically evenly spaced in the time domain. The modulation scheme determines how the reference signal is compared with these carriers and how the switching patterns are generated based on the comparison results. PD modulation offers simplicity and ease of implementation, making it suitable for various power electronic applications. PD modulation can be divided into various subtypes based on the specific modulation scheme used, such as interleaved phase disposition (IPD) and asymmetrical phase disposition (APD), each offering unique advantages and characteristics tailored to specific application requirements.

8.3.2 Phase opposition disposition (POD)

In POD modulation, two carrier signals are generated with opposite polarities, and they are phase-shifted relative to each other by 180 degrees. These carriers are then compared with the reference signal to determine the switching states of the power devices. The modulation scheme seeks to minimize harmonic distortion and enhance the quality of the output waveform. Carriers positioned above the zero reference are 180 degrees out of phase with carriers positioned below the zero reference, as illustrated in Figure 8.5.

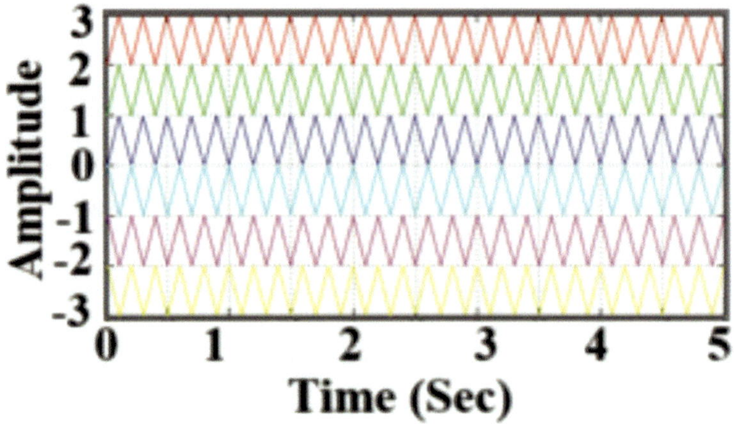

Figure 8.5 MC-PWM comparison signals for POD.

8.3.3 Alternate phase opposition disposition (APOD)

Here, carriers are alternately 180 degrees out of phase of each other as provided in Figure 8.6. In APOD modulation, multiple carrier signals are generated with opposite polarities and are phase-shifted relative to each other. However, unlike traditional phase opposition disposition (POD), where only two carriers are used, APOD employs multiple pairs of carriers with different phase shifts. The phase shifts between the carrier pairs are typically chosen asymmetrically to optimize the modulation scheme for specific application requirements.

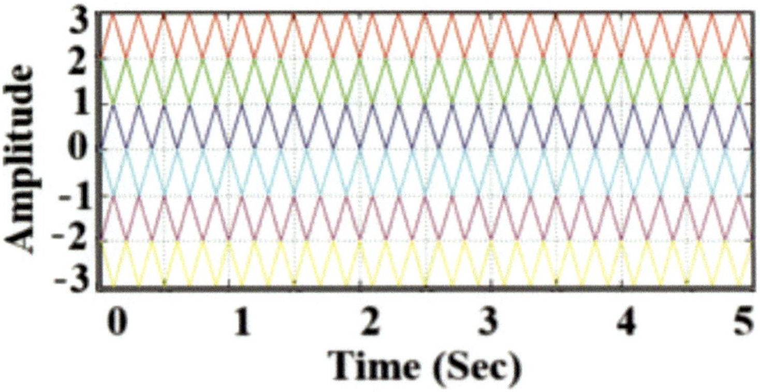

Figure 8.6 MC-PWM comparison signals for APOD.

8.3.4 Phase shifted (PS)

In PS modulation, multiple carrier signals are generated, each with a specific phase shift relative to the others. By adjusting the phase shifts of the carriers, the modulation scheme can be optimized to achieve specific performance objectives, such as reducing harmonic distortion in the output waveform or improving efficiency. In phase-shifted PWM, the carriers have the same amplitude with adjacent carriers being phase displaced, which is given in Figure 8.7.

$$f_{\text{shifted}} = \frac{360^\circ}{m-1} \tag{8.3}$$

where, m = number of output levels.

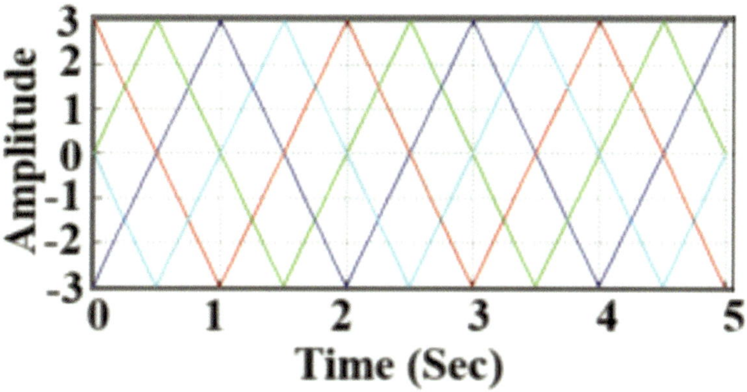

Figure 8.7 MC-PWM comparison signals for PS.

8.4 Simulation Results

All simulations were conducted using Matlab/Simulink software. Figure 8.8 depicts the Matlab diagram of a simulated single-phase ASCMLI, wherein two H-bridges are coupled in series. These H-bridges are supplied by two DC sources with values $V_1 = 70$ V and $V_2 = 140$ V. The resulting output levels are " $0, +70, +140, +210, -70, -140, -210$." It uses $R-L$ load where $R = 100\Omega$ and $L = 0.05$H.

The H-bridge inverter contained in the subsystem is shown in Figure 8.9.a which uses IGBT as switches. Figure 8.9.b. represents the PWM pulse generator for PD, POD, and APOD under the subsystem block. A repeating

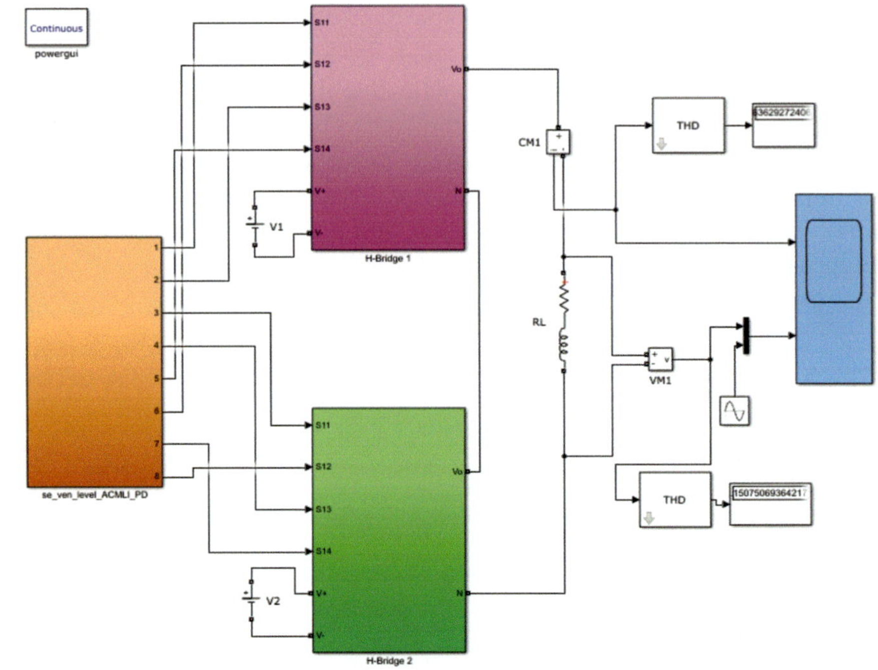

Figure 8.8 Matlab diagram of single-phase ASCMLI.

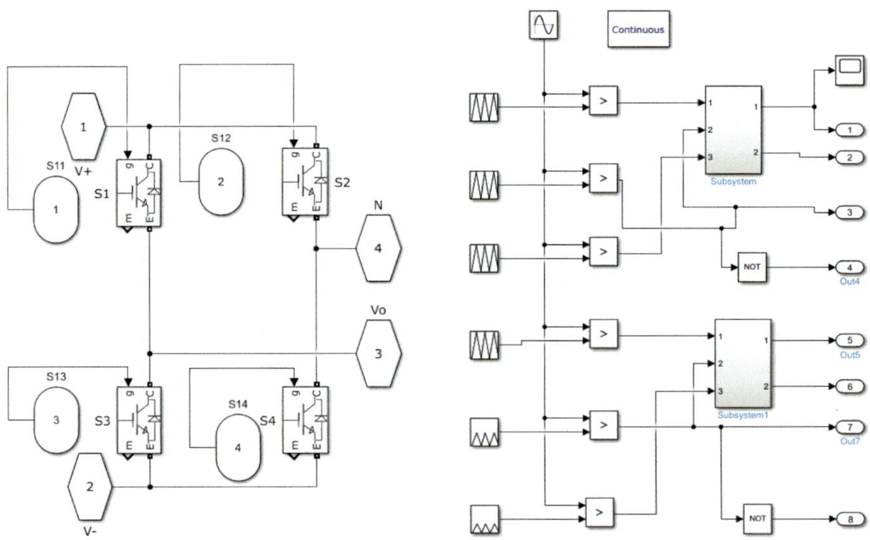

(a) (b)

Figure 8.9 (a) H-bridge inverter. (b) Pulse generator for PD, POD, and APOD.

sequence block has been used to generate the various carrier signals based on the requirement of the number of pulses, and then, these signals are compared with the sinusoidal signal to generate the required pulses for driving the switches.

Figure 8.10.a represents the PWM pulse generator for PS in the subsystem. There are six triangular carriers of 5 kHz compared to a single sinusoidal waveform of 50 Hz by a comparator which is again fed through a logic circuit and the output is PWM pulses given in Figure 8.10.b shows pulses for H-bridges.

Figure 8.11.a shows the current and voltage waveform. Figure 8.11.b represents the current THD% of 6.04% and Figure 8.11.c shows voltage THD% of 13.53% for one-phase POD scheme with $R = 100\ \Omega$, $L = 0.05$ H for $V_1 = 70$, $V_2 = 140$. Figure 8.12.a shows the current and voltage waveform. Figure 8.12.b represents the current THD% of 6.47% and Figure 8.12.c shows voltage THD% of 13.53% for one-phase APOD scheme with $R = 100\ \Omega$, $L = 0.05$ H for $V_1 = 70$, $V_2 = 140$.

Similar waveforms for PD-PWM (ILPDPWM and ASPDPWM) techniques can also be generated for seven-level ASCMLI, considering the same configuration of input and load, whose waveforms are not presented here. Figure 8.13.a shows the current and voltage waveform. Figure 8.13.b represents the current THD% of 10% and Figure 8.13.c shows voltage THD% of 29.36% for one-phase PS scheme with $R = 100\ \Omega$, $L = 0.05$ H for $V_1 = 70$, $V_2 = 140$.

Figure 8.14 represents the voltage and current waveform for a three-phase seven-level ASCMLI with a PD scheme for generating switching pulses. The same single-phase system has been implemented in a three-phase system without modification or changes in the input and output parameters of inverters.

Figure 8.15 depicts the three-phase voltage and current waveform for three-phase seven-level CMLI with pulse generated by using the APOD scheme in order to showcase the working of the respective inverter under different schemes. The input voltage to H-bridges and output load parameters are kept the same as the previous one.

The simulation results for output current and THD% of the same are obtained for a particular load with a carrier frequency of 5 kHz and modulating frequency of 50 Hz for PD, POD, APOD, and PS multicarrier PWM techniques with different modulation indexes are shown in Figure 8.16.a–d.

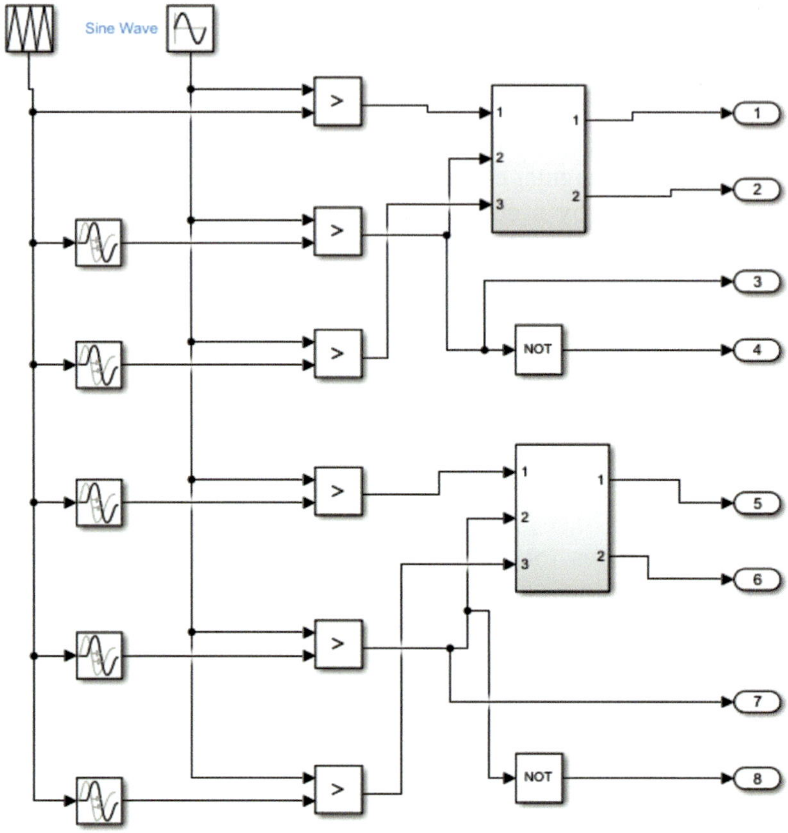

(a)

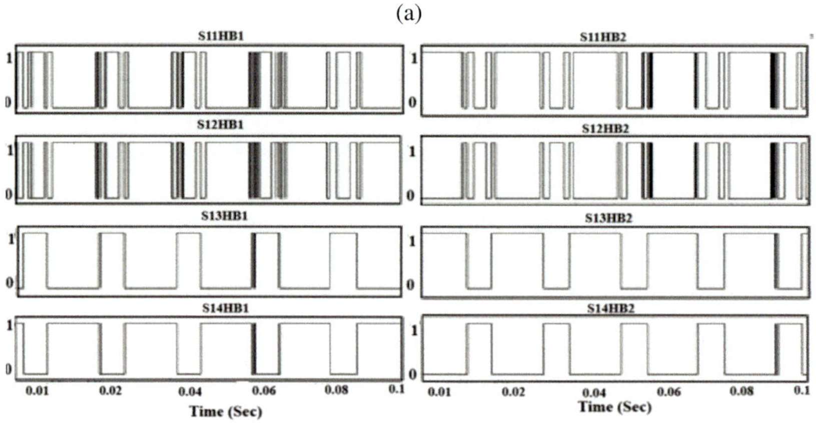

(b)

Figure 8.10 (a) PWM pulse generator for PS. (b) PWM pulses for H-bridges.

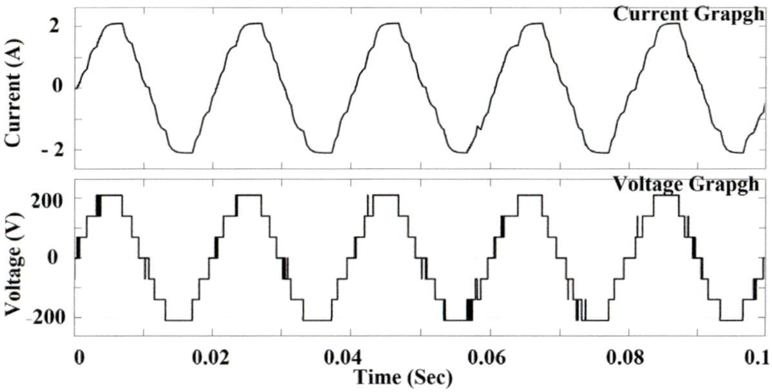

(a)

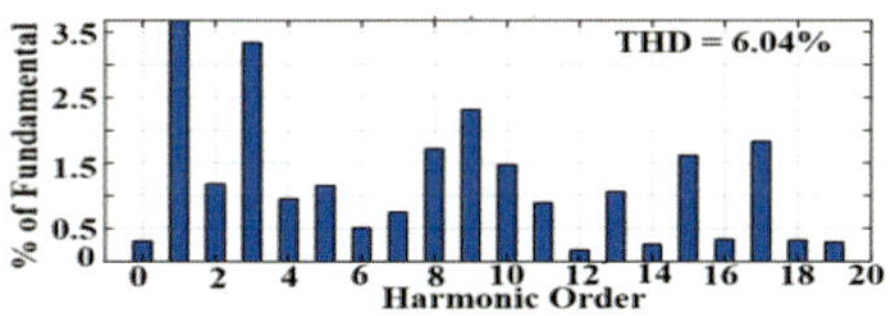

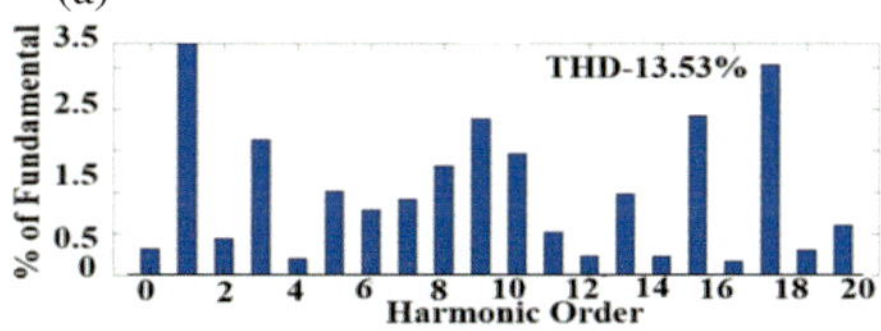

(b) (c)

Figure 8.11 (a) Current and voltage waveform, (b) current THD%, and (c) voltage THD% for one-phase POD scheme with $R = 100\ \Omega$, $L = 0.05$H for $V_1 = 70$, $V_2 = 140$.

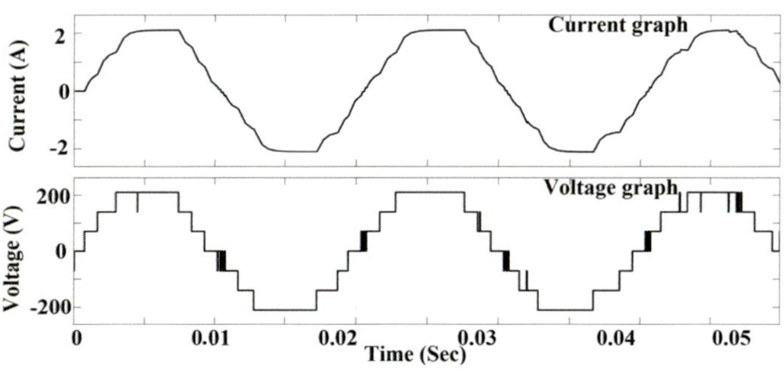

(a)

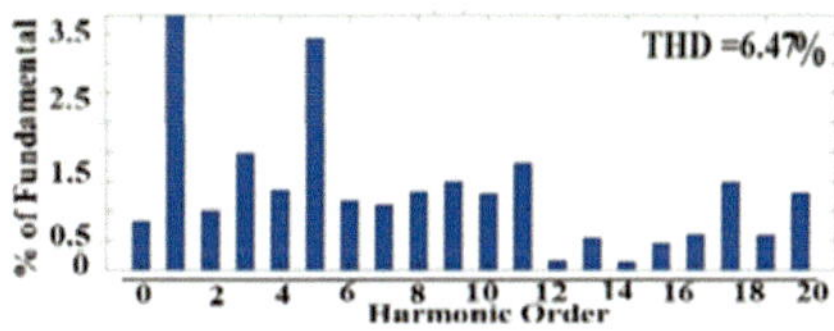

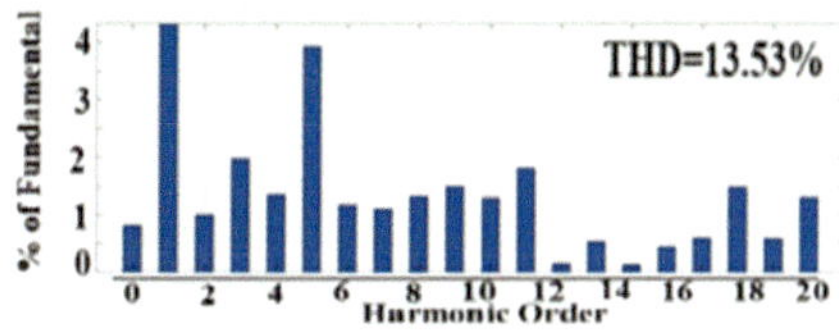

(b) (c)

Figure 8.12 (a) Current and voltage waveform, (b) current THD%, and (c) voltage THD% for one-phase APOD scheme with $R = 100\ \Omega$, $L = 0.05$H for $V_1 = 70$, $V_2 = 140$.

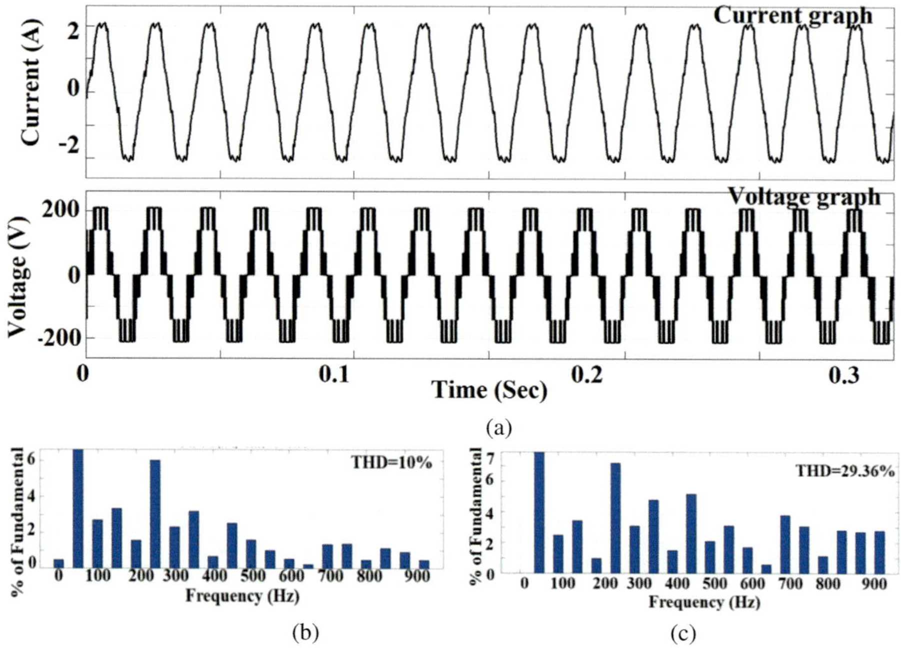

Figure 8.13 (a) Current and voltage waveform, (b) current THD%, and (c) voltage THD% for one-phase PS scheme with $R = 100\ \Omega$, $L = 0.05$ H for $V_1 = 70$, $V_2 = 140$.

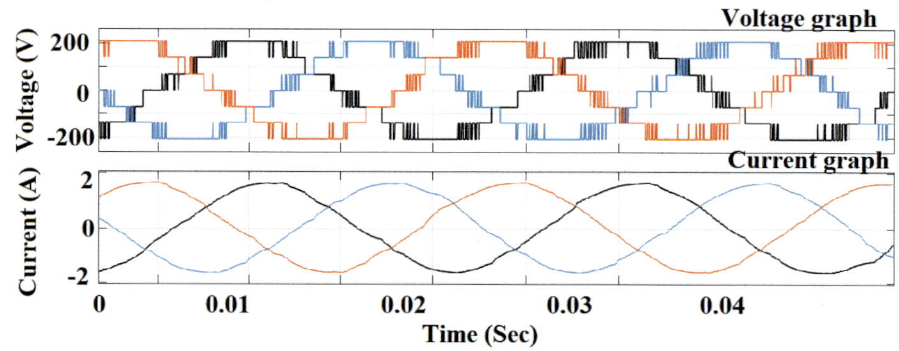

Figure 8.14 Voltage and current waveform for three-phase PD scheme with $R = 100\ \Omega$, $L = 0.05$ H for $V_1 = 70$, $V_2 = 140$.

The simulation results for output voltage and THD% of the same are obtained for PD, POD, APOD, and PS multicarrier PWM techniques with different modulation indexes are shown in Figure 8.17.a–d.

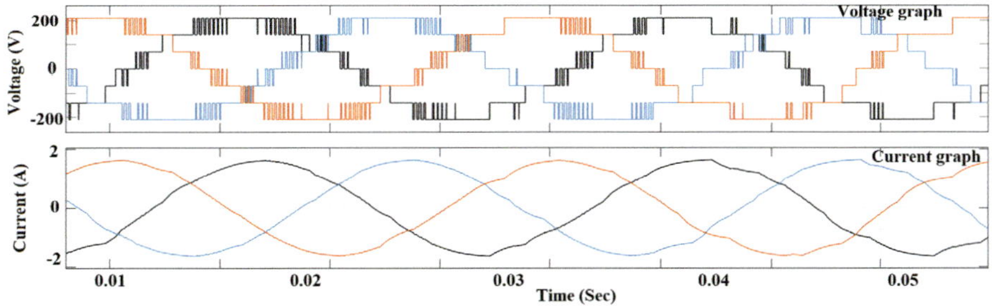

Figure 8.15 Voltage and current waveform for three-phase POD scheme with $R = 100\ \Omega$, $L = 0.05$ H for $V_1 = 70$, $V_2 = 140$.

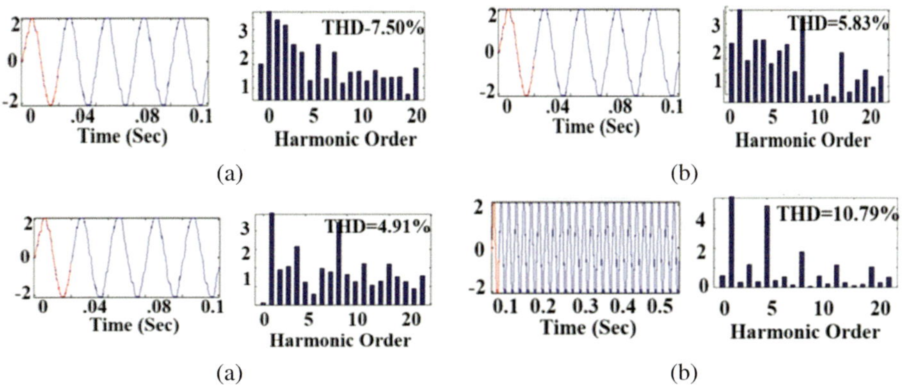

Figure 8.16 Output current and THD for (a) PD, (b) POD, (c) APOD, and (d) PS, respectively, for multicarrier PWM.

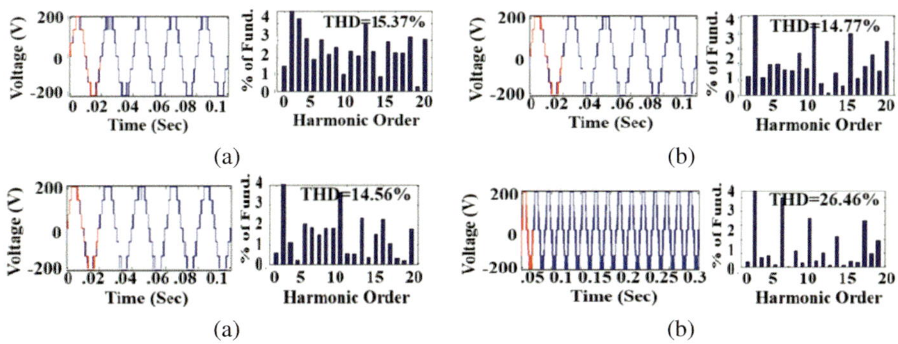

Figure 8.17 Output voltage and THD for (a) PD, (b) POD, (c) APOD, and (d) PS, respectively, for multicarrier PWM.

8.5 Real Time Results

To validate the simulation results, hardware results were provided. Appendix-A shows the microcontroller program to generate pulses. The input DC used

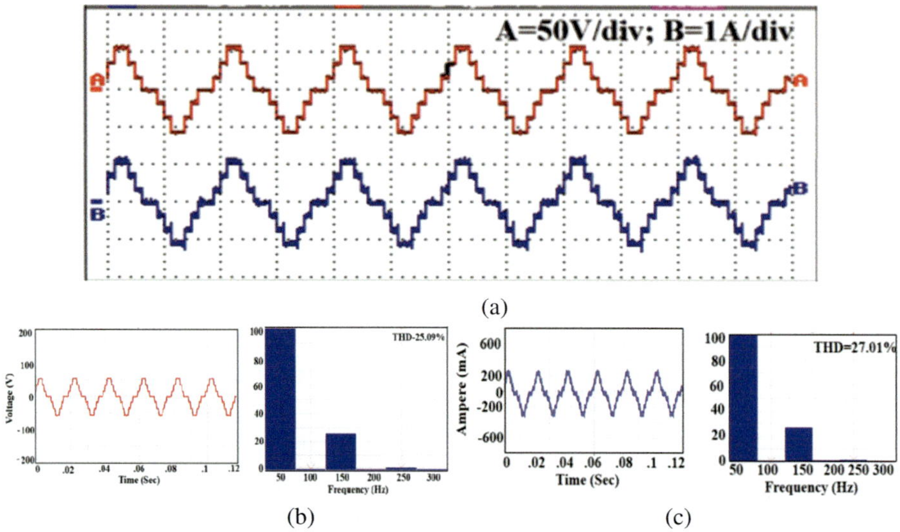

Figure 8.18 (a) Output voltage and current, (b) voltage THD% graph, and (c) current THD% graph for a resistive load of 100 Ω, $V_1 = 20$ V and $V_2 = 40$ V.

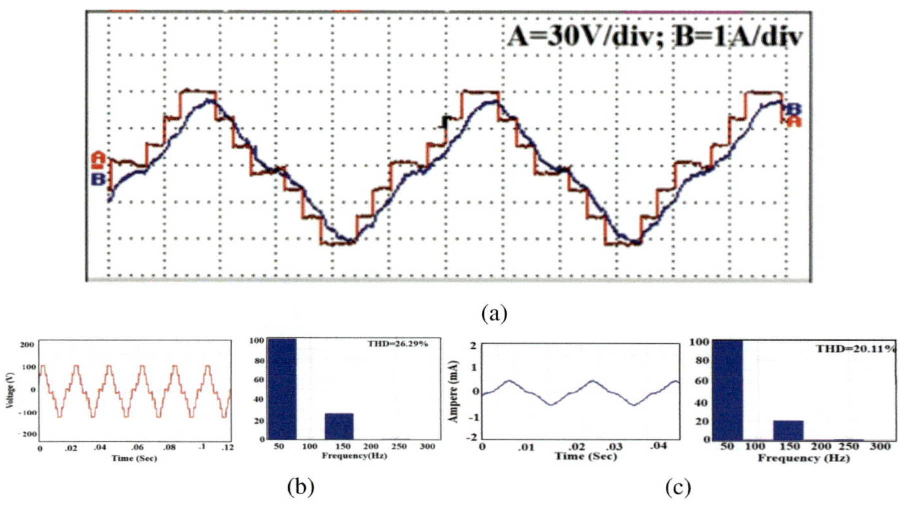

Figure 8.19 (a) Output voltage and current, (b) voltage THD% graph, and (c) current THD% graph for a resistive load of 100 Ω, $V_1 = 20$ V, and $V_2 = 40$ V.

for the inverter was taken from Aplab Make regulated DC power supply. The hardware outputs were taken by four channels, 200 MHz, 2.5 GS/s FLUKE 190-204 Scope meter, and 100 mV/A current probe, respectively.

Figure 8.18.a below shows the hardware result for voltage and current, Figure 8.18.b shows FFT for voltage, and Figure 8.18.c shows the FFT for current with a resistive load of 100 Ω and input voltage in the binary configuration of 20 and 40 V, respectively, to obtain a seven-level output.

Figure 8.19.a below shows the hardware result for voltage and current, Figure 8.19.b shows FFT for voltage, and Figure 8.19.c shows FFT for the current for seven-level ACMLI with a load of 100 Ω, 60 mH, and input DC as previous case.

8.6 Comparison

With all the simulations done by the above methods, the comparison tables have been obtained. Tables 8.1 and 8.2 represent the variation of THD with

Table 8.1 Variation of THDv and THDi with modulation index at a carrier frequency of 5 kHz.

		PD-PWM		POD-PWM		APOD-PWM		PS-PWM	
fc	Ma	V_{THD}	i_{THD}	V_{THD}	i_{THD}	V_{THD}	i_{THD}	V_{THD}	i_{THD}
5 kHz	0.8	20.30	9.4	20.46	9.87	21.03	12.60	37.96	16.91
	1	15.37	7.5	14.77	5.83	14.56	4.91	26.46	10.79
	1.2	15.24	9.68	14.23	8.72	14.35	9.83	23.06	10.36

Table 8.2 Variation of V_{THD} and i_{THD} with modulation index at a carrier frequency of 2.5 kHz.

		PD-PWM		POD-PWM		APOD-PWM		PS-PWM	
fc	Ma	V_{THD}	i_{THD}	V_{THD}	i_{THD}	V_{THD}	i_{THD}	V_{THD}	i_{THD}
2.5 Hz	0.8	20.54	8.63	19.76	9.48	20.32	8.00	41.45	16.41
	1	15.61	6.79	16.20	3.94	14.62	5.95	29.21	10.29
	1.2	14.21	7.72	13.95	7.86	14.02	8.45	24.59	9.66

modulation indices for PD, POD, APOD, and PS MC-PWM techniques at carrier frequencies of 5 and 2.5 kHz, respectively.

V_{THD} and i_{THD} represent distortion for voltage and current, respectively, fc is the carrier frequency, and Ma is the modulation index.

8.7 Conclusion

This chapter explores the analysis of a seven-level binary structure ACMLI topology employing various modulation techniques of MC-PWM. To implement the seven-level inverter with multicarrier techniques, six carriers were utilized, including PD-PWM, POD-PWM, APOD-PWM, and PS-PWM techniques. These techniques were compared in terms of THD on both the output voltage and current, considering different modulation and frequency indexes. On the other hand, some hardware results have been presented here to show the functioning of the ACMLI. It depicts that a modulation index of 1 is preferred for achieving lower THDs. Among the investigated multicarrier PWM techniques, APOD-PWM emerges as the most suitable option, offering lower THDs. Using APOD-PWM at higher carrier frequencies, with a modulation index of 1, results in lower THDs.

References

[1] P. Thongprasri, “A 5-Level Three-Phase Cascaded Hybrid Multilevel Inverter,” International Journal of Computer and Electrical Engineering, Vol. 3, No. 6, pp-789-794, December 2011.

[2] J. Zhao, X. He, R. Zhao, “A Novel PWM Control Method for Hybrid-Clamped Multilevel Inverters,” Industrial Electronics, IEEE Transactions on, vol.57, no.7, pp.2365-2373, July 2010.

[3] T. Prathiba, P.Renuga, “A comparative study of Total Harmonic Distortion in Multilevel inverter topologies,” Journal of Information Engineering and Applications, Vol 2, No.3, pp.26-36, 2012.

[4] M.G.H.Aghdam, S.H.Fathi, G.B.Gharehpetian, "Analysis of multicarrier PWM methods for asymmetric Multi-level inverter," Industrial Electronics and Applications, 2008. ICIEA 2008. 3rd IEEE Conference on, vol., no., pp.2057-2062, 3-5 June 2008.

[5] R. Naderi, A. Rahmati, “Phase-Shifted Carrier PWM Technique for General Cascaded Inverters,” Power Electronics, IEEE Transactions on, vol.23, no.3, pp.1257-1269, May 2008.

[6] P.Palanivel, S.S.Dash, "Analysis of THD and output voltage performance for cascaded multilevel inverter using carrier pulse width modulation techniques," Power Electronics, IET , vol.4, no.8, pp.951-958, September 2011.

[7] I.Colak, E.Kabalci, R.Bayindir, S.Sagiroglu, "The design and analysis of a 5-level cascaded voltage source inverter with low THD," Power Engineering, Energy and Electrical Drives, 2009. POWERENG '09. International Conference on, vol., no., pp.575-580, 18-20 March 2009.

[8] Comparision of Multi-Carrier Techniques in Seven-Level Asymmetric Cascade Multilevel Inverter Ilhami COLAK Faculty of Technology, Department of Electric & Electronic Engineering Gazi University Ankara, Ersan KABALCI Faculty of Engineering & Architecture, Dep. of Electrical & Electronic Eng. Nevsehir University Nevsehir, Turkey Gokhan KEVEN Vocational College of Hac Bekta Veli, Dep. of Computer Programming Nevsehir University Nevsehir, Turkey.

[9] International Journal of Emerging Technology and Advanced Engineering Website: www.ijetae.com (ISSN 2250-2459, Volume 2, Issue 3, March 2012) 230 A Comparative Study of Asymmetric and Symmetric Cascaded Multilevel Inverter employing Variable Frequency Carrier based PWM Dr. R.Seyezhai Associate Professor, Department of EEE, SSN College of Engineering, Chennai, India.

[10] Power Electronics Handbook, M.H.Rashid

[11] Analog control scheme for cascaded multilevel inverters based on multiple carrier PWM technique S.Albert Alexander, Research Scholar Department of Electrical and Electronics Engineering Kongu Engineering College Perundurai, India. T.Manigandan Principal P.A.College of Engineering and Technology Pollachi, India.

[12] A Hybrid Cascaded Multilevel Inverter Nattapat Praisuwanna, Surin Khomfoi Center of Excellence for Innovative Energy Systems, Faculty of Engineering King Mongkut's Institute of Technology Ladkrabang, Chalongkrung Rd. Ladkrabang,, Bangkok, 10520, THAILAND.

APPENDIX-A

The microcontroller program for 7-level Cascaded Inverter is given below;

```
#define F_CPU 1000000UL
#include <avr/io.h>
#include <util/delay.h>
```

```
int main(void)
{
DDRA=0xFF;
DDRC=0xFF;
while(l)
{
PORTA=0xAA;
PORTC=0xAA;
_delay_ms(1.278244/2);
PORTA=0x9A;
PORTC=0xA0;
_delay_ms(1.912522/2);
PORTA=0x99;
PORTC=0xA0;
_delay_ms(3.154345/2);
PORTA=0x99;
PORTC=0x90;
_delay_ms(7.309777/2);
PORTA=0x99;
PORTC=0x50;
_delay_ms(3.154345/2);
PORTA=0x95;
PORTC=0x50;
_delay_ms(1.912522/2);
PORTA=0x55;
PORTC=0x50;
_delay_ms(1.278244/2);
PORTA=0x55;
PORTC=0x50;
_delay_ms(1.278244/2);
PORTA=0x65;
PORTC=0x50;
_delay_ms(1.912522/2);
PORTA=0x66;
PORTC=0x50;
_delay_ms(3.154345/2);
PORTA=0x66;
PORTC=0x60;
_delay_ms(7.309777/2);
```

```
PORTA=0x66;
PORTC=0xA0;
_delay_ms(3.154345/2);
PORTA=0x6A;
PORTC=0xA0;
_delay_ms(1.912522/2);
PORTA=0xAA;
PORTC=0xA0;
_delay_ms(1.278244/2);
}
}
```

9

Cascaded H-Bridge and Modular Multilevel Converters: A Review

P. S. V. Kishore[1], Nakka Jayaram[1], Jami Rajesh[1], Swmay Jakkula[2], and Kusuma Gottapu[3]

[1]Department of Electrical Engineering, NIT Andhra Pradesh, India
[2]Department of Electrical Engineering, IIT Mumbai, India
[3]Department of Electrical and Electronics Engineering, SRKR Engineering College, India
Email: kishorepulavarthi.sclr@nitandhra.ac.in; jayaram@nitandhra.ac.in; Jamirajesh.sclr@nitandhra.ac.in; swamyjakkulanitap@gmail.com; gkusuma9@gmail.com

Abstract

This chapter provides an in-depth exploration of multilevel converter technologies, specifically focusing on the cascade H-bridge multilevel inverter (CHBMLI) and modular multilevel converter (MMC), which are crucial for medium-voltage high-power drives. The CHBMLI is examined in detail, covering its symmetrical and asymmetrical operation modes, comparisons with basic multilevel inverters (MLIs) such as DCMLI and FCMLI, and various pulse width modulation (PWM) techniques. Additionally, its applications in grid integration, FACTS, and motor drives are discussed. The chapter also delves into the MMC, discussing its evolution from basic MLIs, addressing isolated DC sources, capacitor unbalancing, and harmonic profiles. The basic structure of MMC, comprising half-bridge and full-bridge cells, is explained, along with mathematical modeling of energy exchange between arms and phases. Various circuit configurations, including AC-DC, DC-AC, three-phase to one-phase, three-phase to three-phase, three-phase to multiphase, and DC-DC converters, are explored, as well as fault analysis and PWM techniques for balancing capacitor voltage across individual cells. Finally,

DOI: 10.1201/9781003669845-9

the significance of MMC in diverse applications is examined, highlighting its importance in medium-voltage high-power drives.

Keywords: Cascaded H-bridge converters, fault analysis on MMC, modular multilevel converters, MMC circuit configurations, PWM techniques

9.1 Introduction

An inverter is a type of converter that converts direct current (DC) supply voltage into alternating current (AC) output voltage. Inverters can be categorized into several types, as illustrated in Figure 9.1.

A conventional two-level inverter produces output voltages of 0, $\pm V_{\text{in}}$, resulting in a waveform with substantial harmonic distortion. To achieve a nearly sinusoidal output, suitable for AC supply applications, the concept of multilevel inverters (MLIs) was developed. MLIs generate output waveforms with multiple voltage levels, significantly reducing total harmonic distortion (THD) as the number of levels increases. MLIs have created a revolution in the last decades due to the following advantages

- Draws input current with minimal distortion.
- Low harmonic content results in improved power quality.
- Electromagnetic compatibility is better.
- Easy interface with the motor drives.
- Ability to operate without transformers.
- The common mode voltage is less.
- They can operate at higher voltage and current levels.
- Possibility of getting fault tolerant operation.
- Less dV/dt.

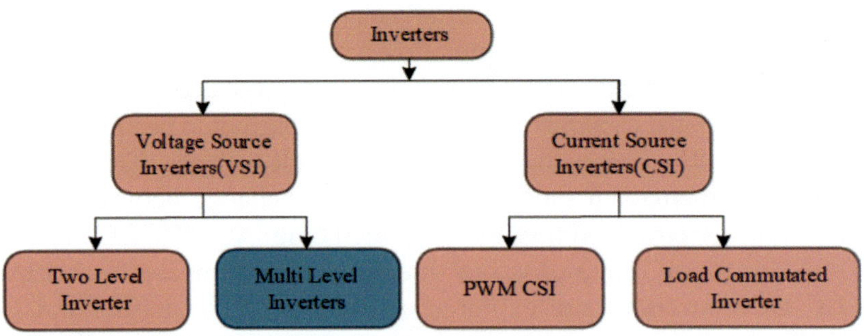

Figure 9.1 Classification of inverters.

9.2 Overview of Multilevel Inverters

Multilevel inverters are inverters having an output voltage with a greater number of levels. That means the output voltage of an MLI resembles a sine wave. The first multilevel inverter was the CHB inverter and it was invented in 1975. The neutral point-clamped or diode-clamped inverter was developed in the 1980s by manipulating the CHB inverter. In this diode-clamped inverter, diodes are used to block the source voltages. Later in the 1990s, capacitor-clamped inverters were developed. Even though the first CHB was invented in 1975, the use of this cascaded inverter was first found in 1990. Therefore, the classical MLI are classified as (1) diode-clamped MLI, (2) flying capacitor MLI, and (3) cascaded MLI. Later in the 2000s, modular multilevel converter (MMC) and active neutral point topologies were developed. Recently, several reduced switch count multilevel inverters were developed by many researchers.

9.2.1 Diode-clamped MLI

The three-level DCMLI circuit is shown in Figure 9.2.a, where the output voltage (V_o) is measured relative to the ground. The two capacitors divide the input voltage (V_{in}) equally, generating three output levels: 0, $+V_{in}/2$, and $-V_{in}/2$. Switching on specific semiconductor pairs S_2 and S_3, S_1 and S_2, or S_3 and S_4 produces these output levels, respectively. Each switch handles a voltage stress of $V_{in}/2$. The circuit has two complementary switch pairs (S_1

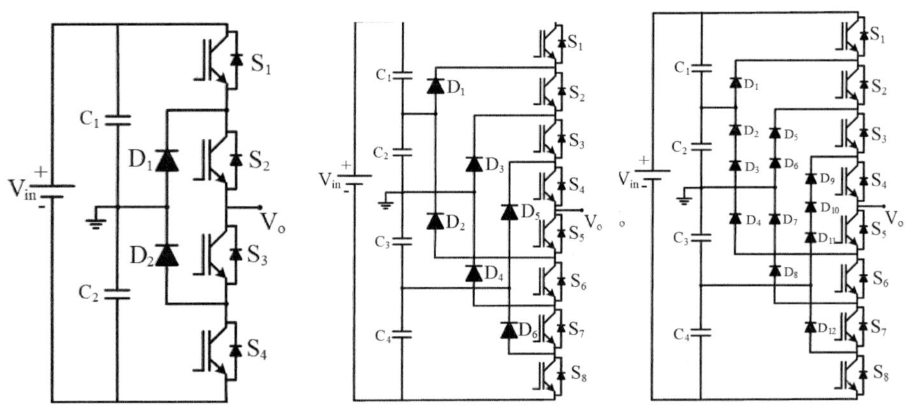

Figure 9.2 (a) Three-level DCMLI, (b) five-level DCMLI, and (c) modified five-level DCMLI.

and S_3 and S_2 and S_4) that operate in an alternating manner. Figure 9.2.b shows the five-level DCMLI circuit, which has four complementary switch pairs (S_1 and S_5, S_2 and S_6, S_3 and S_7, and S_4 and S_8). Note that the clamping diodes have different voltage-blocking capacities, depending on the switch configuration and input voltage division. When all the lower switches (S_5, S_6, S_7, and S_8) are turned on, the clamping diodes experience different voltage-blocking requirements: D_2 blocks 3/4 of the input voltage ($3V_{in}/4$), D_4 blocks half of the input voltage ($V_{in}/2$), and D_6 blocks only 1/4 of the input voltage ($V_{in}/4$), assuming the input voltage is evenly distributed across the DC bus capacitors.

That means different clamping diodes have to be rated for different voltage ratings. This problem can be addressed by connecting three diodes of rating $V_{in}/4$ in place of one diode rated for $V_{in}/4$. The topology of the modified DCMLI is depicted in Figure 9.2.c. In this, all the clamping diodes are rated for $V_{in}/4$ only.

9.2.2 Flying capacitor MLI

The circuit diagrams of three-level and five-level FCMLI are depicted in Figures 9.3.a and 9.3.b, respectively. This FCMLI is also called a capacitor-clamped MLI [2]. The switching arrangement for the three-level FCMLI is

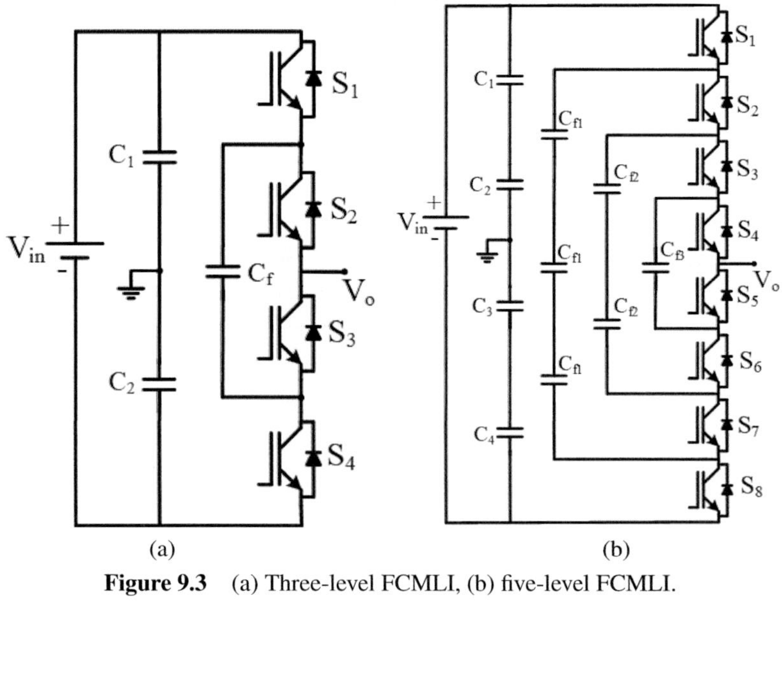

(a) (b)

Figure 9.3 (a) Three-level FCMLI, (b) five-level FCMLI.

the same as the three-level DCMLI. There are four complementary switching pairs in this inverter also, they are (S_1, S_8), (S_2, S_7), (S_3, S_6), and (S_4, S_5). The advantage of FCMLI is that there is a redundancy in the switching pattern. That means, to generate a voltage level of $V_{in}/4$ or $-V_{in}/4$, four switching combinations are possible and six switching combinations are possible to generate a zero voltage level. By properly selecting the switching combinations, it is possible to balance the voltages across the clamping capacitors.

9.2.3 Cascade H-bridge MLI

9.2.3.1 Symmetrical CHBMLI

A cascade H-bridge multilevel inverter (CHBMLI) is symmetrical if the same input voltage magnitude is applied to each H-bridge module; otherwise, it is asymmetrical. The basic building block is a single H-bridge circuit (Figure 9.4.a), which produces three voltage levels (0, $\pm V_{in}$). By switching on specific semiconductor pairs, the circuit generates $+V_{in}$, $-V_{in}$, or zero voltage levels. Connecting multiple H-bridges in cascade mode (Figure 9.4.b) synthesizes a five-level output voltage waveform (Figure 9.4.c). The output voltage is the sum of individual H-bridge outputs. Table 9.1 shows the switching configurations for generating five voltage levels. This modular design allows for easy extension to more levels by adding H-bridges in cascade, with the number of levels (p) related to the number of H-bridges (N) by $p = 2N + 1$. Each switch should have a voltage rating of V_{in}. A comparison of component counts for basic multilevel inverters (MLIs) is provided in Table 9.2.

Advantages of CHBMLI:

- In comparison to the previous two topologies, fewer components are enough to generate the same number of levels.
- No clamping diodes and flying capacitors are needed.
- It has modularity in structure; therefore, it is easy to extend for higher voltage levels.
- By connecting different DC sources to the various bridges, a higher number of voltage levels can be obtained. Thus, asymmetrical CHBMLI is evolved.

Disadvantages of CHBMLI:

- Separate DC supplies are needed in each H-bridge.

Applications: The CHBMLI is used in

- Interfacing with renewable energy resources

Table 9.1 Switching pattern for five-level CHBMLI.

S_1	S_2	S_3	S_4	S_5	S_6	S_7	S_8	V_{o1}	V_{o2}	V_o
1	0	0	1	1	0	0	1	V_{in}	V_{in}	$2V_{in}$
1	0	0	1	1	0	1	0	V_{in}	0	V_{in}
1	0	0	1	0	1	0	1	V_{in}	0	
1	0	1	0	1	0	0	1	0	V_{in}	
0	1	0	1	1	0	0	1	0	V_{in}	
1	0	1	0	1	0	1	0	0	0	0
1	0	1	0	0	1	0	1	0	0	
0	1	0	1	1	0	1	0	0	0	
0	1	0	1	0	1	0	1	0	0	
1	0	0	1	0	1	1	0	V_{in}	$-V_{in}$	
0	1	1	0	1	0	0	1	$-V_{in}$	V_{in}	
0	1	1	0	1	0	1	0	$-V_{in}$	0	$-V_{in}$
0	1	1	0	0	1	0	1	$-V_{in}$	0	
1	0	1	0	0	1	1	0	0	$-V_{in}$	
0	1	0	1	0	1	1	0	0	$-V_{in}$	
0	1	1	0	0	1	1	0	$-V_{in}$	$-V_{in}$	$-2V_{in}$

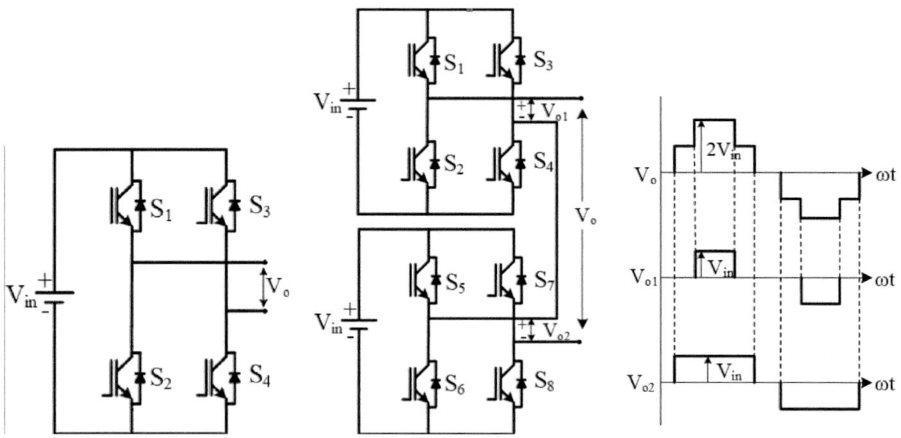

(a) (b) (c)

Figure 9.4 (a) Three-level H-bridge, (b) five-level CHBMLI, (c) output voltage waveform V_o for five-level CHBMLI.

- Back-to-back frequency link systems
- Power factor compensators
- DC power source utilization
- Electric vehicle drives
- Active filters and motor drives

Table 9.2 Comparison of basic MLI topologies.

MLI topology	No. of switches	No. of DC sources	No. of DC bus capacitors	No. of clamping diodes	No. of clamping capacitors
Diode clamped	$2 \times (p-1)$	1	$p-1$	(p-1)*(p-2)	--
Flying capacitor	$2 \times (p-1)$	1	$m-1$	--	$(p-1) \times (p-2)/2$
CHB	$2 \times (p-1)$	$(p-1)/2$	--	--	--

9.2.3.2 Asymmetrical CHBMLI

The CHBMLIs discussed above have source voltages with equal values. Without increasing the number of bridges, source voltages with uneven values can yield a greater number of levels in the output voltage. If the source voltages are in the ratio of 1:2, then it generates a seven-level output voltage and if the ratio is 1:3, then it generates a nine-level output voltage. The circuit diagrams of these asymmetrical CHBMLIs are depicted in Figure 9.5.a,b. The

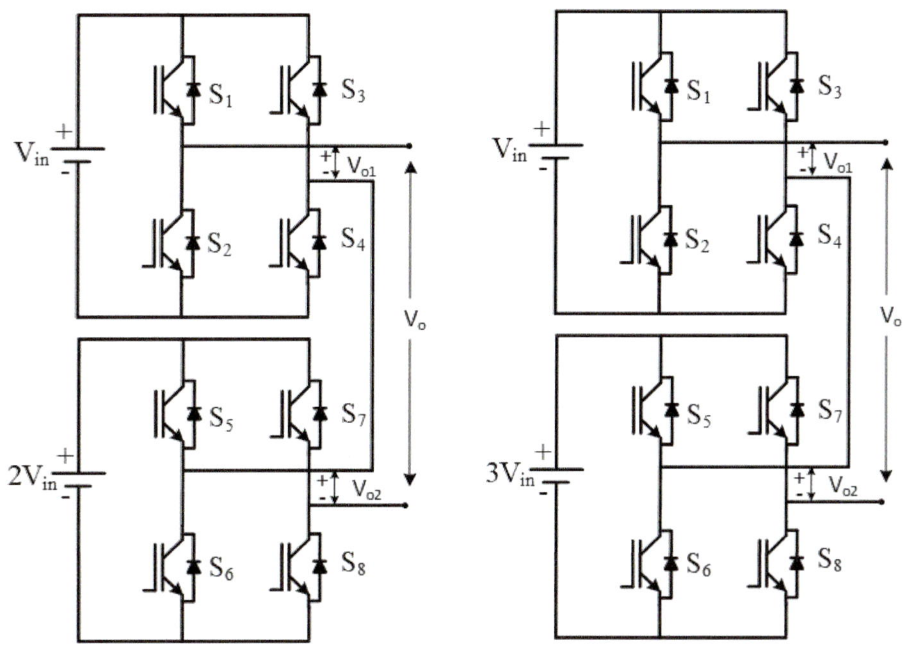

Figure 9.5 (a) Seven-level asymmetrical CHBMLI, and (b) nine-level asymmetrical CHBMLI.

Table 9.3 Switching pattern for seven-level asymmetrical CHBMLI.

S_1	S_2	S_3	S_4	S_5	S_6	S_7	S_8	V_{o1}	V_{o2}	V_o
1	0	0	1	1	0	0	1	V_{in}	$2V_{in}$	$3V_{in}$
1	0	1	0	1	0	0	1	0	$2V_{in}$	$2V_{in}$
0	1	0	1	1	0	0	1	0	$2V_{in}$	
1	0	0	1	1	0	1	0	V_{in}	0	V_{in}
1	0	0	1	0	1	0	1	V_{in}	0	
0	1	1	0	1	0	0	1	$-V_{in}$	$2V_{in}$	
1	0	1	0	1	0	1	0	0	0	0
1	0	1	0	0	1	0	1	0	0	
0	1	0	1	1	0	1	0	0	0	
0	1	0	1	0	1	0	1	0	0	
1	0	0	1	0	1	1	0	V_{in}	$-2V_{in}$	$-V_{in}$
0	1	1	0	1	0	1	0	$-V_{in}$	0	
0	1	1	0	0	1	0	1	$-V_{in}$	0	
1	0	1	0	0	1	1	0	0	$-2V_{in}$	$-2V_{in}$
0	1	0	1	0	1	1	0	0	$-2V_{in}$	
0	1	1	0	0	1	1	0	$-V_{in}$	$-2V_{in}$	$-3V_{in}$

switching arrangement to generate seven levels for Figure 9.5.a is shown in Table 9.3. The switching pattern to get nine levels in the output voltage for the circuit of Figure 9.5.b can be obtained by replacing the output voltage of the second H-bridge (V_{o2}) of $\pm 2V_{in}$ with $\pm 3V_{in}$ in Table 9.3. The main disadvantage of this asymmetrical CHBMLI is that the modularity in the structure is no more and the switches in each H-bridge are rated for different voltage ratings.

9.3 Different Control Techniques

In the previous section, an overview of the multilevel converters is presented. To control these MLIs, different control techniques were proposed by the researchers. These control strategies are divided into two categories based on the value of switching frequency. (1) Fundamental switching frequency, and (2) high switching frequency.

The control strategies based on fundamental switching frequency are again classified into four types as follows.

- Nearest level control
- Space vector control
- Optimal switching angle
- Selective harmonic elimination

The control strategies based on high switching frequency are classified in the flowchart shown below.

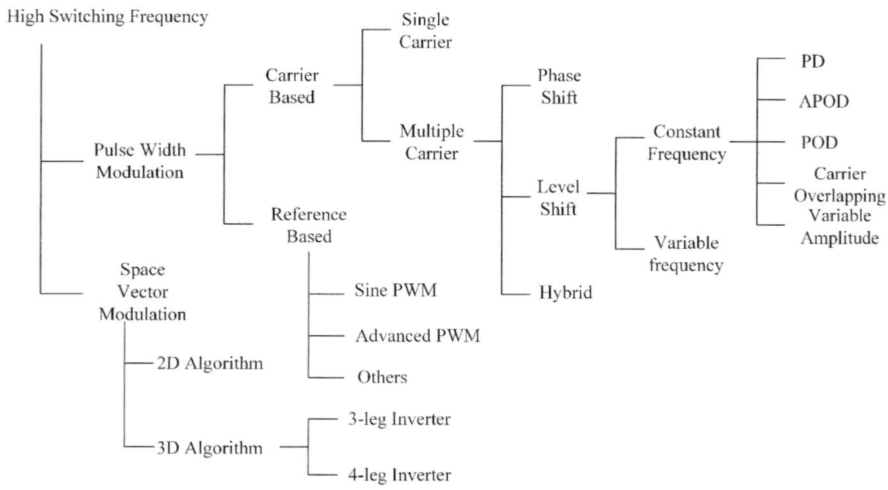

9.3.1 Fundamental switching frequency

This low switching frequency category of controlling has one or two commutations per cycle. This group mainly has four different techniques as shown in the classification.

9.3.1.1 Selective harmonic elimination (SHE)

The SHE technique for a 1-ϕ five-level inverter is described using Fourier coefficients, and the solution of the equations is derived using two switching angle variables.

$$4\pi\left[\cos\delta_1 + \cos\delta_2\right] = m. \tag{9.1}$$

$$\cos\left(3\delta_1\right) + \cos\left(3\delta_2\right) = 0. \tag{9.2}$$

Eqn (9.1) represents the fundamental component amplitude, and to make third-order harmonic to zero, eqn (9.2) is made equal to zero. The other harmonics 5th and 7th also eliminated. There may be multiple solutions for higher-order harmonic elimination, but the best method is one amongst them, using THD as the criterion. This method can be generalized to any number of switching angles and levels easily. The major challenge of this technique is finding the solutions of transcendental equations [4, 5]. The major drawback is the solution found from this technique changes for every modulation index (m).

9.3.1.2 Space vector control (SVC)

Nearest vector control (NVC), also known as space vector control (SVC), is a control method that operates at a low switching frequency. NVC does not generate the required load voltage average in every switching cycle. Instead, it selects the vector closest to the reference vector, minimizing the error between them [6]. By choosing the nearest vector, NVC reduces switching losses and commutations, making it an efficient control strategy [7].

9.3.1.3 Nearest level control (NLC)

The round modulation technique is another name for this. This approach generates a staircase voltage with a lower output voltage quality. This method is also known as space vector control in the time domain. By comparing the intended reference waveform to the output voltage level, the nearest voltage level is computed. Figure 9.6 shows a control diagram for the nearest level control.

$$\text{Closest voltage level} = xf_{\text{round}}, \tag{9.3}$$

where x is the difference between actual and reference signals. Figure 9.7 depicts that the maximum error is $x/2$ [7].

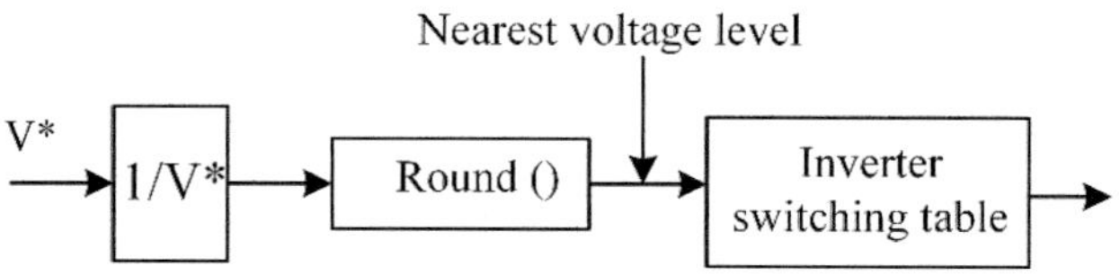

Figure 9.6 Nearest level control block diagram.

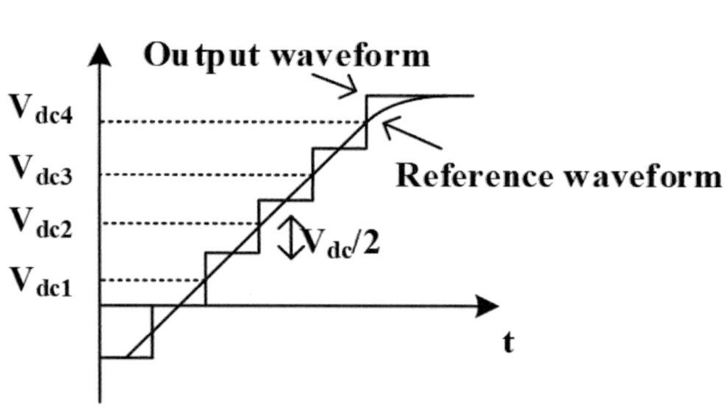

Figure 9.7 Generation of voltage in nearest level control.

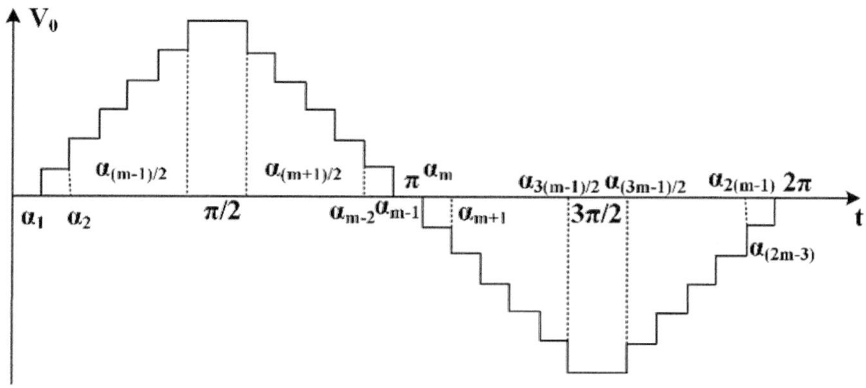

Figure 9.8 13-level MLI output waveform.

9.3.1.4 Switching angle (SA) calculation techniques

Calculating the SA is critical for resolving power quality issues. The THD in output voltage and current waveforms can be increased if the SA is not appropriately taken. Appropriate SA is necessary to create optimum output voltage [8] in order to improve power quality. The shape of the waveform depicted in Figure 9.8 represents the generalized MLI output voltage.

The waveform is centrally symmetrical. SA calculated in the first quadrant is enough for calculating the remaining SA in other quadrants. Switching angle calculation technique is divided into four subtypes, they are feed-forward (FF), half-height (HH), half-equal-phase (HEP), and equal-phase (EP) method.

9.3.1.5 EP method

This is the simplest technique to generate the SA in the wide range 0°–180° [9]. The formula for generating the SA is as follows:

$$\alpha_{\mathrm{k}} = k\,(180^{\circ}/m)\,, \tag{9.4}$$

here, $k = 1,2....(m-1)/2$. The waveform generated using this method is like a triangle and the angles calculated are very narrow. A higher percentage of THD is produced in this method, which is a major drawback.

9.3.1.6 HEP method

To alleviate the disadvantages of the EP method, the half-equal method was developed. To obtain a quality voltage waveform, this approach estimates the primary switching angles. In this procedure, the output voltage does not have a sinusoidal shape. The SA in the first quadrant is given in eqn (9.5).

$$\alpha_{\mathrm{k}} = k\left(180^{\circ}/m+1\right), \tag{9.5}$$

here, $k = 1,2\ldots(m-1)/2$.

9.3.1.7 HH method

To deal with the challenges of the previous methods, the half-height method [9] is used to compute the switching angle for MLIs. This method was developed based on the sinusoidal value. When the function's value reaches half height, the switching angle is generated. The formula for calculating the angle is as follows:

$$\alpha_{\mathrm{k}} = \sin^{-1}\left[\frac{2k-1}{m-1}\right], \tag{9.6}$$

here, $k = 1,2\ldots(m-1)/2$.

9.3.1.8 FF method

This method overcomes the drawbacks of the above methods in the output voltage quality. This method generates the SA and quality output voltage. The key issue in this method is the elimination of gaps between the positive and negative waveforms, as well as the waveform shape. The switching angle is determined by using the formula given below.

$$\alpha_{\mathrm{k}} = \frac{\sin^{-1}\left[\frac{2k-1}{m-1}\right]}{2}, \tag{9.7}$$

here, $k = 1,2\ldots(m-1)/2$.

This method reduces the percentage of THD when compared with all other methods. Researchers have used the modified switching angle calculation formula in the literature.

$$\alpha_{\mathrm{k}} = \sin^{-1}\left[\frac{k-0.5}{m}\right], \tag{9.8}$$

here, $k = 1,2\ldots(m-1)/2$.

9.3.2 High switching frequency

9.3.2.1 Pulse width modulation (PWM)

This switching process of different switches is called modulation. PWM methods are divided into two categories based on whether they are reference-based or carrier-based, as discussed below.

9.3.2.2 Reference-based modulation

This modulation method is classified as sinusoidal PWM, discontinued reference PWM, hybrid reference, staircase, third harmonic injection (THI), trapezoidal PWM, and 60° PWM. The modulating signal in open-loop control is sinusoidal, but in closed-loop control, the same modulating signal is employed with a compensating signal. Because the needed output voltage has a sine wave pattern, this approach is known as SPWM [10]. The 60° PWM approach [11] is a modified PWM technique. In both line and phase voltage waveforms, this approach has lower switching losses and a greater fundamental output voltage. The trapezoidal waveform may be created by clipping the magnitude of the triangle waveform at a specific level. This form of modulation can be used to boost RMS voltage [12]. The number of steps in staircase modulation is determined by the output voltage level necessary. The disadvantage of this modulation is that it should not be used for more than 15 pulses each cycle [13]. This staircase modulation has a quarter wave symmetry waveform as a reference. Each step's magnitude is controlled individually and each step is divided into 30° intervals. When compared with other PWM techniques, this method gives a more fundamental RMS value. By injecting the selected harmonics into the reference sinusoidal waveform, the THI reference signal is generated. Reference signal composed by

$$V_{\mathrm{r}} = -0.029\sin 9\omega t + 0.27\sin 3\omega t + 1.15\sin\omega t. \tag{9.9}$$

In the case of only the 3rd harmonic injection, V_r can be written as

$$V_{\mathrm{r}} = 0.19\sin 3\omega t + 1.15\sin\omega t. \tag{9.10}$$

The 9th harmonic injection does not affect the output waveform quality. The output voltage is free from the triplen harmonics. The reference voltage is calculated by subtracting the instantaneous average of the three reference voltages' maximum and minimum values from each individual reference voltage. The offset value is calculated using the below formula:

$$V_{\text{offset}} = \text{Avg}\left[\max\left(V_{\mathrm{a}}, V_{\mathrm{b}}, V_{\mathrm{c}}\right),\ \min\left(V_{\mathrm{a}}, V_{\mathrm{b}}, V_{\mathrm{c}}\right)\right]. \tag{9.11}$$

Eqn (9.12) is used to generate the THI waveform for each phase.

$$V_{\mathrm{k}}\left(\text{THI}\right) = V_{\mathrm{k}} - V_{\text{offset}}, \tag{9.12}$$

where $k = a, b, c$.

In the THI method, 15% more fundamental output voltage can be attained than the SPWM. Hybrid pulse generation is used as a PWM technique. Two reference signals make the hybrid reference signal. One of the cycles is used in the first half cycle and another signal in the other half cycle.

9.3.2.3 Carrier-based modulation

This method is mainly divided into two categories, level shifted PWM (LSPWM) and phase shifted PWM (PSPWM) technique [14]. The LSPWM is divided into two categories: equal frequency and unequal frequency. PD, POD, APOD, variable amplitude, carrier overlapping, and other hybrid techniques come under the equal frequency category. Some of the equal frequency modulation techniques, PD, POD, and APOD, are shown in Figure 9.9. Carrier arrangement is two types, unipolar and bipolar. For an m-level inverter, in a unipolar configuration, "$(m-1)/2$" carriers are needed, but in a bipolar setup, "$m-1$" carriers are required. [15]. Other than traditional triangular

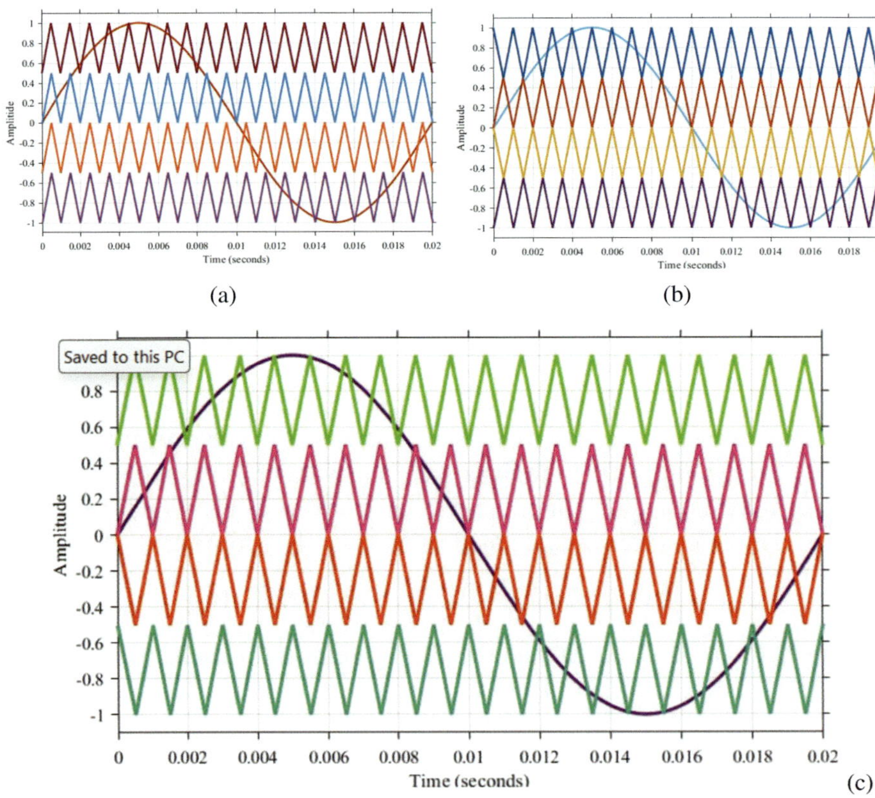

Figure 9.9 Equal frequency modulation for (a) PD, (b) APOD, and (c) POD.

carriers, modified triangular, ramp signal, unipolar sinusoidal carriers, and others are used as carriers [16]. Losses in MLIs mainly depend on the frequency range and carrier type. The PSPWM method is used to diminish the distortion (carriers shifted by 120°) and evenly power distribution among all the cells for any number of levels.

9.3.2.4 Space vector-based modulation (SVM)

The reference waveform, which includes a huge number of vector states, is modulated using space vector modulation. PWM load line voltages identical to a specified reference load line voltage are generated using this modulating approach [17]. Control algorithm variables are extracted directly from the control employed in it, and all switching vectors in complex space are identified [18]. For an m-level converter, $(m-1)^2$ and n^3 switching sequences are required per sector [19]. By selecting the right switching combinations and modulation vector while utilizing floating capacitors, switching losses, common mode voltage, and managing the DC link voltages are decreased. The voltage vectors are defined as

$$\vec{V} = \frac{2}{3}\left[V_{\mathrm{aN}} + e^{\mathrm{j}(2\pi/3)}V_{\mathrm{bN}} + e^{-\mathrm{j}(2\pi/3)}V_{\mathrm{cN}}\right]. \tag{9.13}$$

The closest vector used to calculate the reference vector is shown in Figure 9.10.

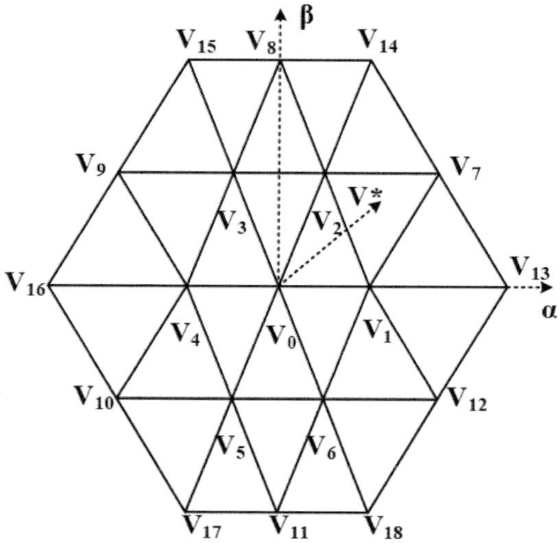

Figure 9.10 Reference vector generation in SVM method.

9.4 Applications of Multilevel Inverters

9.4.1 Grid integration of renewable energy sources (RES)

Multilevel inverters are widely employed in the grid integration of RES because they improve power quality and lower harmonic distortion. This lower total harmonic distortion (THD) reduces the overall cost and size of equipment. MLIs also have high output voltage capability without a step-up circuit, high efficiency, and low electromagnetic interface (EMI). In [20], CHBMLI is used to transfer the power from the two solar panels into the single-phase grid network as depicted in Figure 9.11. Different types of multilevel inverters are used in grid-connected PV systems [21–22]. A new MLI is developed for the grid integration of PV systems to include MPPT, and to improve the THD and the efficiency [23].

9.4.2 FACTS devices

Among the several types of FACTS devices available, STATCOM is regarded as a critical controller. Among all the multilevel inverters, CHB is best suitable topology for the operation of STATCOM. Figure 9.12 represents the

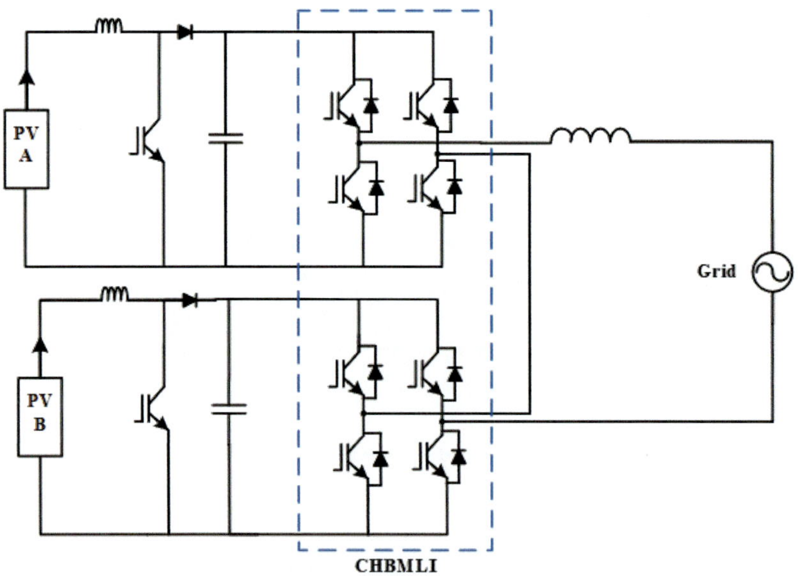

Figure 9.11 Single-phase grid connected system using CHBMLI.

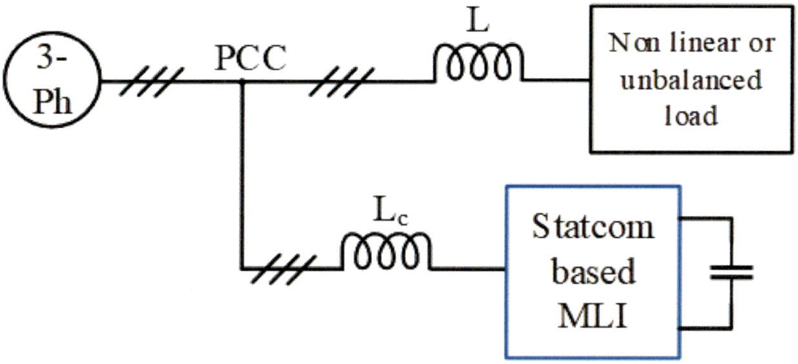

Figure 9.12 Block diagram of STATCOM using MLI.

basic block diagram of STATCOM using MLI. In [24], a multilevel CHB inverter is used in distribution level STATCOM. For harmonic mitigation, the authors suggest a five-level DCMLI-based distribution STATCOM with SRF-based control as in [25].

9.4.3 Motor drives

The ripple contents in currents or torque are particularly damaging in high-power applications, lowering the drive's performance. In high-power applications, the MLI is essential because it reduces torque and current ripples. Further, the use of MLIs in motor drives offers the following advantages.

- MLIs can provide the high power required for large drives.
- Compared to PWM inverters, MLIs may switch at a considerably lower frequency, which can result in large voltage change rates and common-mode voltage to the motor windings.
- The output wave's harmonic distortion diminishes as the number of levels grows, eventually nearing zero.
- MLIs reduce the stress of the motor and do not damage it.
- MLIs produce a high-quality waveform
- MLIs have an increased power rating.

A five-level reduced switch MLI is used in the DTFC-SVM-based control of interior-type PMSM drive [26] as shown in Figure 9.13. Analysis of the single-phase induction motor drive was done using a step-up converter integrated with a voltage multiplier circuit with a different number of levels [27].

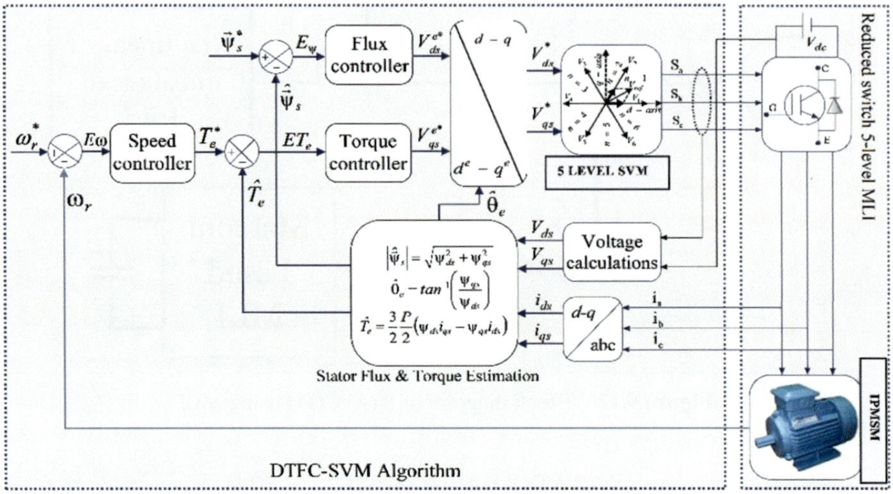

Figure 9.13 Block diagram of PMSM drive.

9.5 Modular Multilevel Converters

The main advantages of MMC [28] are as follows.

- Modular and simple converter modules.
- Low-voltage IGBTs are used in each module.
- Easy scalability of current and voltage.
- Bypass of faulty modules is easy in case of a fault.
- Multipulse rectifiers, or AFE, can provide the DC side, while the grid can supply the AC side.
- Applications involving AC-DC or DC-AC can be handled by the same converter.

However, it has some limitations as follows.

- There is more availability but less reliability.
- The capacitance is significantly higher than in other topologies.
- Too many components make control difficult. However, the control approach is the same for every module. It costs extra.
- Vector-controlled drives demand full load current at nearly zero speed, making control difficult. But appropriate for drives that have pumps or fans.

The circuit configuration of the MMC is depicted in Figure 9.14.a. It has a DC source separated into two parts by a neutral point. It has three legs/phases,

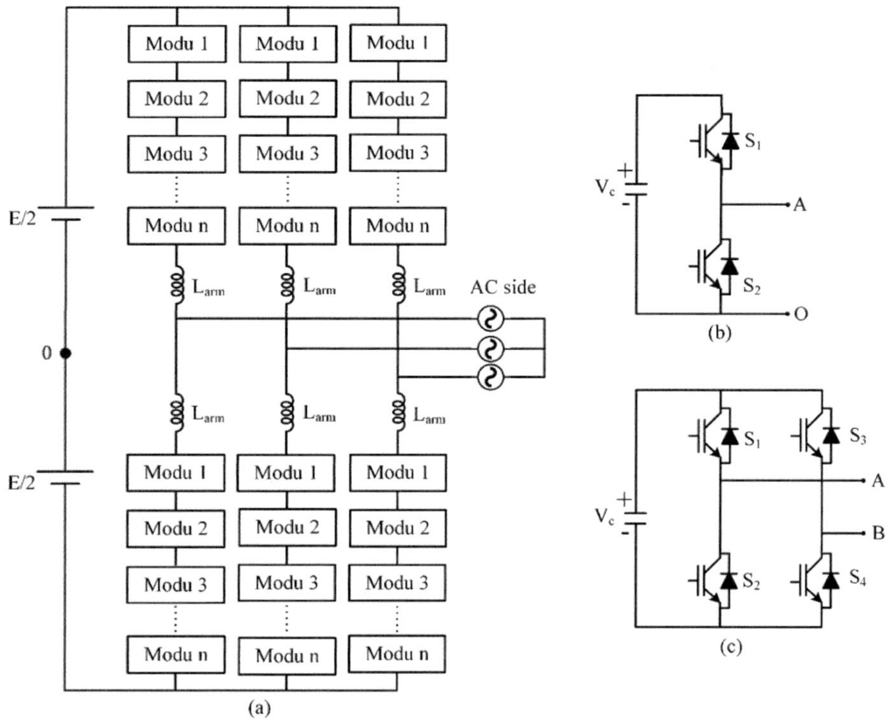

Figure 9.14 (a) Modular multilevel converter, (b) half-bridge, and (c) full-bridge.

each leg having two arms and each arm consists of n number of modules. Here, n may be more than 100. Each module can be replaced by a half-bridge or full-bridge as depicted in Figures 9.14.b and 9.14.c, respectively. Each arm has one inductor to reduce the components of high frequency in the arm current.

If n number of half-bridges are connected in cascade as depicted in Figure 9.15.a and Figure 9.15.b depicts the voltage waveform between points A and O, the peak voltage is seen to be nV_C. It can be seen that it has both AC and DC components. The cascade connection of n full-bridge modules in an arm is depicted in Figure 9.15.c and the voltage V_{AB} is given in Figure 9.15.d. This voltage V_{AB} has only an AC component, and does not contain a DC component. The output becomes very smooth with many numbers of modules [29]. As illustrated in Figure 9.16, the modules forming an arm can be considered as a configurable voltage source.

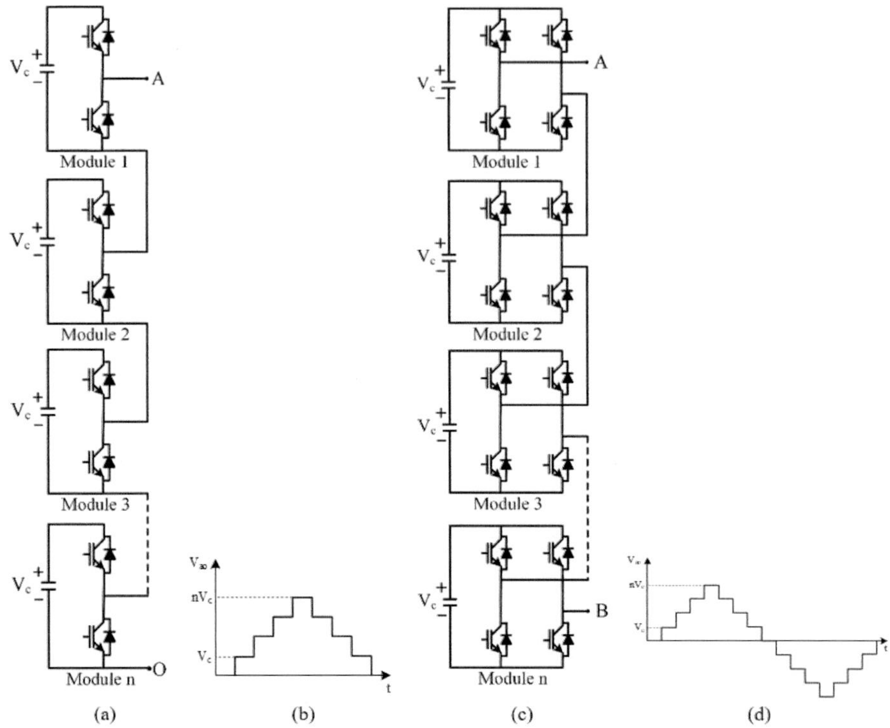

Figure 9.15 (a) Cascade connection of half-bridge modules, (b) its voltage waveform. (c) Cascade connection of full-bridge modules, and (b) its voltage waveform.

9.5.1 Converter voltage equations

The KVL equation for a phase upper arm can be written as

$$\frac{E}{2} - v_{\mathrm{U}}(t) - L\frac{di_{\mathrm{U}}}{dt} = v_{\mathrm{AO}}(t) = V_{\mathrm{a}} = V_{\mathrm{m}}\sin\omega t. \tag{9.14}$$

Neglecting the drop across the inductor

$$\frac{E}{2} - v_{\mathrm{U}}(t) = V_{\mathrm{m}}\sin\omega t. \tag{9.15}$$

Therefore, the voltage generated by the upper arm should be

$$v_{\mathrm{U}}(t) = \frac{E}{2} - V_{\mathrm{m}}\sin\omega t. \tag{9.16}$$

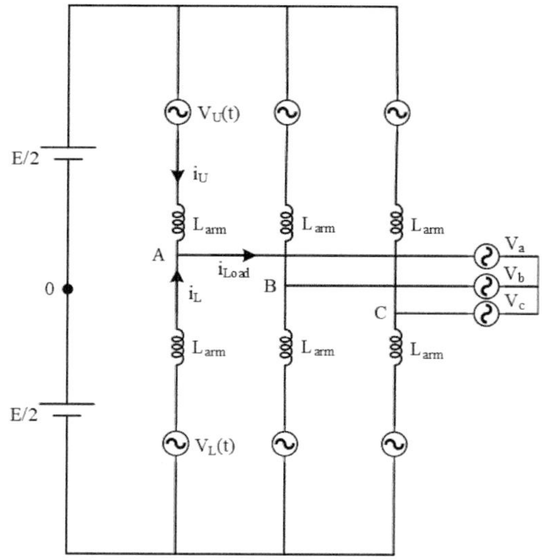

Figure 9.16 MMC with the arms as controllable voltage sources.

Similarly, applying KVL for the lower arm gives

$$-\frac{E}{2} + v_{\mathrm{L}}(t) - L\frac{di_{\mathrm{U}}}{dt} = v_{\mathrm{AO}}(t) = V_{\mathrm{a}} = V_{\mathrm{m}}\sin\omega t. \tag{9.17}$$

The voltage generated by the lower arm should be

$$v_{\mathrm{L}}(t) = \frac{E}{2} + V_{\mathrm{m}}\sin\omega t. \tag{9.18}$$

The converter arm voltages can be changed to vary the output AC voltages as follows:

$$v_{\mathrm{U}}^{*}(t) = \frac{E}{2}(1 - m\sin\omega t),\ \text{where } 0 \leq m \leq 1 \text{ and } m = \frac{V_{\mathrm{m}}}{(E/2)}. \tag{9.19}$$

Similarly

$$v_{\mathrm{L}}^{*}(t) = \frac{E}{2}(1 + m\sin\omega t), \tag{9.20}$$

where m is the modulation index which gives the flexibility to change the value of V_m. The upper and lower arms produce the voltages in such a way as to get the required value of AC voltage. From Figure 9.17.a,b, the instantaneous voltages produced by the upper arm are represented by the red-colored arrows, and the instantaneous voltages produced by the lower arm are represented by the green-colored arrows.

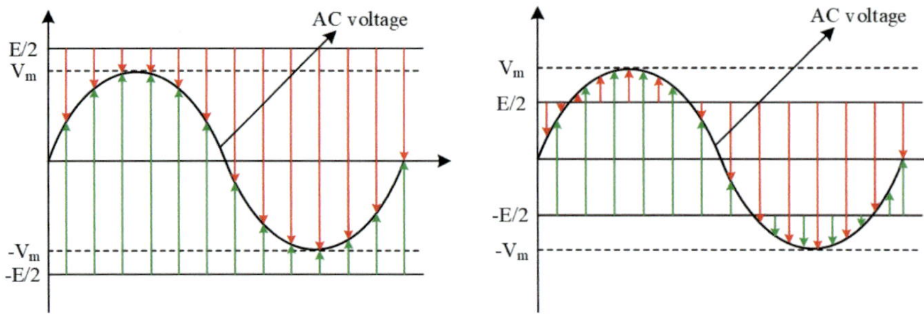

Figure 9.17 MMC operation when (a) $E/2 > V_m$, (b) $E/2 < V_m$.

If $E/2$ is greater than V_m, then the upper and lower arms need to produce the voltages only in one direction as shown in Figure 9.17.a. Therefore, only half-bridges are sufficient to produce these voltages. However, if $E/2$ is less than V_m, then the upper and lower arms need to produce the voltages in both directions as depicted in Figure 9.17.b. Therefore, full-bridges are necessary to produce these voltages [30].

9.5.2 Arm voltage rating

From eqn (9.19) and (9.20)

$$\text{If } V_{\mathrm{m}} \leq \frac{E}{2}, \text{ then } 0 \leq v_{\mathrm{U}} \leq E. \tag{9.21}$$

Similarly,

$$0 \leq v_{\mathrm{L}} \leq E.$$

Therefore, the voltage rating of the lower and upper arms equals the total DC link voltage. Each module's voltage rating is expressed as E/N, where N is the total number of submodules in an arm. Capacitor stored energy is also correlated with submodule count. Each switch's and capacitor ripple's voltage rating will rise to a low value of N. The cost will go up with a high value of N.

9.5.3 Arm currents

Three components make up the current flowing through the arms: (1) DC component to maintain power balance, (2) half of the load current at fundamental frequency, and (3) higher harmonic currents as a result of capacitor voltage ripple.

Under steady state, arm currents can be represented as

$$i_{\mathrm{U}}(t) = \frac{1}{2} i_{\mathrm{L}}(t) + i_{\mathrm{C}}(t). \tag{9.22}$$

Similarly

$$i_{\mathrm{L}}(t) = \frac{1}{2} i_{\mathrm{L}}(t) - i_{\mathrm{C}}(t), \tag{9.23}$$

where $i_L(t)$ is the load current and $i_C(t)$ is the circulating current.

The sum of both arm currents gives the load current

$$i_{\mathrm{L}}(t) = I_{\mathrm{m}} \sin(\omega t - \varphi). \tag{9.24}$$

$i_C(t)$ has higher harmonics (mostly fundamental and second harmonics) and a DC component.

9.5.4 Arm energy balancing at steady state

To keep arm voltages constant, the power flow through the module should be zero for a certain amount of time. [31]. Therefore

$$\int_0^{\mathrm{T}} v_{\mathrm{cell}} i_{\mathrm{cell}} \mathrm{d}t = 0. \tag{9.25}$$

This means arm energy should also be zero.

$$W_{\mathrm{arm}} = \int_0^{\mathrm{T}} p_{\mathrm{arm}} \mathrm{d}t = \int_0^{\mathrm{T}} v_{\mathrm{arm}} i_{\mathrm{arm}} \mathrm{d}t = 0. \tag{9.26}$$

The instantaneous power for the upper arm is given as

$$p_{\mathrm{U}}(t) = v_{\mathrm{U}}(t) \times i_{\mathrm{U}}(t). \tag{9.27}$$

$$p_{\mathrm{U}}(t) = \left(\frac{E}{2}(1 - m\sin\omega t)\right) \times \left(\frac{1}{2} i_{\mathrm{L}}(t) + i_{\mathrm{C}}(t)\right). \tag{9.28}$$

$$p_{\mathrm{U}}(t) = \left(\frac{E}{2} - \frac{E}{2} m\sin\omega t\right) \times \left(\frac{1}{2} I_{\mathrm{m}} \sin(\omega t - \varphi) + i_{\mathrm{C}}(t)\right). \tag{9.29}$$

$$p_{\mathrm{U}}(t) = \frac{E}{2} i_{\mathrm{C}}(t) - \frac{mEI_{\mathrm{m}}}{8} \cos\varphi + \frac{EI_{\mathrm{m}}}{4} \sin(\omega t - \varphi) - \frac{mE}{2} i_{\mathrm{C}}(t) \sin\omega t + \frac{mEI_{\mathrm{m}}}{8} \cos(2\omega t - \varphi). \tag{9.30}$$

The DC and AC components of the instantaneous arm power can be written as

$$p_{\mathrm{U}}(t) = p_{\mathrm{U\_DC}}(t) + p_{\mathrm{U\_AC}}(t). \tag{9.31}$$

$$p_{\mathrm{U\_DC}}(t) = \frac{E}{2} i_{\mathrm{C}}(t) - \frac{mEI_{\mathrm{m}}}{8} \cos\varphi. \tag{9.32}$$

$$p_{\mathrm{U\_AC}}(t) = \frac{EI_{\mathrm{m}}}{4} \sin(\omega t - \varphi) - \frac{mE}{2} i_{\mathrm{C}}(t) \sin\omega t + \frac{mEI_{\mathrm{m}}}{8} \cos(2\omega t - \varphi). \tag{9.33}$$

Each arm's average DC power must equal zero, to maintain the steady-state arm energy balance.

$$p_{\mathrm{U\_DC}}(t) = 0. \tag{9.34}$$

$$i_{\mathrm{C}}(t) = \frac{mI_{\mathrm{m}}}{4} \cos\varphi. \tag{9.35}$$

This gives the DC component of circulating current, which flows through both upper and lower arms.

9.5.5 Relation between AC and DC side current

Let I_{d} be the current that is drawn from DC source (E). The power that is then drawn from or flowing into the DC side is equal to EI_{d}. If there are no converter losses, it should match AC power.

Thus

$$EI_{\mathrm{d}} = 3 \frac{V_{\mathrm{m}}}{\sqrt{2}} \frac{I_{\mathrm{m}}}{\sqrt{2}} \cos\varphi. \tag{9.36}$$

Hence

$$I_{\mathrm{d}} = \frac{3}{4} m I_{\mathrm{m}} \cos\varphi. \tag{9.37}$$

I_{d} is therefore three times the circulating current. Ideally, I_{d} is pure DC.

An alternative notation for the arm currents is

$$i_{\mathrm{U}}(t) = \frac{1}{2} I_{\mathrm{m}} \sin(\omega t - \varphi) + \frac{I_{\mathrm{d}}}{3}. \tag{9.38}$$

9.5.6 Arm energy balancing at steady state

Energy variations in each arm are caused by the AC power components of arm power [32]. The energy in the upper arm can be expressed as

$$W_{\mathrm{U}}(t) = \int p_{\mathrm{U\_AC}}(t)\,\mathrm{d}t. \tag{9.39}$$

$$W_{\mathrm{U}}(t) = \int \left(\frac{EI_{\mathrm{m}}}{4}\sin(\omega t - \varphi) - \frac{mE}{2} i_{\mathrm{C}}(t)\sin\omega t + \frac{mEI_{\mathrm{m}}}{8}\cos(2\omega t - \varphi) \right) \mathrm{d}t. \tag{9.40}$$

Substituting $i_{\mathrm{C}}(t) = \frac{I_{\mathrm{d}}}{3}$, and solving the eqn (9.40), then

$$W_{\mathrm{U}}(t) = -\frac{EI_{\mathrm{m}}}{4\omega}\cos(\omega t - \varphi) + \frac{mEI_{\mathrm{d}}}{6\omega}\cos\omega t + \frac{mEI_{\mathrm{m}}}{16\omega}\sin(2\omega t - \varphi) + \mathrm{C}, \tag{9.41}$$

where C is the integration constant and can be obtained by assuming $W_U(t) = W_{U0}$ at $t = 0$

$$\mathrm{C} = W_{\mathrm{U0}} + \frac{EI_{\mathrm{m}}}{4\omega}\cos\varphi - \frac{mEI_{\mathrm{d}}}{6\omega} + \frac{mEI_{\mathrm{m}}}{16\omega}\sin\varphi. \tag{9.42}$$

Substituting the value of C, the upper arm energy obtained is

$$W_{\mathrm{U}}(t) = W_{\mathrm{U0}} + \frac{EI_{\mathrm{m}}}{4\omega}\cos\varphi - \frac{mEI_{\mathrm{d}}}{6\omega} + \frac{mEI_{\mathrm{m}}}{16\omega}\sin\varphi - \frac{EI_{\mathrm{m}}}{4\omega}\cos(\omega t - \varphi) + \frac{mEI_{\mathrm{d}}}{6\omega}\cos\omega t + \frac{mEI_{\mathrm{m}}}{16\omega}\sin(2\omega t - \varphi). \tag{9.43}$$

Similarly, the lower arm energy can be written as

$$W_{\mathrm{L}}(t) = W_{\mathrm{L0}} + \frac{EI_{\mathrm{m}}}{4\omega}\cos\varphi - \frac{mEI_{\mathrm{d}}}{6\omega} - \frac{mEI_{\mathrm{m}}}{16\omega}\sin\varphi - \frac{EI_{\mathrm{m}}}{4\omega}\cos(\omega t - \varphi) + \frac{mEI_{\mathrm{d}}}{6\omega}\cos\omega t - \frac{mEI_{\mathrm{m}}}{16\omega}\sin(2\omega t - \varphi). \tag{9.44}$$

Based on the circuit diagram and the sign convention that was utilized, which states that the current directions in the upper and lower arms are opposite to one another. Therefore

$$W_{\mathrm{L}}(t) = -W_{\mathrm{L0}} - \frac{EI_{\mathrm{m}}}{4\omega}\cos\varphi + \frac{mEI_{\mathrm{d}}}{6\omega} + \frac{mEI_{\mathrm{m}}}{16\omega}\sin\varphi + \frac{EI_{\mathrm{m}}}{4\omega}\cos(\omega t - \varphi) - \frac{mEI_{\mathrm{d}}}{6\omega}\cos\omega t + \frac{mEI_{\mathrm{m}}}{16\omega}\sin(2\omega t - \varphi). \tag{9.45}$$

It is evident that the upper and lower arms share the same fundamental component. Nonetheless, their aggregate cancels each other out, suggesting that the upper and lower arms exchange energy at a fundamental frequency. It does not go out of the arms.

9.5.7 Phase energy balancing at steady state

Each phase energy variation can be represented as

$$W(t) = W_{\mathrm{U}}(t) + W_{\mathrm{L}}(t). \tag{9.46}$$

$$W(t) = \frac{2mEI_{\mathrm{m}}}{16\omega}\sin\varphi + \frac{2mEI_{\mathrm{m}}}{16\omega}\sin(2\omega t - \varphi). \tag{9.47}$$

$$W(t) = W_0(t) + \Delta W(t). \tag{9.48}$$

It should be noticed that the phase energy expression has two terms: a constant and a second harmonic. The energy held in the capacitors in phase is given by the constant phase. The energy that is transferred between the three phases is shown by the second harmonic component. The result of adding the second harmonic component of three phases is zero.

9.6 Circuit Configurations Possible With Modular Multilevel Converters

MMCs are used as DC-AC, AC-DC, DC-DC, and AC-AC converters in different applications. Merits of MMC like modularity and low voltage ratings of components are present in all the above-mentioned applications [30]. Some of the major circuit configurations of MMC are given below.

9.6.1 Double star half-bridge cell

It consists of two star-connected modular cells. Each leg of the configuration consists of a cascade connection of a large number of half-bridge cells as depicted in Figure 9.18.b,d. It is similar to the circuit shown in Figure 9.16.

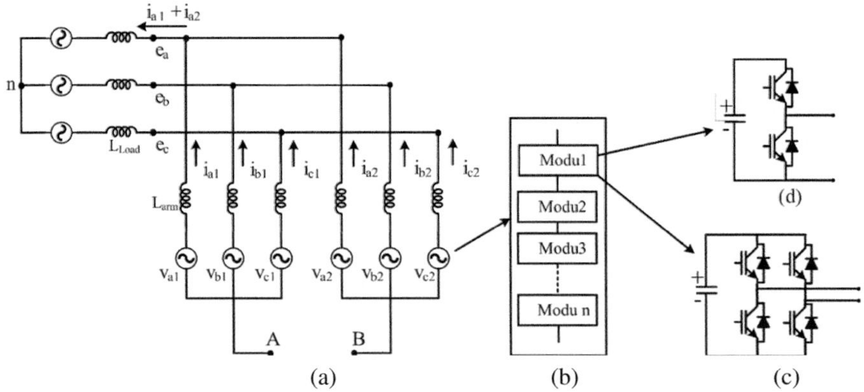

Figure 9.18 (a) Double star half-bridge cell, (b) arm, (c) full-bridge, and (d) half-bridge.

This configuration is also known as a "double star chopper cell." The configuration shown in Figure 9.18.a is operated in a DC-AC converter by connecting a DC voltage source between A and B terminals and drives the three-phase AC load. By connecting three three-phase supply at the load terminal, and operating in AC-DC conversion mode the output DC is generated at terminals A and B. This MMC configuration is mainly used in HVDC and motor drive applications [33].

9.6.2 Double star full-bridge cell

In the above MMC configuration shown in Figure 9.18.a, by considering a full-bridge in place of each cell as shown in Figure 9.18.c, the configuration is called a "double star full-bridge cell." In this configuration, by taking one star connected module, then it is a CHB multilevel inverter or it is also called a "single star full-bridge cell." In a double-star full-bridge by connecting a single-phase supply between A and B terminals, this can be operated as a single-phase to three-phase converter, and reverse cab also be operated (three-phase to single-phase). This type of configuration is mainly used in traction drives, where the supply is 25 kv single phase and three phase drives are used.

9.6.3 Triple star full-bridge

Triple star full-bridge cell is used for applications where conversion is needed from three-phase AC to three-phase AC as shown in Figure 9.19. The frequency and voltage at load terminals are varied while the supply voltage

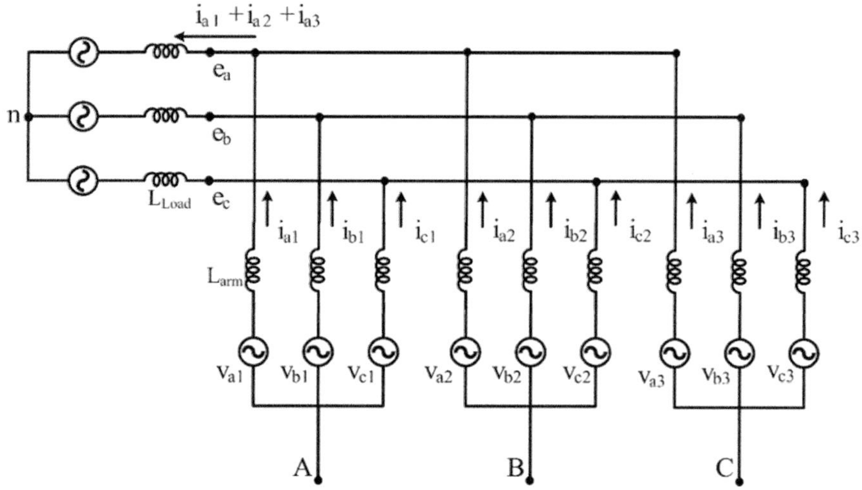

Figure 9.19 Triple star full-bridge cell.

and frequency are constant. Here, a three-phase supply is connected at ABC terminals, full-bridges are used as MMC cells, and the output load voltage is a controlled three-phase voltage. This MMC configuration is mainly used in industrial drive applications.

9.6.4 M-phase to N-phase MMC configuration

By increasing the number of star modules in triple star bridge cell configuration, the converter can be operated as an M-phase to N-phase converter as shown in Figure 9.20. This MMC configuration is used for applications that require multiphase supply. In Figure 9.20, N-phase voltage is obtained at AB...N terminals by connecting the three-phase supply at the load terminals. Here, full-bridges are used as cells in the star-connected modules.

9.6.5 Single delta bridge cell

The modular cell can also be connected in a delta configuration as shown in Figure 9.21, then it is called a "single delta bridge cell." In delta configuration, the circulating currents flow in the delta loop, this is useful during the unbalanced loading conditions in the system. This MMC configuration is mostly used in STATCOM and BESS applications.

9.6.6 Double delta bridge cell or Hexverter

In the above triple-star bridge configuration, by using only two modules in place of each star-connected module, a three-phase to three-phase converter can be obtained as shown in Figure 9.22.a. It is called a "double delta bridge

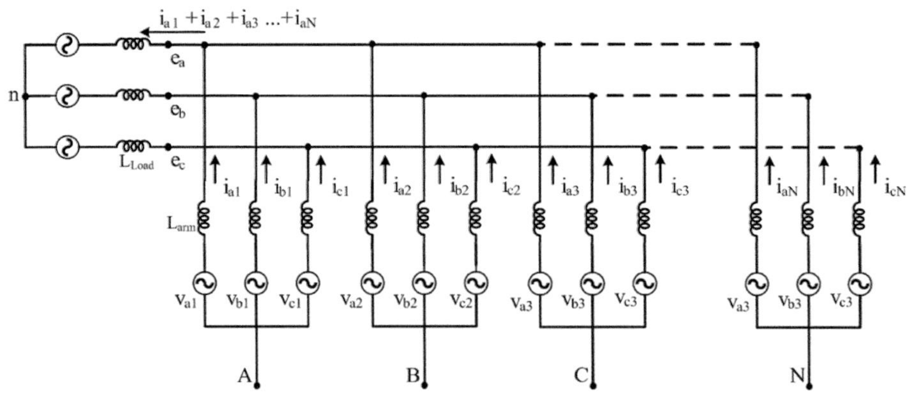

Figure 9.20 M-phase to N-phase MMC configuration.

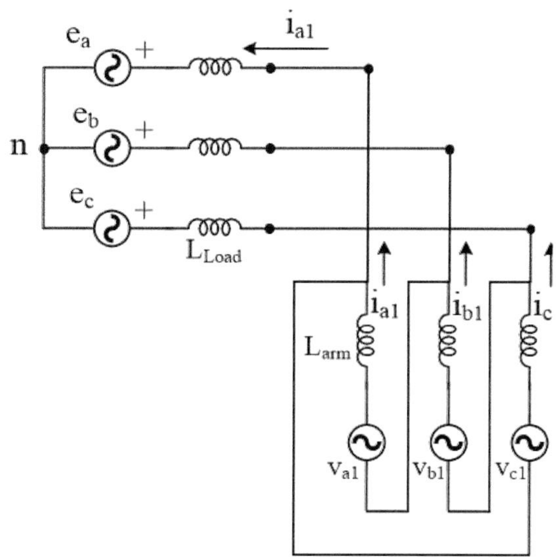

Figure 9.21 Single delta bridge cell.

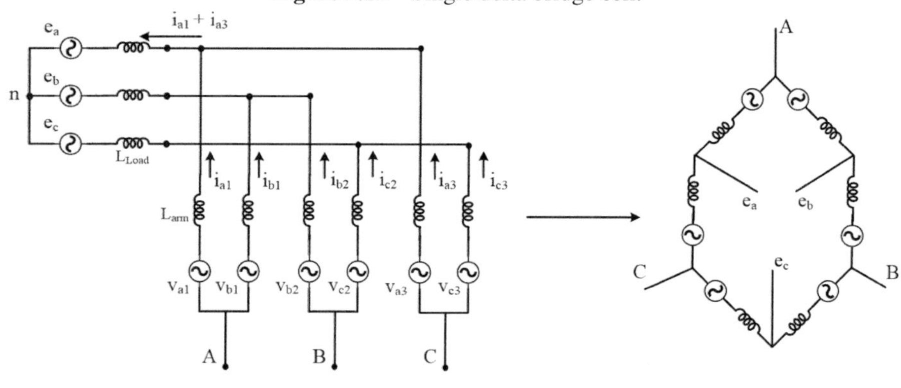

Figure 9.22 Double delta bridge cell.

cell." This configuration forms a hexagonal loop with the modular structures as shown in Figure 9.22.b, so it is also called a "Hexverter." The hexagonal loop provides the path for circulating currents which is useful in the converter for the energy balance.

9.6.7 DC-DC converters with MMC

The MMC configuration in Figure 9.23.a is a nonisolated DC-DC converter. Here, the input is a high-voltage DC and it can be converted into low voltage

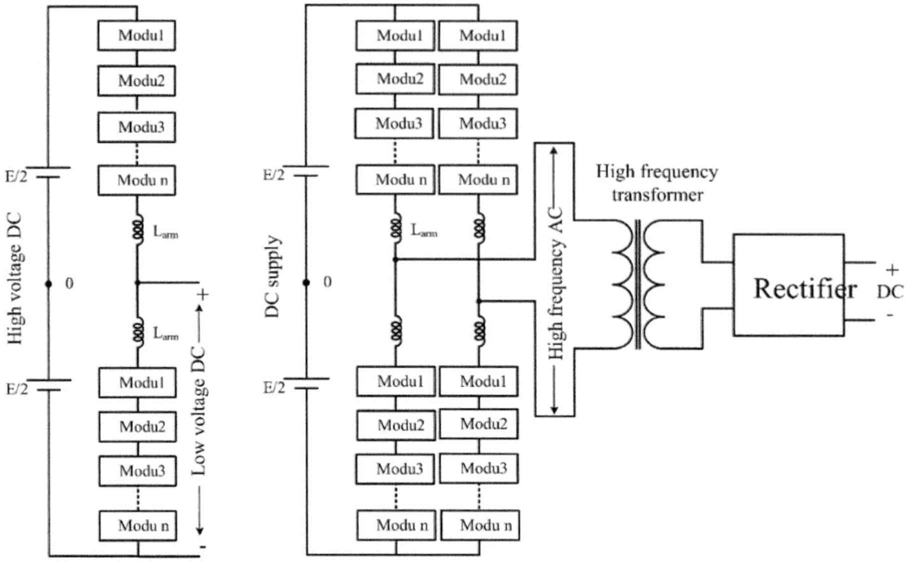

(a) (b)

Figure 9.23 (a) Nonisolated DC-DC converter with MMC and (b) isolated DC-DC converter with MMC.

DC. The cells used in the modules can be half-bridges or full-bridges. These configurations are mainly used in HVDC applications where power has to be tapped from the HVDC transmission lines at remote locations in the middle of the lines, where an AC grid is not available. These configurations are also used for injecting the power from renewable energy into the HVDC lines at remote locations. The configuration shown in Figure 9.23.b is the isolated DC-DC converter formed from a modular structure. In this configuration, two modular legs are used to convert high-voltage DC to AC (high frequency and low voltage), and a high-frequency transformer is used for isolation. This high-frequency AC is converted into DC with the help of a rectifier and the rectifier may be an MMC or any other rectifier as it is low voltage.

9.7 Fault Analysis on Modular Multilevel Converters

The faults in converters are mainly thermal faults, open-short switch faults, DC source short circuit faults, and capacitors faults. The serious faults in

MMC are short circuit faults, they must be isolated within 10 μ sec to avoid further damage. During a fault, all the IGBT/MOSFETs are turned off in a leg to stop the capacitor short circuit. Here, the large arm inductors are also used to limit the fault currents. To avoid the saturation of arm inductors, air core inductors are used [34].

9.7.1 Analysis with half-bridges in MMC

For analysis, a severe pole-to-pole fault in the HVDC line of MMC is considered as shown in Figure 9.24.a. During the fault, the current circulates through the antiparallel diode of the semiconductor switch as shown in Figure 9.24.b. However, the antiparallel diodes are inverter class diodes that have high switching frequency and low power handling nature. Rectifier class diodes are needed to withstand these high surge currents, therefore, a thyristor is connected across the IGBT which bypasses the surge currents during fault as shown in Figure 9.24.c. This thyristor is turned on until the circuit breaker opens to clear the fault. The thyristor has large i^2t ratings, these thyristors along with arm inductors reduce the severity of pole-pole short-circuit fault.

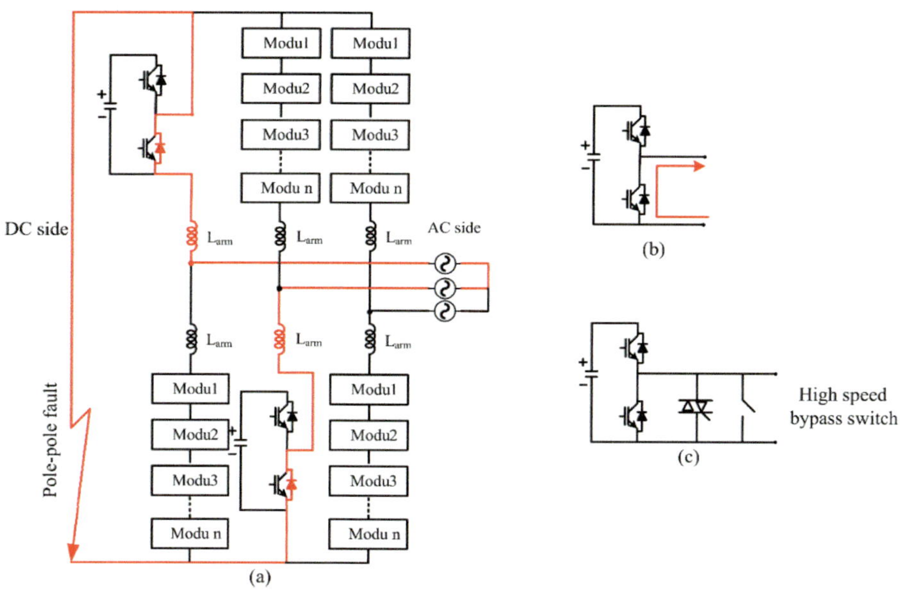

Figure 9.24 Fault analysis of MMC with half-bridge cells.

9.7.2 Analysis with full-bridges in MMC

During the pole-pole short circuit fault on the DC side, initially, the gate signals are stopped for the power semiconductors. The fault currents will circulate through antiparallel diodes. In this case, currents always pass through DC-link capacitors as shown in Figure 9.25.a,b. During the process, the capacitors are charged to a peak of the AC voltage, and the diode acts in reverse blocking mode and currents stop flowing. So, in full-bridge MMC, the currents are automatically blocked on their own during faults. Therefore, full-bridge MMC gives better performance than half-bridge MMC during faults. Hence, full-bridge MMC is more popularly used commercially.

The full-bridge MMC is operated in a better manner by controlling the voltage of full-bridge cells with switching, instead of turning off the power semiconductor switches. In this technique, the full-bridge MMC cells are controlled to produce a voltage similar to back emf in the circuit and this back emf reduces the short circuit currents. Therefore, the severity of the fault can be reduced with the help of full-bridge cells.

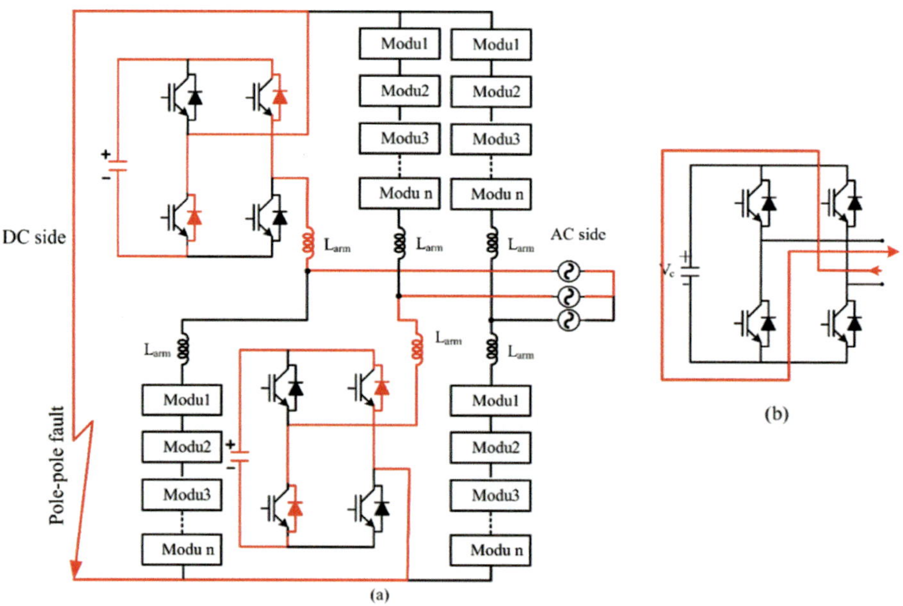

Figure 9.25 Fault analysis of MMC with full-bridge cells.

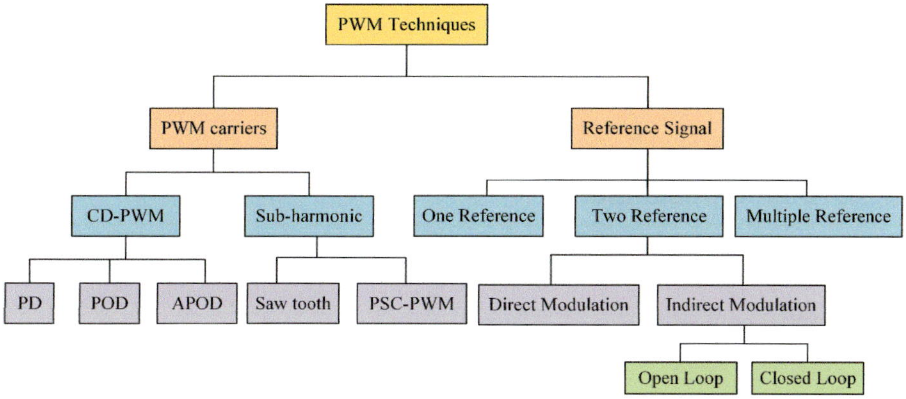

Figure 9.26 Classification of the PWM techniques for MMC.

9.8 PWM Techniques

PWM techniques are mainly classified into two types. The classification of the PWM techniques is based on PWM carriers and reference signals as shown in Figure 9.26.

9.8.1 Carrier disposition PWM (CD-PWM) techniques

To produce gate pulses for "*N*" cell architectures, these approaches need "*N*" triangular waveforms. This approach uses a separate switching pattern for each cell. The x-axis is symmetrically shifted in these triangular waveforms. The output phase voltage is determined by comparing the voltage reference waveform with triangular carriers. This approach is further categorized as PD, APOD, and POD based on the phase shift. The operation of these techniques is very similar to the PWM techniques explained in the CHB operation.

In phase shifted PWM (PSPWM), each cell loss is identical, however, the capacitor voltage balancing requires an additional PI controller. In the case of LSPWM, uneven losses take place in each cell. To balance the capacitor voltages, a sorting algorithm is used in this technique.

9.8.2 Subharmonic

In this technique, "$2N$" identical carrier waveforms per phase-leg with a phase shift of $\Theta = 360°/2N$, either triangular or sawtooth waveforms are used. The

PD PWM approach delivers superior line-to-line voltage by considering the same number of switching transitions as the subharmonic technique.

Based on the number of reference signals, modulation techniques are subclassified.

9.8.3 Direct modulation

Two complementary sinusoidal reference signals are employed to regulate both the upper and lower arms voltages of phase-i, as shown in eqn (9.49) and (9.50).

$$n_{\mathrm{p,i}} = N\frac{\frac{V_{\mathrm{dc}}}{2} - v_{\mathrm{i,ref}}}{V_{\mathrm{dc}}}. \tag{9.49}$$

$$n_{\mathrm{n,i}} = N\frac{\frac{V_{\mathrm{dc}}}{2} + v_{\mathrm{i,ref}}}{V_{\mathrm{dc}}}. \tag{9.50}$$

These reference signals are compared with carrier waveforms to determine the required number of cells in the upper and lower arms. The existence of circulating currents, which increases total losses, is a major disadvantage of this technology.

9.8.4 Indirect modulation

Two sinusoidal reference signals are utilized to regulate the upper and lower arms, similar to direct modulation. The waveforms of reference provided by

$$n_{\mathrm{p,i}} = N\frac{\frac{V_{\mathrm{dc}}}{2} - v_{\mathrm{i,ref}} - \sum_{\mathrm{v_{reg,i}}} - v_{\mathrm{reg,i}}^{\mathrm{circ}}}{\sum_{\mathrm{j=0}}^{\mathrm{N}} v_{\mathrm{cn,j,i}}}. \tag{9.51}$$

$$n_{\mathrm{n,i}} = N\frac{\frac{V_{\mathrm{dc}}}{2} + v_{\mathrm{i,ref}} - \sum_{\mathrm{v_{reg,i}}} - v_{\mathrm{reg,i}}^{\mathrm{circ}}}{\sum_{\mathrm{j=0}}^{\mathrm{N}} v_{\mathrm{cn,j,i}}}. \tag{9.52}$$

Similar to the direct approach for determining the number of cells put in the lower and upper arms, these reference signals were compared to carrier waveforms. This approach is divided into the following subcategories.

9.8.4.1 Closed-loop

The reference signal equation value $\sum_{\mathrm{j=0}}^{\mathrm{N}} v_{\mathrm{cx,j,i}}$ is computed using the observed capacitor voltages in this closed-loop approach. The terms $\sum v_{\mathrm{reg,i}}$ and $v_{\mathrm{reg,i}}^{\mathrm{circ}}$ are derived by controlling the total energy stored in phase-i leg capacitors and the energy balance between the arms in a closed loop. The benefits of this technology are (1) control of the upper and lower arm's energy imbalance, and (2) regulation of the average cell capacitor voltage, allowing operation with a wide range of voltage levels at low output voltage.

9.8.4.2 Open-loop

In this closed-loop method, the reference signal equation value $\sum_{j=0}^{N} v_{cx,j,i}$ is computed based on the estimated capacitor voltages. The terms $\sum_{v_{reg,i}} = 0$ and $v_{reg,i}^{circ}$ are estimated so as to eliminate the harmonics in circulating currents. The advantage of this technique is the absence of voltage sensors and the drawback is that accurate estimation is necessary to describe the system dynamics [35].

9.9 Capacitor Balancing Technique

In MMC architecture, like in other topologies such as NPC, it is necessary to balance and keep the submodule capacitor voltages at V_{dc}/N. Researchers suggested a number of capacitor voltage-balancing solutions, however, the sorting algorithm is the most often used voltage-balancing strategy. This method involves monitoring and sorting the capacitor voltages of each arm. If the currents in the upper arm are positive, the "*N*" cells with the lowest voltages in the corresponding arm are located and inserted. The voltages of these inserted cell capacitors rise as they are charged. If the currents in the top area are negative, the "*N*" cells with the higher voltages in the associated arm cells are identified and inserted. The voltages of the inserted cell capacitors fall as they are discharged. This sorting technique ensures that the capacitor voltage is balanced under all MMC working situations, however, it results in extra switching transitions between the cells. Unless the number of on-state cells required during two successive control intervals is not changed, cell bypassing/insertion may occur. As a result, switching frequency rises, resulting in unwanted power losses in high-power systems.

9.10 MMC Applications

9.10.1 HVDC systems

HVDC transmission is the main application of MMC although it is used in other applications like distribution STATCOM and low voltage inverter application. The absence of short circuit management capacity on the DC-side half-bridge cells is a drawback in this application.

9.10.2 Variable-speed drives

Other classic converters, such as NPC, have been shown to be inferior to the usage of medium-voltage variable speed drives. This application presents its own set of control issues. At lower frequencies, the ripple magnitude of capacitor voltage in cell capacitors is considerable.

9.10.3 Dynamic braking chopper

In order to absorb and distribute energy, some applications require a DC braking chopper. For example, during a breakdown in an onshore wind plant, the onshore plant should absorb and disperse the wind farm's rated power without halting the wind farm's operation.

9.10.4 Other applications

The other applications of MMC include distribution STATCOM, propulsion system of ships, electric traction system, DC-DC converter topologies, and low voltage inverter applications reported by the researchers.

9.11 Conclusion

Multilevel inverters have emerged as a viable solution for medium-voltage high-power applications, driven by advancements in self-commutated switching devices. This chapter provides a comprehensive overview of classical multilevel inverters, covering their operation principles, advantages, disadvantages, and applications. Additionally, various modulation schemes for MLIs are explored, along with their uses in grid-connected renewable energy systems, FACTS devices, and motor drives. To address the limitations of cascade H-bridge multilevel inverters, modular multilevel converters are discussed in depth, featuring half-bridge and full-bridge cells, mathematical analysis, energy balancing, and circuit configurations for specific applications. The chapter also delves into fault analysis, tolerant capabilities, and modulation schemes with capacitor balancing conditions for MMC. Finally, the applications of MMC in various fields are examined.

References

[1] X. Liang, N. C. Kar, and J. Liu, "Load Filter Design Method for Medium-Voltage Drive Applications in Electrical Submersible Pump

Systems," IEEE Trans. Ind. Appl., vol. 51, no. 3, pp. 2017–2029, 2015, doi:10.1109/TIA.2014.2369814.

[2] K. K. Gupta, A. Ranjan, P. Bhatnagar, L. K. Sahu, and S. Jain, "Multilevel inverter topologies with reduced device count: A review," IEEE Trans. Power Electron., vol. 31, no. 1, pp. 135–151, 2016, doi:10.1109/TPEL.2015.2405012.

[3] L. Sun, W. Zhenxing, M. Weiming, F. Xiao, X. Cai, and L. Zhou, "Analysis of the DC-link capacitor current of power cells in cascaded H-bridge inverters for high-voltage drives," IEEE Trans. Power Electron., vol. 29, no. 12, pp. 6281–6292, 2014, doi:10.1109/TPEL.2014.2302081.

[4] C. Buccella, C. Cecati, M. G. Cimoroni, and K. Razi, "Analytical method for pattern generation in five-level cascaded H-bridge inverter using selective harmonic elimination," IEEE Trans. Ind. Electron., vol. 61, no. 11, pp. 5811–5819, Nov. 2014.

[5] M. S. A. Dahidah, G. Konstantinou, and V. G. Agelidis, "A review of multilevel selective harmonic elimination PWM: Formulations, solving algorithms, implementation and applications," IEEE Trans. Power Electron., vol. 30, no. 8, pp. 4091–4106, Aug. 2015.

[6] J. Rodriguez, P. Correa and L. Moran, "A vector control technique for medium voltage multilevel inverters," APEC 2001. Sixteenth Annual IEEE Applied Power Electronics Conference and Exposition (Cat. No.01CH37181), 2001, pp. 173-178 vol.1.

[7] S. Kouro et al., "Recent Advances and Industrial Applications of Multilevel Converters," in IEEE Transactions on Industrial Electronics, vol. 57, no. 8, pp. 2553-2580, Aug. 2010.

[8] A. Edpuganti and A. K. Rathore, "Optimal low switching frequency pulse width modulation of medium voltage seven-level cascade-5/3H inverter," IECON 2013 - 39th Annual Conference of the IEEE Industrial Electronics Society, 2013, pp. 6287-6292.

[9] Jinghua Z, Zhengxi L. Investigations on best switching angles to obtain lowest THD for multilevel DC/AC inverters. In: Proceedings of the 8th IEEE conf on ind. electr and app; 2013, p. 1814–18.

[10] Perez M, Rodriguez J, Pontt J, Kouro S. Power distribution in hybrid multi-cell converter with nearest level modulation. In: Proceedings of the IEEE Inter sympon ind electronics; 2007, p. 736–41.

[11] Rashid MH. Power electronics circuits, devices, and applications, 3rd ed..Pearson: Prentice Hall; 2011.

[12] Ayob SM, Salam Z, Jusoh A. Trapezoidal pwm scheme for cascaded multilevel inverter. In: Proceedings of the IEEE International power and energy conference;2006, p. 368–72.
[13] Thorborg K, Nystrom A. Staircase PWM: an uncomplicated and efficient modulation technique for AC motor drives. IEEE Trans Power Electron1988;3(4):391–8.
[14] Tsunoda A, Hinago Y, Koizumi H. Level- and phase-shifted PWM for seven-level switched-capacitor inverter using series/parallel conversion. IEEE Trans Ind Electron 2014;61(8):4011–21.
[15] Salam Z, Bahari N. Selective harmonics elimination PWM (SHEPWM) using differential evolution approach. In: Proceedings of the Joint International conference on power electronics, drives and energy systems; 2010, p. 1–5.
[16] Alexander SA. Development of solar photovoltaic inverter with reduced harmonic distortions suitable for Indian sub-continent. Renew Sustain Energy Rev2016; 56:694–704.
[17] C. A. Silva, L. A. Cordova, P. Lezana and L. Empringham, "Implementation and Control of a Hybrid Multilevel Converter with Floating DC Links for Current Waveform Improvement," in IEEE Transactions on Industrial Electronics, vol. 58, no. 6, pp. 2304-2312, June 2011.
[18] S. Mekhilef, M. N. Abdul Kadir and Z. Salam, "Digital Control of Three Phase Three-Stage Hybrid Multilevel Inverter," in IEEE Transactions on Industrial Informatics, vol. 9, no. 2, pp. 719-727, May 2013.
[19] Massoud A, Finney S, Williams B. Mapped hybrid spaced vector modulation formultilevel cascaded-type voltage source inverters. IET Power Electronics 2008;1(3):318–35.
[20] Bhupender Sharma, Jayaram Nakka, "Single-phase cascaded multilevel inverter topology addressed with the problem of unequal photovoltaic power distribution in isolated dc links", IET Power Electronics, 2019, Vol. 12 Iss. 2, pp. 284-294
[21] D. K. Singh, S. Manna and A. K. Akella, "Grid Connected PV System Using Multilevel Inverter," 2021 7th International Conference on Electrical Energy Systems (ICEES), 2021, pp. 346-351, doi:10.1109/ICEES51510.2021.9383721.
[22] A. Kumar, R. K. Mandal, R. Raushan and P. Gauri, "Grid Connected Photovoltaic Systems with Multilevel Inverter," 2020 International Conference on Emerging Frontiers in Electrical and Electronic Technologies (ICEFEET), 2020, pp. 1-9, doi:10.1109/ICEFEET49149.2020.9187010.

[23] Muhammad Bilal Satti, Ammar Hasan, and Mian Ilyas Ahmad, "A New multilevel Inverter Topology for Grid-Connected Photovoltaic Systems", International journal of photoenergy, 2018.

[24] L. K. Haw, M. S. A. Dahidah and H. A. F. Almurib, "A New Reactive Current Reference Algorithm for the STATCOM System Based on Cascaded Multilevel Inverters," in IEEE Transactions on Power Electronics, vol. 30, no. 7, pp. 3577-3588, July 2015, doi:10.1109/TPEL.2014.2341318.

[25] V. R. Kota and S. Vinnakoti, "Synchronous Reference Frame based control of MLI-STATCOM in power distribution network," 2016 IEEE Power and Energy Conference at Illinois (PECI), 2016, pp. 1-6, doi: 10.1109/PECI.2016.7459247.

[26] R. Pothuraju, T. Ramesh and P. D. Kumar, "DTFC-SVM based Five-level cascaded Inverter with reduced switches fed Interior Type Permanent Magnet Synchronous Motor Drive," 2020 IEEE 9th Power India International Conference (PIICON), 2020, pp. 1-6, Doi:10.1109/PIICON49524.2020.9113067.

[27] Kumar, N. Bharath. Design and Analysis of Different Multi-Level Inverter Topologies for Single Phase Im Drive. Turk. J. Comput. Math. Educ. (TURCOMAT) 2021, 10, 4104–4117.

[28] Martinez-Rodrigo, F.; Ramirez, D.; Rey-Boue, A.B.; De Pablo, S.; Herrero-de Lucas, L.C. Modular Multilevel Converters: Control and Applications. *Energies* 2017, *10*, 1709. https://doi.org/10.3390/en10111709.

[29] Wang, Y.; Aksoz, A.; Geury, T.; Ozturk, S.B.; Kivanc, O.C.; Hegazy, O. A Review of Modular Multilevel Converters for Stationary Applications. *Appl. Sci.* 2020, *10*, 7719. https://doi.org/10.3390/app10217719.

[30] Anandrup Das, "High Power Multilevel Converters- Analysis design and operational issues", NPTEL.

[31] M. Utvić and D. Dujić, "Generalized theory on direct arm energy control in modular multilevel converters," in *CPSS Transactions on Power Electronics and Applications*, vol. 5, no. 4, pp. 388-399, Dec. 2020, doi:10.24295/CPSSTPEA.2020.00032.

[32] Marzo, A. Sanchez-Ruiz, J. A. Barrena, G. Abad and I. Muguruza, "Power Balancing in Cascaded H-Bridge and Modular Multilevel Converters Under Unbalanced Operation: A Review," in *IEEE Access*, vol. 9, pp. 110525-110543, 2021, doi:10.1109/ACCESS.2021.3103337.

[33] P. Sun, Y. Tian, J. Pou and G. Konstantinou, "Beyond the MMC: Extended Modular Multilevel Converter Topologies and Applications," in IEEE Open Journal of Power Electronics, vol. 3, pp. 317-333, 2022, doi:10.1109/OJPEL.2022.3175714.

[34] G. A. Reddy and A. Shukla, "The Enhanced Modular Multilevel Converter With DC Fault Blocking Capability," in *IEEE Transactions on Power Electronics*, vol. 38, no. 7, pp. 8571-8582, July 2023, doi: 10.1109/TPEL.2023.3263004.

[35] S. Debnath, J. Qin, B. Bahrani, M. Saeedifard and P. Barbosa, "Operation, Control, and Applications of the Modular Multilevel Converter: A Review," in IEEE Transactions on Power Electronics, vol. 30, no. 1, pp. 37-53, Jan. 2015, doi:10.1109/TPEL.2014.2309937.

10

Low-frequency Analysis of an Asymmetric Inverter

Lakshmi Prasanna and T. R. Jyothsna

Electrical Engineering, Andhra University, India
Email: lakshmiprasanna.rs@andhrauniversity.edu.in;
thummalajyothsna@gmail.com

Abstract

As a viable technology for numerous applications, including battery-powered vehicles and sources of alternative energy, multilevel inverters (MLIs) are showing interest from both academia and industry. In high, as well as medium-voltage applications, MLIs are frequently used in complex converter designs. One of the primary goals of the current chapter is to construct minimal switch MLI structures. The work proposes the design of an asymmetrical cascaded H-bridge multilevel inverter with fundamental switching frequency. This method facilitates the design of multiple voltage levels utilizing an identical number of switches. This chapter offers low-frequency analysis related to several variants of harmonic mitigation techniques, as well as effective switching angle improvements to create various levels. Moreover, the design's efficacy is assessed by lowering THD as a result of the low-order harmonics receding. THD is calculated theoretically for the suggested configurations and contrasted with simulation results. OPAL-RT test environment is employed to analyze the configuration's simulation findings.

Keywords: Asymmetric configuration, mitigation methods, low-frequency methods, total harmonic distortion (THD), OPAL-RT(OP4510)

DOI: 10.1201/9781003669845-10

10.1 Introduction

The reliability of the power network has been greatly enhanced by utilizing the integration of renewable energy sources. The two most important environmentally friendly sources of electricity are solar and wind power. A DC-to-AC converter is one type of converter needed for connections to solar energy systems. Due to their distinctive features, including minimal harmonic distortion, low disruption actions, and high dependability, multilevel inverters are frequently utilized for these kinds of applications [1, 2].

Traditional MLI designs, like flying capacitor MLI (FCMLI), cascaded H-bridge MLI (CHBMLI), and neutral point clamped MLI (NPCMLI) are frequently used for practical applications. For the same level of output, the CHB inverter structure requires fewer components than other traditional inverter designs. Due to characteristics such as a high level of flexibility and the capacity to securely interface with medium-power appliances, CHB inverters are transitioning from a classical approach to real-world applications [3, 4].

To give the same number of levels as in the classic configuration, MLIs with fewer components have been used in recent investigations. Despite this, there are still certain difficulties for academics using the component count reduction technique.

Challenges that researchers have encountered in their attempt to reduce the number of components include the need for multiple sources, complex control techniques, extensibility losses, fewer concurrent modes, higher rated voltage of switching devices, and fewer bidirectional switches in order to achieve the desired level from the current topologies [5, 6, 7, 8, 9]. Nevertheless, this issue can be solved by incorporating renewable energy sources that can power the cascade inverters on their own.

Compared to other traditional inverter architectures, the CHBMLI requires fewer switches and consideration of all other variables in order to generate an equivalent number of voltage levels in the output [10]. Depending on the output voltage generation, CHBMLI is often classified as either symmetric or asymmetric. These are: (1) equal voltage cascaded MLI (ECHBMLI), where the amplitude of each input voltage is the same; (2) the natural sequence cascaded MLI (NSCHBMLI) model has all sources that vary by one and progress arithmetically; (3) binary cascaded MLI (BCHBMLI), wherein the sources are double in turn; (4) triplen cascaded MLI (TCHBMLI), input voltage are triplen ratio; and (5) quasi-linear cascaded MLI (QLCHBMLI), where all sources have equal value until the target output voltage is obtained.

The essential parameter of MLI structure is the modulation techniques. The technique of modulating voltage waveform to accomplish preset objectives. Two categories of MLI modulation techniques: high frequency (HF) and low frequency (LF). A number of noteworthy characteristics of the LF scheme are as follows: (1) it is very effective due to its low-frequency operation; (2) it has wide capacity and excellent voltage boosting; (3) it requires minimal filter component count; (4) it eliminates lower harmonics completely; (5) it has minimal switching losses with limited harmonic control; and (6) it also optimizes the performance index for further superior aspects. Indeed, proposing new modulation techniques that may be used in any kind of design to satisfy certain needs might be considered an entirely distinct area of investigation [9, 13]. Various modulation schemes for distinct topologies are presented [23, 24]. Utilizing HF Scheme MLI configuration is developed in [25]. In [26, 28], the topologies switching function is based on a novel HF scheme which is related to a novel arrangement of carrier shapes. Selective harmonic elimination (SHE), space vector control (SVC), and nearest level control (NLC) are frequently used LF approaches [11, 12]. Employing the NLC method, a novel configuration is implemented in [27]. SVC is a sensible method if the number of levels is higher, the main drawback is that it does not get rid of a particular harmonic. Using the SHE technique, the previously noted problem is lowered by regulating the switch angles listed [10, 14]. An outline preview of the algorithms described in the SHE approaches [18, 19, 20]. An approach to a symmetrical MLI setup that minimizes harmonics using particle swarm optimization (PSO) is presented [15, 17]. Moreover, switch angle calculations for certain modulation indices (ma) are absent from the PSO based on numerical results produced by the genetic algorithm (GA) technique [21, 22]. Obtaining solutions through SHE implementation becomes more difficult as output levels increase. Thus, harmonic mitigation which involves a significant reduction in lower-order harmonics rather than their total elimination has been employed in [16] as a basic SHE technique to overcome the aforementioned issue.

This work proposes an asymmetrical cascaded bridge MLI architecture along with mitigating strategies. The offered work is set up as follows: all switching states are encountered by the proposed topology for the 19-level in Section 10.2. Analysis of losses, voltage stress, and mitigation strategies are covered in Section 10.3. Simulation outcomes, thermal analysis, and OPAL-RT results are discussed in Section 10.4; the suggested configuration is contrasted with alternative topologies in Section 10.5, and the work's conclusion is discussed in Section 10.6.

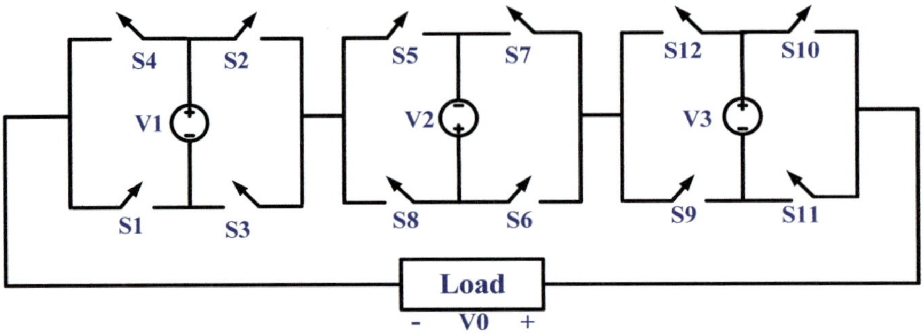

Figure 10.1 Proposed topology.

10.2 Proposed Topology

Figure 10.1 represents the planned H-bridge configuration, it features 12 IGBT switches that are antiparallel with diodes, along with three DC sources. The following subsection describes the operating modes of the 19-level configuration.

10.2.1 Operating modes

The switching states of operation for this setup, which fall under QCHBMLI are depicted in Figure 10.2. The third voltage source (V_3) is six times the base source (first source (V_1)), and the second source (V_2) is double the value of the base source (V_1). Twelve switches are employed as mentioned above, however, each H-bridge switch ratings are different. Table 10.1 describes the switching modes of operation. Eqn (10.1) represents the peak voltage (V_{opeak}) of this topology. The following is an explanation of how output voltage levels are generated:

$$V_{\text{opeak}} = \sum_{n=1}^{\infty} \left(\frac{n(n+1)}{2} + (4n-6) \right) V_{\text{dc}}. \tag{10.1}$$

0 V_{dc}: To attain zero output voltage, either S_1, S_3, S_6, S_8, S_9, and S_{11} conducts or S_2, S_4, S_5, S_7, S_{10}, and S_{12} activated as indicated in Figure 10.2.a,b.

± 1 V_{dc}: S_1, S_2, S_6, S_8, S_9, and S_{11} conducts which is represented in Figure 10.2.c for getting $+1$ V_{dc} output. S_3, S_4, S_5, S_7, S_{10}, and S_{12} conduct to create -1 V_{dc} output voltage which is displayed in Figure 10.2.

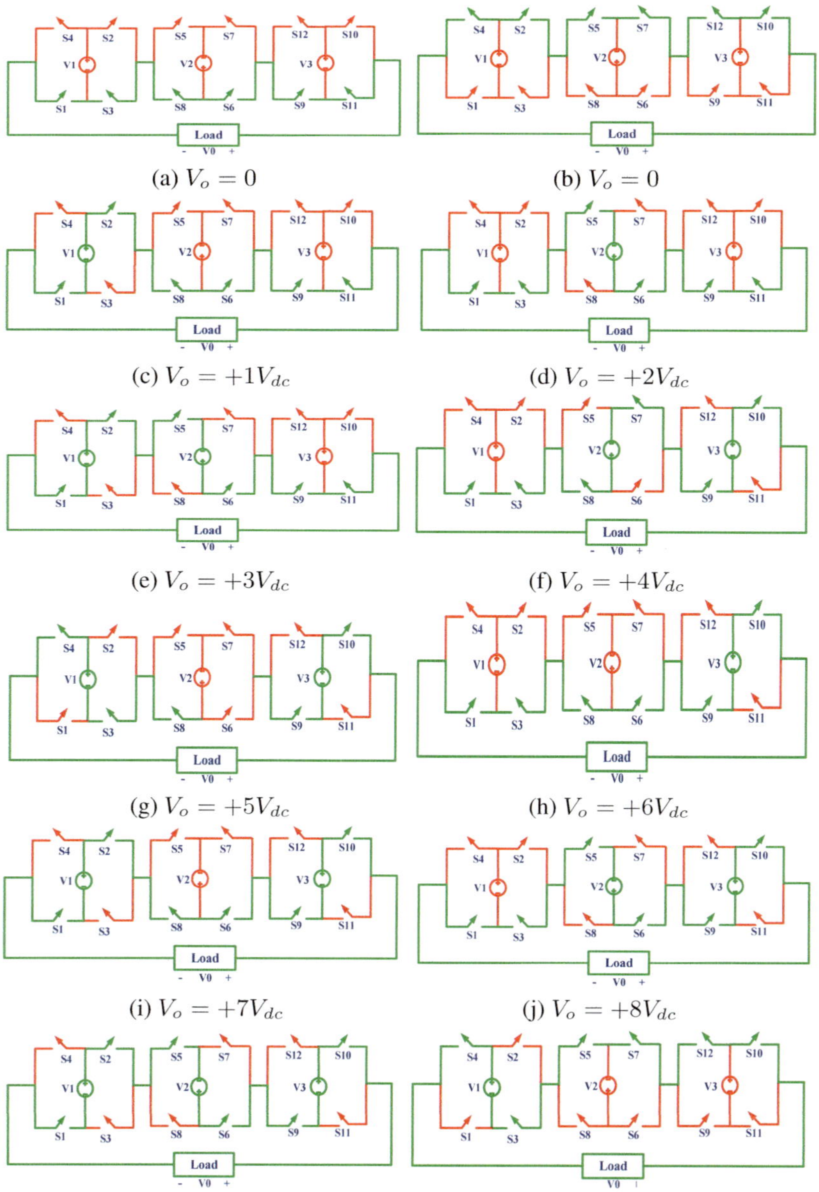

(a) $V_o = 0$

(b) $V_o = 0$

(c) $V_o = +1V_{dc}$

(d) $V_o = +2V_{dc}$

(e) $V_o = +3V_{dc}$

(f) $V_o = +4V_{dc}$

(g) $V_o = +5V_{dc}$

(h) $V_o = +6V_{dc}$

(i) $V_o = +7V_{dc}$

(j) $V_o = +8V_{dc}$

(k) $V_o = +9V_{dc}$

(l) $V_o = --1V_{dc}$

Figure 10.2 *Continued*

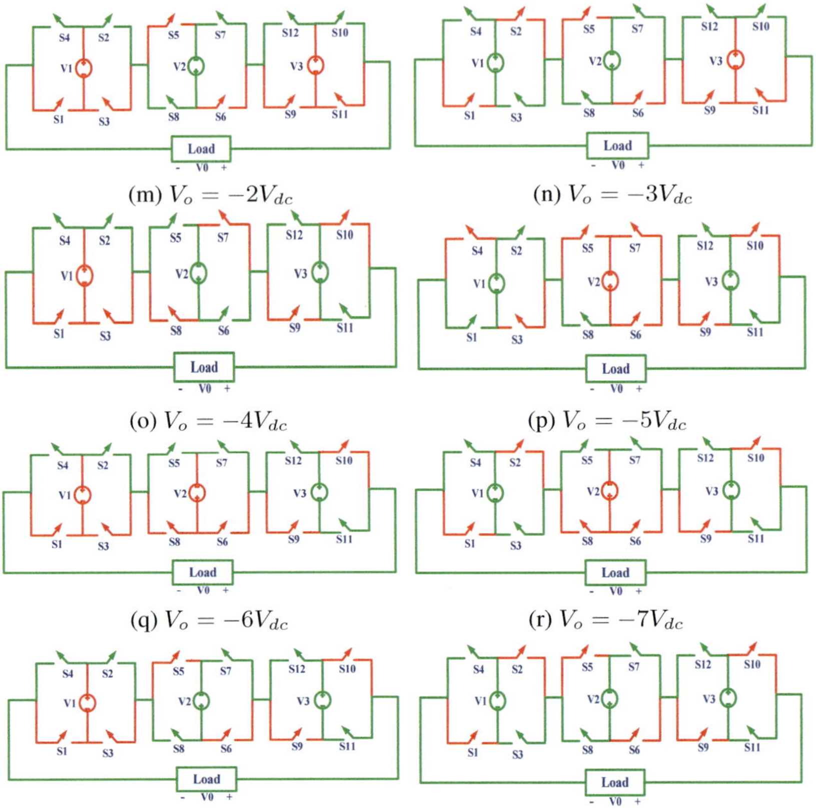

(m) $V_o = -2V_{dc}$ (n) $V_o = -3V_{dc}$

(o) $V_o = -4V_{dc}$ (p) $V_o = -5V_{dc}$

(q) $V_o = -6V_{dc}$ (r) $V_o = -7V_{dc}$

(s) $V_o = -8V_{dc}$ (t) $V_o = -9V_{dc}$

Figure 10.2 Operating modes.

±2 V_{dc}: To develop +2 V_{dc} as output voltage, the conducting path of switches S_3, S_4, S_5, S_6, S_9, and S_{11} are shown in Figure 10.2.d. To provide an output of −2 V_{dc}, the conduction path pattern is S_1, S_2, S_7, S_8, S_{10}, and S_{12}, as portrayed in Figure 10.2.m and Table 10.1 (12) mode.

±3 V_{dc}: Switch conduction paths are presented in Figure 10.2.e and Table 10.1 (7) mode, for the purpose of producing an output voltage of +3 V_{dc}. Using the sequence shown in Figure 10.2.n, −3 V_{dc} output is produced.

±4 V_{dc}: By using the conductivity of switches S_1, S_3, S_7, S_8, S_9, and S_{10}, +4 V_{dc} voltage is created represented in Figure 10.2.f and Table 10.1 (6) mode. The conductivity path of devices S_2, S_4, S_5, S_6, S_{11}, and S_{12} are used to produce an output voltage of −4 V_{dc} and is indicated in Figure 10.2.o and Table 10.1 (14) mode.

Table 10.1 Switching modes.

Switching action	Mode	Output voltage
S_1-S_2-S_5-S_6-S_9-S_{10}	1	+9 V_{dc}
S_1-S_3-S_5-S_6-S_9-S_{10}	2	+8 V_{dc}
S_1-S_2-S_6-S_8-S_9-S_{10}	3	+7 V_{dc}
S_1-S_3-S_6-S_8-S_9-S_{10}	4	+6 V_{dc}
S_3-S_4-S_6-S_8-S_9-S_{10}	5	+5 V_{dc}
S_1-S_3-S_7-S_8-S_9-S_{10}	6	+4 V_{dc}
S_1-S_2-S_5-S_6-S_9-S_{11}	7	+3 V_{dc}
S_1-S_3-S_5-S_6-S_9-S_{11}	8	+2 V_{dc}
S_1-S_2-S_5-S_7-S_9-S_{11}	9	+1 V_{dc}
S_1-S_3-S_6-S_8-S_9-S_{11}	10	0 V_{dc}
S_3-S_4-S_6-S_8-S_{10}-S_{12}	11	−1 V_{dc}
S_2-S_3-S_7-S_8-S_{10}-S_{12}	12	−2 V_{dc}
S_3-S_4-S_7-S_8-S_{10}-S_{12}	13	−3 V_{dc}
S_2-S_3-S_5-S_6-S_{11}-S_{12}	14	−4 V_{dc}
S_1-S_2-S_6-S_8-S_{11}-S_{12}	15	−5 V_{dc}
S_2-S_3-S_6-S_8-S_{11}-S_{12}	16	−6 V_{dc}
S_3-S_4-S_6-S_8-S_{11}-S_{12}	17	−7 V_{dc}
S_2-S_3-S_7-S_8-S_{11}-S_{12}	18	−8 V_{dc}
S_2-S_3-S_7-S_8-S_{11}-S_{12}	19	−9 V_{dc}

±5 V_{dc}*:* Figure 10.2.g and Table 10.1 (5) mode require the $+5\ V_{dc}$ output voltage to be provided using S_1, S_2, S_5, S_6, S_9, and S_{11} conducts respectively. S_3, S_4, S_7, S_8, S_{10}, and S_{12} conduct to get an output of $-5\ V_{dc}$ which is depicted in Figure 10.2.p.

±6 V_{dc}: In accordance with the conductivity of the switches indicated in Figure 10.2.h, an output voltage of $+6\ V_{dc}$ is produced and shown in Table 10.1 (4) mode. By following the conduction path displayed in Figure 10.2.q, the output $-6\ V_{dc}$, as in Table 10.1 (16) mode is obtained.

±7 V_{dc}: As seen in Table 10.1 (3) mode and Figure 10.2.i accordingly, S_1, S_3, S_5, S_6, S_9, and S_{10} conduct for developing $+7\ V_{dc}$ as output voltage level. According to Figure 10.2.r and Table 10.1 (17), respectively, S_2, S_4, S_7, S_8, S_{11}, and S_{12} conduct to produce $-7\ V_{dc}$ as the output voltage level.

$\pm 8\ V_{dc}$: As can be observed from the switches conduction paths in Figure 10.2.j, it generates $+8\ V_{dc}$ as the output voltage. It creates an output voltage level of $-8\ V_{dc}$, as illustrated in Table 10.1 (18) in the same manner as the conductivity path indicated in Figure 10.2.s.

$\pm 9\ V_{dc}$: It follows the conductivity path shown in Figure 10.2.k, and generates an output voltage of $+9\ V_{dc}$, displayed in Table 10.1 (1) mode. A similar way to the conductivity path depicted in Figure 10.2.t, these proceeds as shown in Table 10.1 (19) mode, yield an output voltage level of $-9\ V_{dc}$.

10.3 Mitigation Techniques, Voltage Stress, and Power Loss Analysis

10.3.1 Mitigation techniques

In consideration of a quarter wave symmetry, primary angles are presented in Figure 10.3. Figure 10.3 shows the division of switching angles. During the interval of 1st quadrant (0–π/2), the primary switching angles are displayed as

$$\alpha_1, \alpha_2, \ldots\ldots, \alpha_{(m-1)/2}. \tag{10.2}$$

The switching angles are shown as in the 2nd quadrant interval (π/2–π).

$$\alpha_{(m+1)/2} = \pi - \alpha_{(m-1)/2}, \ldots., \alpha_{(m-1)} = \pi - \alpha_1. \tag{10.3}$$

The switching angles in the 3rd quadrant interval ($\pi - 3\pi/2$), are shown as

$$\alpha_m = \pi + \alpha_1, \ldots., \alpha_{3(m-1)/2} = \pi + \alpha_{(m-1)/2}. \tag{10.4}$$

When examining the switching angles in the 4th quadrant interval (3π/2–2π)

$$\alpha_{(3m-1)/2} = 2\pi - \alpha_{(m-1)/2}, \ldots., \alpha_{2(m-1)} = 2\pi - \alpha_1. \tag{10.5}$$

Equal phase method: this method is used to generate an evaluation of switching angles using eqn (10.6). The switching angles have an average spectrum of 0–π/2 and are dispersed uniformly. The number of levels is indicated by the number *m*.

$$\alpha_i = i \times \frac{180}{m}, \tag{10.6}$$

where i=1, 2, ……., $(m-1)/2$

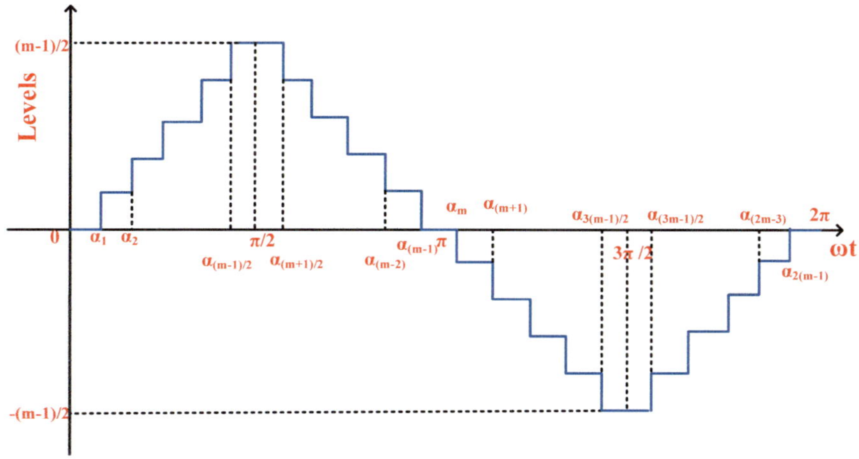

Figure 10.3 Multilevel inverter output voltage waveform.

Half-equal phase method: since this approach produces an EPM waveform that resembles a triangle and produces an output that is too narrow, this was developed to extract effective output from the MLIs. Using the following formula, switching angles can be determined

$$\alpha_{\mathrm{i}} = i \times \frac{180}{(m+1)}. \tag{10.7}$$

Half-*height method*: the aforementioned strategies are able to organize switching angles in a smooth manner, the output waveform does not seem to be sinusoidal. Using the sine function as a basis, this method calculates the switching angles in the first quadrant. The primary switching angles are represented as

$$\alpha_{\mathrm{i}} = \sin^{-1}\left(\frac{2i-1}{m-1}\right). \tag{10.8}$$

Feed Forward method: In contrast to earlier methods, this strategy was developed to minimize the width of two half cycles, as shown by eqn (10.9)

$$\alpha_i = \frac{1}{2}\sin^{-1}\left(\frac{2i-1}{m-1}\right). \tag{10.9}$$

Consequently, utilizing eqn (10.6)–(10.9), the principal switching angles are effectively calculated for the "*m*" number of levels. Eqn (10.6)–(10.9) demonstrate that the computation of the switching angle is independent of

Table 10.2 Switching angles.

Method	Main switching angle								
	α_1	α_2	α_3	α_4	α_5	α_6	α_7	α_8	α_9
EPM	9.47	18.95	28.42	37.89	47.37	56.84	66.32	75.79	85.26
HEPM	9	18	27	36	45	54	63	72	81
HHM	3.18	9.59	16.13	22.89	30	37.67	46.24	56.44	70.81
FFM	1.59	4.79	8.06	11.44	15	18.83	23.12	28.22	35.40

other techniques. Table 10.2 displays the principal switching angles for the 19-level topology, which are computed based on the four techniques.

In terms of fundamental maximum voltage (V_m), the modulation index (MI) formula is represented by eqn (10.10). According to eqn (10.11), the modulation index appears to be altered as a result of the primary switching angle variations. Thus, for different main switching angles, the optimal value of MI varies based on the inverter configuration.

$$\mathrm{MI} = \frac{V_{\mathrm{m}}}{V_1 + V_2 + \ldots.. + V_{\mathrm{n}}}. \tag{10.10}$$

$$V_{\mathrm{m}} = \frac{4}{\pi}\left[V_1 \cos\alpha_1 + V_2 \cos\alpha_2 + \ldots. + V_{\mathrm{n}} \cos\alpha_{(m-1)/2}\right]. \tag{10.11}$$

Theoretical calculation of THD: the calculation of total harmonic distortion (THD) is done using eqn (10.12). In this case, *P* represents the switching angle count and α denotes the primary switching angle. For a 19-level inverter, for instance, *P* becomes 9 and the numerator's first term becomes $\pi^2 \times 81/8$. Eqn (10.13) and (10.14), respectively, express the remaining two terms. Based on the asymmetric arrangement, Table 10.3 expresses the determined value of %THD for all four approaches.

$$\mathrm{THD} = \frac{\sqrt{\left(\left(\frac{\pi^2 p^2}{8}\right) - \left(\frac{\pi}{4}\left(\sum_{i=0}^{p-1}(2i-1)\,\alpha_{\mathrm{i}}\right)\right) - \left(\sum_{i=1}^{p}\cos\alpha_i\right)^2\right)}}{\left(\sum_{i=1}^{p}\cos\alpha_i\right)}. \tag{10.12}$$

$$\text{2nd } term = \frac{\pi}{4}(\alpha_1 + 3\alpha_2 + 5\alpha_3 + 7\alpha_4 + 9\alpha_5 + 11\alpha_6 + 13\alpha_7 + 15\alpha_8). \tag{10.13}$$

$$\text{3rd } term = \cos\alpha_1 + \cos\alpha_2 + \cos\alpha_3 + \cos\alpha_4 + \cos\alpha_5 + \cos\alpha_6 + \cos\alpha_7 + \cos\alpha_8 + \cos\alpha_9. \tag{10.14}$$

Table 10.3 Calculated % THD.

Mitigation method	%THD
EPM	17.04
HEPM	16.22
HHM	4.31
FFM	20.52

Table 10.3 indicates that the computed value of % THD for HHM is lesser than other methods. As a result, the output waveform is maximized for HHM and the cost of filter design is reduced. Because HHM simulation has advanced above previous methods, HHM results are reported in this work.

10.3.2 Voltage stress analysis

The proportion of the total value of the switch blocking voltages to the peak output voltage is known as total standing voltage (TSV). The peak voltage stress of complimentary switches possesses identical values in the proposed topology. Eqn (10.15)–(10.17) show the peak voltage stress of switching devices for 19-level configuration.

$$V_{\mathrm{s1}} = V_{\mathrm{s2}} = V_{\mathrm{s3}} = V_{\mathrm{s4}} = V_{\mathrm{dc}}. \tag{10.15}$$

$$V_{\mathrm{s5}} = V_{\mathrm{s6}} = V_{\mathrm{s7}} = V_{\mathrm{s8}} = 2V_{\mathrm{dc}}. \tag{10.16}$$

$$V_{\mathrm{s9}} = V_{\mathrm{s10}} = V_{\mathrm{s11}} = V_{\mathrm{s12}} = 6\,V_{dc}. \tag{10.17}$$

$$\mathrm{TSV} = 4\times\left(V_{\mathrm{s1}} + V_{\mathrm{s5}} + V_{\mathrm{s9}}\right) = 36V_{\mathrm{dc}}, \tag{10.18}$$

where V_{sn} stands for the peak voltage appearance of the switch during the turn-off process. Therefore, TSV is expressed as eqn (10.18). For the 19-level configuration, the voltage stress of devices is presented in Figure 10.4. As mentioned in Table 10.1, at +6 V_{dc} level S_1, S_3, S_6, S_7, S_9, and S_{10}, the remaining switches seem like a nonconductive element whereas the remaining switches are conducting. Given this, the blocking voltage of switches S_2 and S_4 is 1 V_{dc}, S_5 and S_8 are 2 V_{dc}, and S_{11} and S_{12} are 6 V_{dc}, respectively. In a similar fashion, other levels are depicted in Figure 10.4.

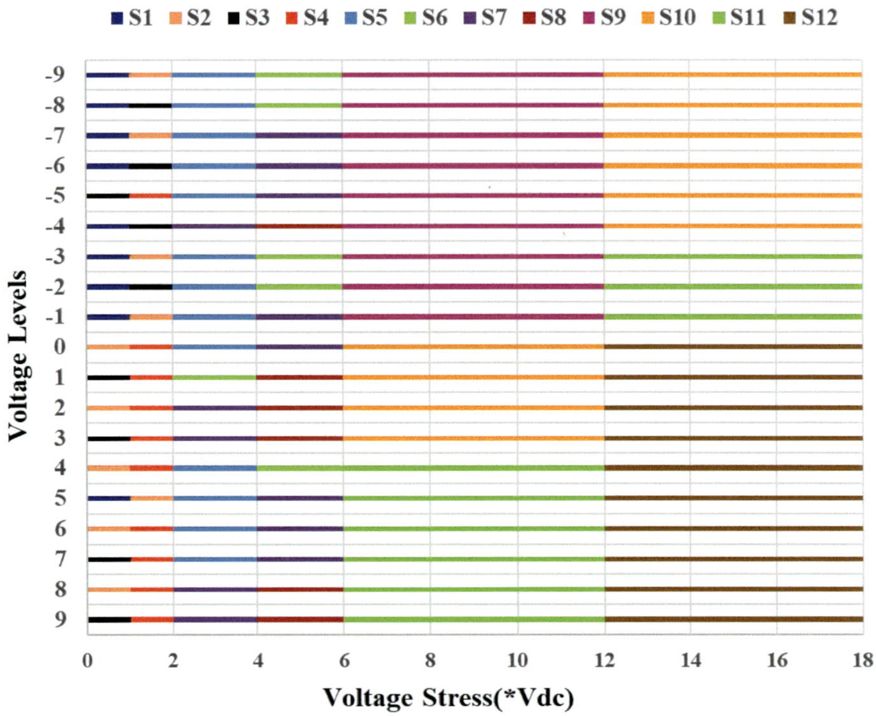

Figure 10.4 Bar chart representation of voltage stress analysis.

10.3.3 Power loss analysis

Switching devices encounter two distinct forms of power losses: switching and conduction. Switching losses are due to interruptions in the switch's on/off processes, whereas conduction losses are caused by on-state resistance. The switching loss that occurs during the process of turning on can be expressed as

$$\begin{aligned} P_{\text{swlturnon}}(i) &= f_{\text{carrier}} \int_0^{t_{\text{on}}} v(t)\, i(t)\, \mathrm{d}t \\ &= f_{\text{carrier}} \int_0^{t_{\text{on}}} \left(\left(\frac{V_{\text{swoff,i}}}{t_{\text{on}}} (t_{\text{on}} - t) \right) \left(\frac{i_{\text{on,i}}}{t_{\text{on}}} t \right) \right) \mathrm{d}t \\ &= \frac{1}{6} f_{\text{carrier}} \times V_{\text{swoff,i}} \times i_{\text{on,i}} \times t_{\text{on}}. \end{aligned} \tag{10.19}$$

$$P_{\text{swlturnoff}}(\mathrm{i}) = f_{\text{carrier}} \int_0^{t_{\text{off}}} v(t) i(t) \mathrm{d}t$$

$$= f_{\text{carrier}} \int_0^{t_{\text{off}}} \left(\left(\frac{V_{\text{swoff,i}}}{t_{\text{off}}} t \right) \left(\frac{i_{\text{off,i}}}{t_{\text{off}}} (t_{\text{off}} - t) \right) \right) \text{d}t$$

$$= \frac{1}{6} f_{\text{carrier}} \times V_{\text{swoff,i}} \times i_{\text{off,i}} \times t_{\text{off}}, \tag{10.20}$$

where $P_{\text{swlturnon}}(i)$, $P_{\text{swltunoff}}(i)$, and V_{swoff} indicate the off-state switching voltage, turn-on loss, and turn-off voltage of the *i*th switch, respectively. The current that flows via the switch is referred to as I_{on} and I_{off}, related to the active or inactive position. Turn-on and turn-off losses are added together to determine total switching losses (P_{sw}).

$$P_{\text{sw}}(\text{Total}) = \sum_{i=1}^{N_{\text{sw}}} \left(\begin{array}{c} \sum_{j=1}^{N_{\text{on}}(i)} P_{\text{swlon}}(ij) \\ + \sum_{j=1}^{N_{\text{off}}(i)} P_{\text{swloff}}(ij) \end{array} \right), \tag{10.21}$$

where N_{sw} is the switch count.

Because of on-state resistance and voltage drop of devices, they experience losses during the whole conduction phase. The equations are expressed as

$$P_{\text{Dcon}} = V_{\text{Don}} \times i_{\text{Davg}} + R_{\text{Don}} \times i_{\text{Drms}}^2. \tag{10.22}$$

$$P_{\text{swcon}} = V_{\text{swon}} \times i_{\text{swavg}} + R_{\text{swon}} \times i_{\text{swrms}}^2, \tag{10.23}$$

where P_{Dcon} and P_{swcon} are conduction losses (diode, switch), V_{Don} and V_{swon} are voltage drop (diode, switch), R_{Don} and R_{swon} are resistances (diode, switch) and i_{Davg}, i_{swavg}, i_{Drms}, and i_{swrms} are average and RMS currents (diode, switch).

10.4 Results and Discussions

10.4.1 Simulation results

The recommended configuration is constructed and simulated using MATLAB/Simulink. Quarter-wave symmetry estimations are used to estimate corresponding angles of the next quadrants once the switching angles of all methods are established with respect to the first quadrant. The QCHBMLI technique is used to construct a 19-level inverter. The variables which are considered are shown in Table 10.4. Figure 10.5 depicts the switching pattern for a single HHM cycle. The complete H bridge output for HHM is shown in Figure 10.6, which also shows that bridges 1, 2, and 3 have peak output voltages of 100, 200, and 600 volts, respectively.

Table 10.4 Simulating parameters.

Components	Values
DC source	100 V, 200 V, 600 V
Resistance	100 Ω
Inductances	100 mH,150 mH, 200 mH
Reference frequency	50 Hz

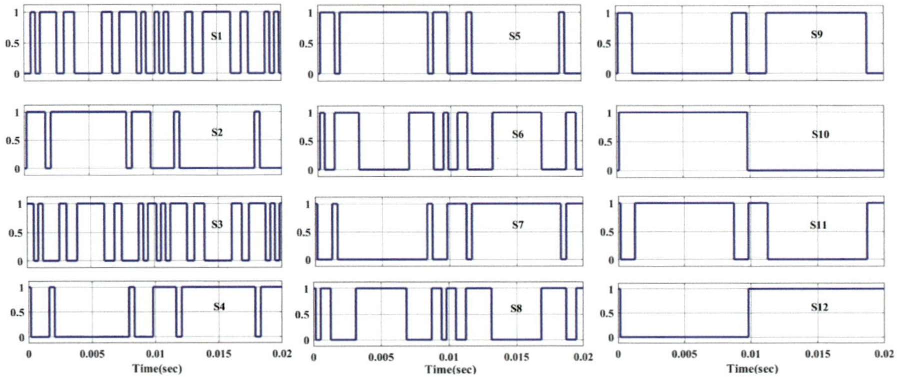

Figure 10.5 Switching pattern.

The inverter operation under R-load circumstances is shown in Figure 10.7.a. It seems that output voltage and current are unique. The performance of the inverter under different load situations for 19-level operation is displayed in Figure 10.7.b. While the load voltage remains constant under all loading conditions, as seen in Figure 10.7.b, the current is zero when there is no load, showing an identical voltage pattern for R-load and lagging for RL-load. The waveform variations associated with inductive load fluctuations are seen in Figure 10.7.c. Figure 10.7.c illustrates how a high inductive load causes the current to fall, which causes greater current lagging. In Figure 10.7.d, the behavior of the output waveforms at various source voltages is displayed. Regardless of variations in the source voltage, RL-load causes the output current to always follow the output voltage with a lagging characteristic. Figure 10.8 displays the harmonic spectra of outputs with the THD values. When the load is RL, resultant voltage and current % THD values are 2.72 and 4.27, respectively.

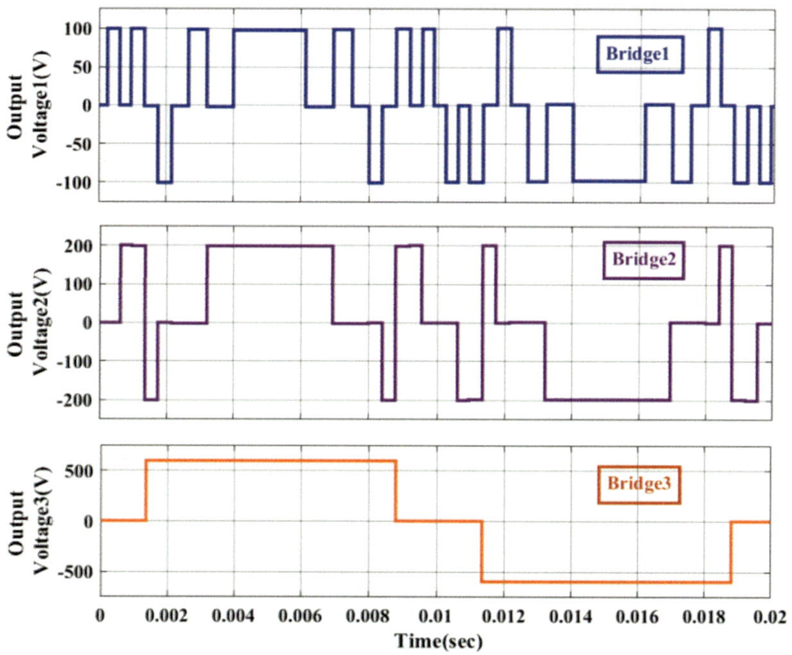

Figure 10.6 Resultant of each bridge for HHM.

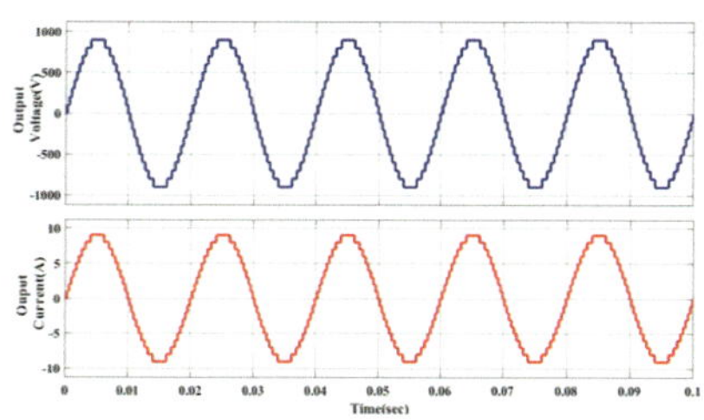

(a) Resistive Load

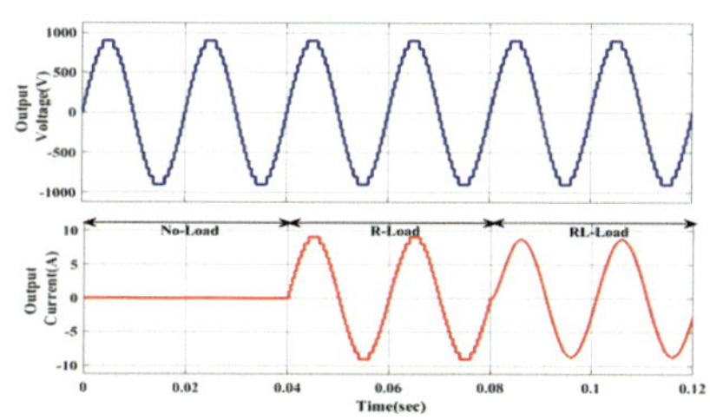

(b) Distinct load situations

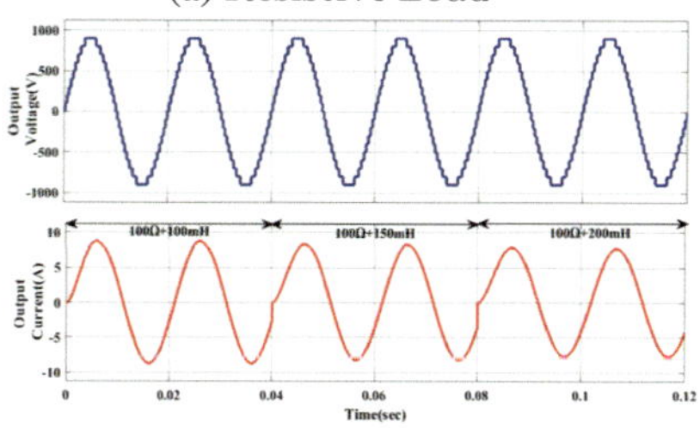

(c) Distinct Impedances

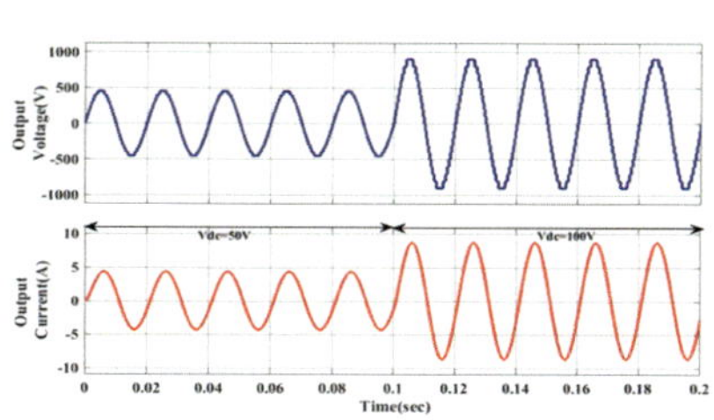

(d) Regulation of Input voltage

Figure 10.7 Resultant waveforms.

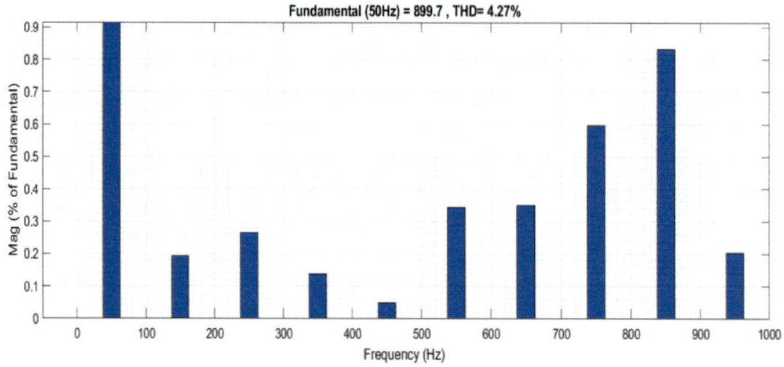

(a) Resultant Voltage

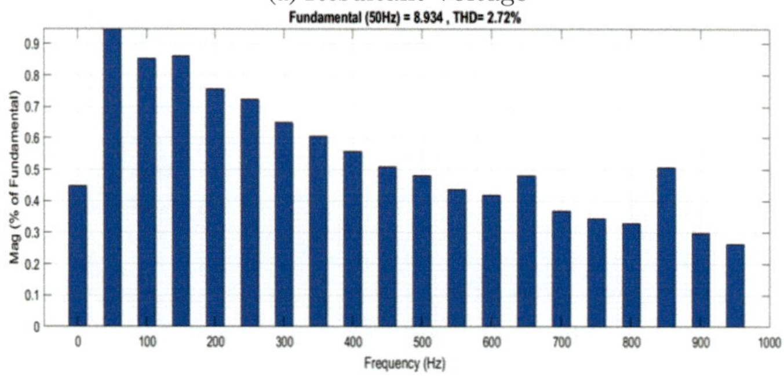

(b) Resultant Current

Figure 10.8 % THD.

Figure 10.8 shows that the fundamental component with very few low-order harmonics. Using the same process, simulation was used to obtain the % THD values for additional harmonic mitigation techniques. When R load is taken into account, Figure 10.9 compares the THD of each method. Figure 10.9 shows that the % THD for EPM, HEPM, HHM, and FFM are 17.1, 16.25, 4.27, and 20.52. Consequently, when compared to alternative approaches, the HHM approach is the most optimal harmonic mitigation strategy.

10.4.2 Thermal modeling

The recommended configuration power semiconductor components are thermally modeled using PLECS software. The loss analysis approach used is shown in Figure 10.10. Figure 10.11 displays the switching losses (%) of switches for various load scenarios. Even though individual switches may experience losses, the net effect of all system parameters' performance

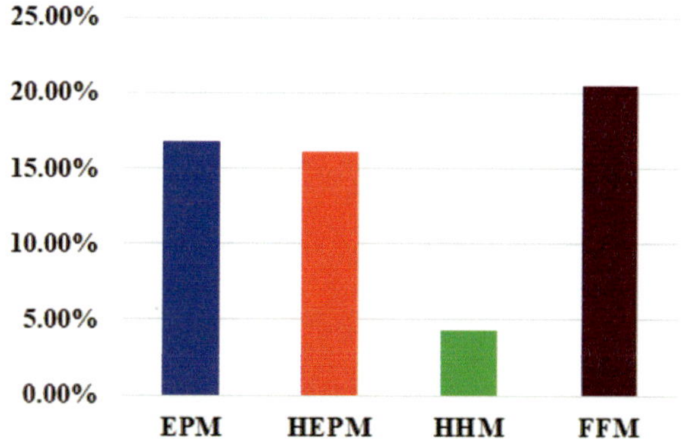

Figure 10.9 % THD comparison of all methods.

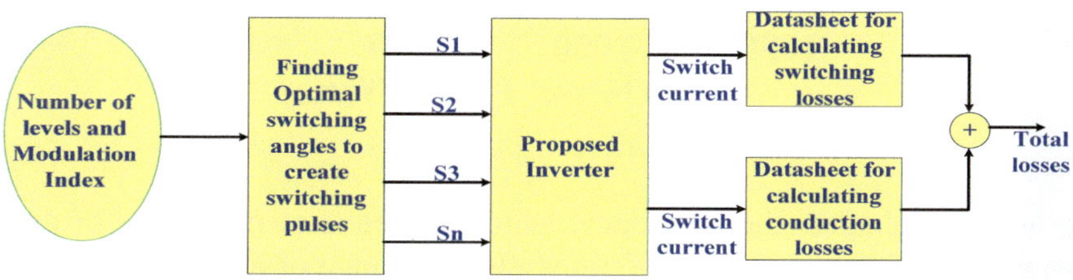

Figure 10.10 Power loss analysis.

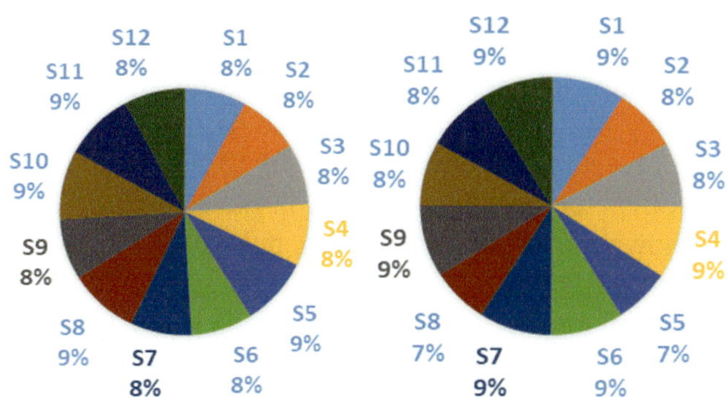

(a) R-Load (b) RL-Load

Figure 10.11 Percentage losses.

determines the system's overall efficiency. This implies that despite the possibility of switch losses, other elements or factors in the system may offset these losses, leading to the overall efficiency shown by the curve. R loads

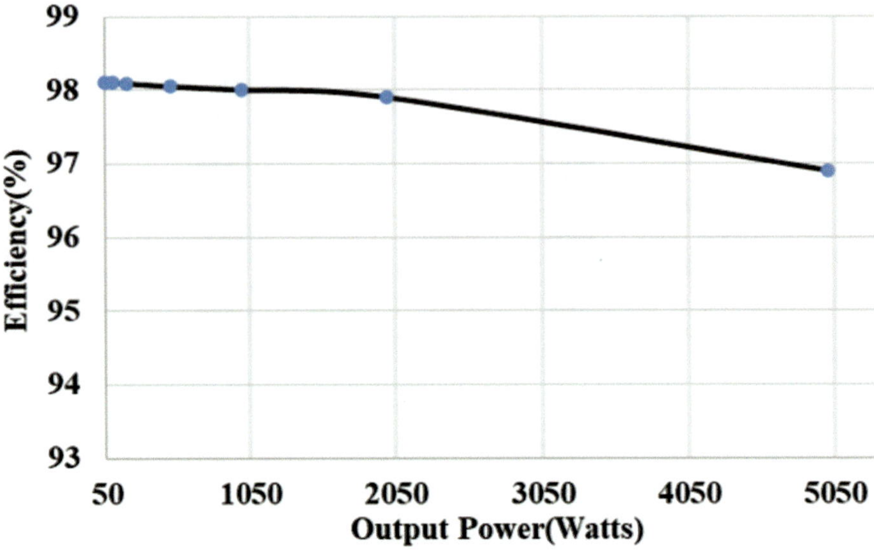

Figure 10.12 Efficiency curve.

are used to calculate the efficiency, which is then shown in favor of output power (0–5000W). Figure 10.12 illustrates the range of effectiveness, which is 98.2% to 97.0%.

10.4.3 HIL implementation

A real-time simulator is essential for both creating and validating the efficacy and accuracy of systems since the models it creates run at the same speed as those of actual systems. Figure 10.13.a indicates waveforms for R-Load. The voltage across three bridges is illustrated in Figure 10.13.b,c. Figure 10.14 shows the 19-level switching pulses for HHM. In comparison with other methods, HHM results are given as the best results for the specified configuration, taking into account both the simulated and OPAL-RT(OP4510) outcomes.

10.5 Comparison

Table 10.5 contrasts the proposed configuration with all other relevant configurations. To compare, the switch count (N_{sw}), modulation scheme, THD (%), and switching loss parameters are used. In [25], the switches count is 16, and the modulation scheme applied is hybrid pulse width modulation related to

HF Scheme. %THD is not mentioned in the presented chapter and switching losses are high. In [26], the number of switches is 14, and the working method of the switching scheme is modified LSPWM, and power losses are high. In this chapter, two transformers are utilized for designing the layout, and the %THD is 4.5. In [28], the switches count is similar to the proposed topology but switching pulses are provided through modified LSPWM, the %THD for the designed structure is 6.22, and switching losses are high. In [27], the number of switches used is 13, and LF scheme (NLC) is employed to send gate pulses, %THD is 4.5, and power losses are low. The suggested arrangement offers sophisticated results in contrast to other topologies listed in Table 10.5.

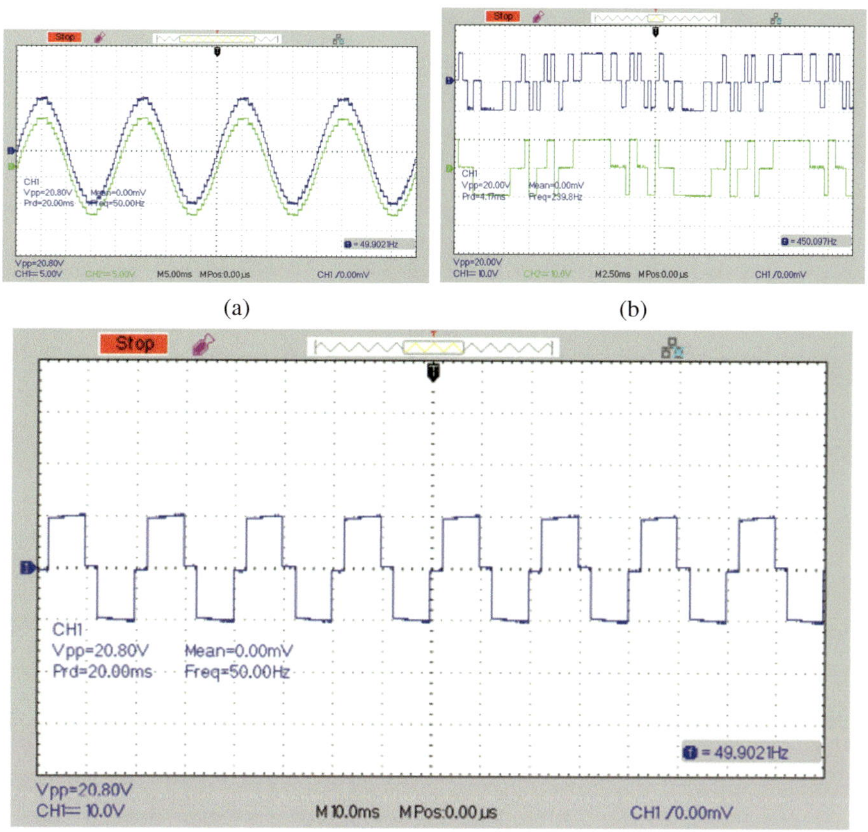

(a) (b) (c)

Figure 10.13 (a) Output waveforms. (b) and (c) Resultant of bridges.

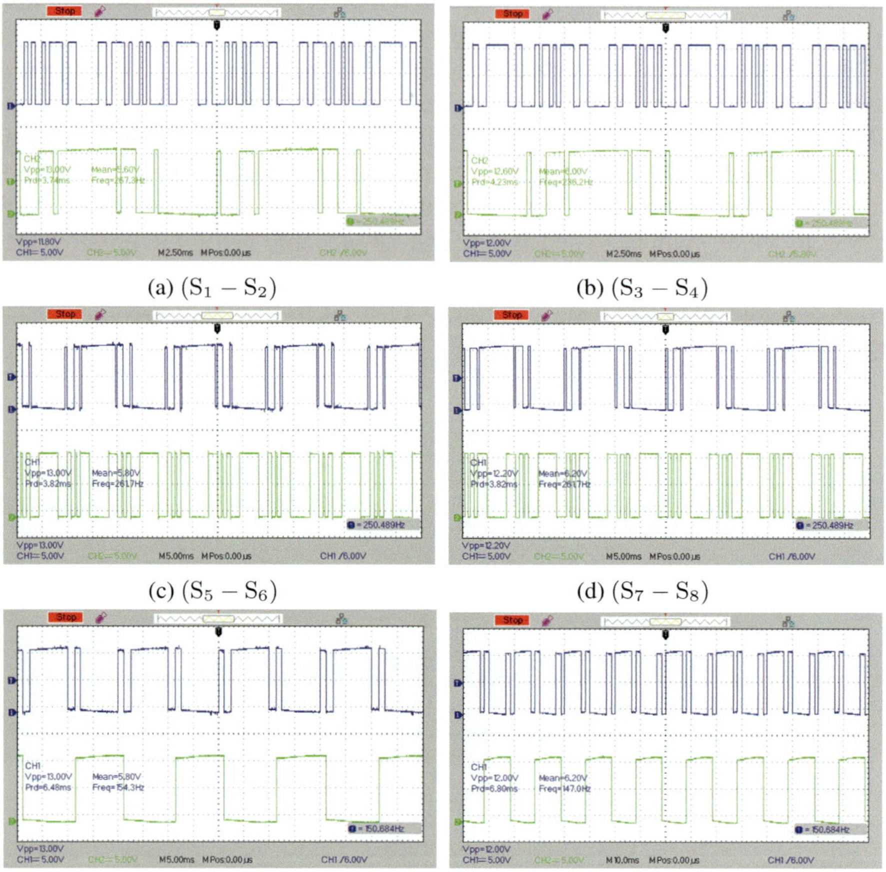

(a) $(S_1 - S_2)$ (b) $(S_3 - S_4)$

(c) $(S_5 - S_6)$ (d) $(S_7 - S_8)$

(e) $(S_9 - S_{10})$ (f) $(S_{11} - S_{12})$

Figure 10.14 Switching pulses of HHM for 19-level.

Table 10.5 Comparison with other topologies.

	N_{sw}	**THD (%)**	**Modulation scheme**	**Switching loss**
[25]	16	–	HPWM	High
[26]	14	4.3	Modified LSPWM	High
[27]	13	4.5	NLC	Low
[28]	12	6.22	Modified LSPWM	High
Proposed	12	4.27	HHM	Low

10.6 Conclusion

The suggested asymmetrical cascaded bridge inverter employs three DC sources and minimum devices to provide the output voltage of a 19-level structure. Using a mathematical switching technique, the current chapter developed a straightforward control and modulation strategy. The suggested bridge configuration leads to a smaller size, minimum component losses, along with low installation cost. Total losses are calculated with PLECS software. Additionally, the recommended configuration achieves unique criteria in terms of efficacy and cost aspects with lower conduction and switching losses. Similarly, the recommended cascading bridge structure can be applied to different levels in the modified module that uses uneven DC sources. Additionally, as per IEEE 519-2014 requirements, the suggested inverter output has the lowest possible THD. The simulation results show that HHM experiences the least THD. It is possible to use the suggested topology for both standalone and grid-connected renewable energy applications.

10.7 Acknowledgments

OPAL-RT Test bed Simulator (OP-4510) is supported by Raghu Engineering College, Dakamarri, Visakhapatnam, Andhra Pradesh, India.

References

[1] J. Rodriguez, J. S. Lai, and F. Z. Peng, "Multilevel inverters: A survey of topologies, controls, and applications," IEEE Transactions on Industrial Electronics, vol. 49, no. 4, pp. 724–738, Aug. 2002, doi:10.1109/TIE.2002.801052.

[2] S. Kouro et al., "Recent advances and industrial applications of multilevel converters," IEEE Transactions on Industrial Electronics, vol. 57, no. 8, pp. 2553–2580, Aug. 2010, doi:10.1109/TIE.2010.2049719.

[3] S. De, D. Banerjee, K. Siva Kumar, K. Gopakumar, R. Ramchand, and C. Patel, "Multilevel inverters for low-power application," IET Power Electronics, vol. 4, no. 4, pp. 384–392, Apr. 2011, doi:10.1049/iet-pel.2010.0027.

[4] H. Abu-Rub, J. Holtz, J. Rodriguez, and G. Baoming, "Medium-voltage multilevel converters State of the art, challenges, and requirements in Industrial applications," IEEE Transactions on Industrial Electronics, vol. 57, no. 8, pp. 2581–2596, Aug. 2010, doi:10.1109/TIE.2010.2043039.

[5] K. K. Gupta, A. Ranjan, P. Bhatnagar, L. K. Sahu, and S. Jain, "Multilevel inverter topologies with reduced device count: A review," IEEE Transactions on Power Electronics, vol. 31, no. 1. Institute of Electrical and Electronics Engineers Inc., pp. 135–151, Jan. 01, 2016. doi:10.1109/TPEL.2015.2405012.

[6] P. R. Bana, K. P. Panda, R. T. Naayagi, P. Siano, and G. Panda, "Recently Developed Reduced Switch Multilevel Inverter for Renewable Energy Integration and Drives Application: Topologies, Comprehensive Analysis and Comparative Evaluation," IEEE Access, vol. 7, pp. 54888–54909,2019, doi:10.1109/ACCESS.2019.2913447.

[7] M. Vijeh, M. Rezanejad, E. Samadaei, and K. Bertilsson, "A General Review of Multilevel Inverters Based on Main Submodules: Structural Point of View," IEEE Trans Power Electron, vol. 34, no. 10, pp. 9479–9502, Oct. 2019, doi:10.1109/TPEL.2018.2890649.

[8] H. P. Vemuganti, D. Sreenivasarao, S. K. Ganjikunta, H. M. Suryawanshi, and H. Abu-Rub, "A survey on reduced switch count multilevel inverters," IEEE Open Journal of the Industrial Electronics Society, vol. 2, pp. 80–111, 2021, doi:10.1109/OJIES.2021.3050214.

[9] V. K. N. V. and G. T, "A comprehensive survey on reduced switch count multilevel inverter topologies and modulation techniques," Journal of Electrical Systems and Information Technology, vol. 10, no. 1, Jan. 2023, doi:10.1186/s43067-023-00071-8.

[10] M. Malinowski, K. Gopakumar, J. Rodriguez, and M. A. Perez, "A survey on cascaded multilevel inverters," IEEE Transactions on Industrial Electronics, vol. 57, no. 7. pp. 2197–2206, Jul. 2010. doi:10.1109/TIE.2009.2030767.

[11] M. S. A. Dahidah, G. Konstantinou, and V. G. Agelidis, "A Review of Multilevel Selective Harmonic Elimination PWM: Formulations, Solving Algorithms, Implementation and Applications," IEEE Trans Power Electron, vol. 30, no. 8, pp. 4091–4106, 2015, doi:10.1109/TPEL.2014.2355226.

[12] A. Rakesh Kumar, T. Deepa, S. Padmanaban, and J. Bo Holm-Nielsen, "Low-Switching Frequency Modulation Schemes for Multi-level Inverters", CRC Press, Taylor& Francis Group, First Edition ,2021.

[13] A. Poorfakhraei, M. Narimani and A. Emadi, "A Review of Modulation and Control Techniques for Multilevel Inverters in Traction Applications," in IEEE Access, vol. 9, pp. 24187–24204, 2021, doi: 10.1109/ACCESS.2021.3056612.

[14] J. Chiasson, L. M. Tolbert, K. McKenzie and Z. Du, "Elimination of harmonics in a multilevel converter using the theory of symmetric polynomials and resultants," 42nd IEEE International Conference on Decision and Control (IEEE Cat. No.03CH37475), Maui, HI, USA, 2003, pp. 3507–3512 vol.4, doi:10.1109/CDC.2003.1271691.

[15] M. Srndovic, A. Zhetessov, T. Alizadeh, Y. L. Familiant, G. Grandi, and A. Ruderman, "Simultaneous Selective Harmonic Elimination and THD Minimization for a Single-Phase Multilevel Inverter with Staircase Modulation," IEEE Trans Ind Appl, vol. 54, no. 2, pp. 1532–1541, Mar. 2018, doi:10.1109/TIA.2017.2775178.

[16] J. Napoles, J. I. Leon, L. G. Franquelo, R. Portillo and M. A. Aguirre, "Selective harmonic mitigation technique for multilevel cascaded H-bridge converters," 2009 35th Annual Conference of IEEE Industrial Electronics, Porto, Portugal, 2009, pp. 806–811, doi:10.1109/IECON.2009.5415023.

[17] C.A.L. Espinosa, I. Portocarrero, and M. Izquierdo, "Minimization of THD and angles calculation for multilevel inverters," Int. Journal of Eng. and Technology IJET-IJENS, vol. 12, no. 5, pp. 83–86, Oct. 2012.

[18] M. S. A. Dahidah, G. Konstantinou, and V. G. Agelidis, "A Review of Multilevel Selective Harmonic Elimination PWM: Formulations, Solving Algorithms, Implementation and Applications," IEEE Trans Power Electron, vol. 30, no. 8, pp. 4091–4106, 2015, doi:10.1109/TPEL.2014.2355226.

[19] J. Napoles et al., "Selective Harmonic Mitigation Technique for Cascaded H-Bridge Converters with Nonequal DC Link Voltages," in IEEE Transactions on Industrial Electronics, vol. 60, no. 5, pp. 1963–1971, May 2013, doi:10.1109/TIE.2012.2192896.

[20] T. Kato, "Sequential homotopy-based computation of multiple solutions for selected harmonic elimination in PWM inverters," in IEEE Transactions on Circuits and Systems I: Fundamental Theory and Applications, vol. 46, no. 5, pp. 586–593, May 1999, doi:10.1109/81.762924.

[21] S. Barkat, E. M. Berkouk, and M. S. Boucherit, "Particle swarm optimization for harmonic elimination in multilevel inverters," Electrical Engineering, vol. 91, no. 4–5, pp. 221–228, Dec. 2009, doi:10.1007/s00202-009-0135-9.

[22] J. Pontt, J. Rodríguez, and R. Huerta, "Mitigation of noneliminated harmonics of SHEPWM three-level multipulse three-phase active front end converters with low switching frequency for meeting standard

IEEE-519-92," IEEE Trans Power Electron, vol. 19, no. 6, pp. 1594–1600, Nov. 2004, doi:10.1109/TPEL.2004.836616.

[23] S. X. Zhou, Z. X. Sang, J. Zhang, L. Jing, Z. Du, and Q. T. Guo, "Comparison on modulation schemes for 15-level cascaded H-bridge multilevel inverter," IOP Conf Ser Earth Environ Sci, vol. 188, p. 012039, Oct. 2018, doi:10.1088/1755-1315/188/1/012039.

[24] N. Vishwajith, S. Nagaraja Rao, and S. Sachin, "Performance analysis of reduced switch ladder type multilevel inverter using various modulation control strategies," J Phys Conf Ser, vol. 1706, no. 1, p. 012092, Dec. 2020, doi:10.1088/1742-6596/1706/1/012092.

[25] R. Barzegarkhoo, E. Zamiri, M. Moradzadeh, and H. Shadabi, "Symmetric hybridised design for a novel step-up 19-level inverter," IET Power Electron., Vol. 10, pp. 1377–91, sep2017, doi:10.1049/iet-pel.2016.0558.

[26] J. Venkataramanaiah, Y. Suresh and A. K. Panda, "Design and Development of a Novel 19-Level Inverter Using an Effective Fundamental Switching Strategy," in IEEE Journal of Emerging and Selected Topics in Power Electronics, vol. 6, no. 4, pp. 1903–1911, Dec. 2018, doi: 10.1109/JESTPE.2017.2776294.

[27] C. Dhanamjayulu, P. Kaliannan, S. Padmanaban, P. K. Maroti and J. B. Holm-Nielsen, "A New Three-Phase Multi-Level Asymmetrical Inverter with Optimum Hardware Components," in IEEE Access, vol. 8, pp. 212515–212528, 2020, doi:10.1109/ACCESS.2020.3039831.

[28] V. Kubendran, Y. Mohamed Shuaib, S. Vidyasagar, V. Kalyanasundaram, K. Saravanan, "The development of a generalized multilevel inverter for symmetrical and asymmetrical dc sources with a minimized-ON state switch", Ain Shams Engineering Journal, Volume 15, Issue 2, 2024,102358, doi:10.1016/j.asej.2023.102358.

11

Multilevel Converters for Renewable Energy Systems

Lei Zhang

School of Electrical and Electronic Engineering, Nanyang Technological University, Singapore
Email: lei.zhang@ntu.edu.sg

Abstract

Modular multilevel converters (MMCs) are preferred in high-power applications such as renewable energy sources (RESs), medium-voltage motor drives, energy storage systems (ESSs), and solid-state transformers (SSTs). One major challenge in MMC-based renewable systems is coordinating energy flow between capacitors, batteries, and distributed energy sources. The integration of energy storage components and renewable sources into MMC submodules significantly alters operational behavior, adding complexity to energy management. A sophisticated control strategy is essential to optimize system efficiency and stability.

High-gain DC-DC converters also empower the development of renewable systems, particularly in large-scale offshore wind farms where significant voltage step-up is needed. SSTs have been explored to reduce transformer size and weight while achieving high voltage gains. MMCs are ideal for these applications, but maintaining voltage balance across capacitors remains a challenge. Current voltage-balancing and modulation methods have not been fully evaluated.

This chapter provides a comprehensive analysis of various voltage-balancing methods and modulation techniques for MMCs in SSTs, proposing two improved algorithms. Simulation and experimental results confirm their

DOI: 10.1201/9781003669845-11

effectiveness. A novel MMC-based configuration for hybrid energy storage is also introduced, with an optimized control strategy managing diverse components to enhance system performance. These optimized algorithms address low-frequency ripple in submodule capacitor voltage, improving MMC reliability and longevity, as demonstrated by detailed simulations and experimental validations.

Keywords: Modular multilevel converters, solid-state transformer, hybrid energy storage system

11.1 Introduction

The modular multilevel converter (MMC) was initially introduced by R. Marquardt in 2003 [1], which gained popularity in high-power applications due to its numerous advantages. As a result, it has become a key component in multiterminal direct current (MTDC) systems and DC grids [2]. Recently, MMCs have been employed in various high-power applications, including grid integration for renewable energy sources (RESs), medium-voltage motor drive systems, and energy storage systems (ESSs), as well as solid-state transformers (SSTs), as illustrated in Figure 11.1.

Figure 11.2.a is the schematic of an MMC. Each phase leg contains two arms. In each arm, there are multiple submodules (SMs) and an arm inductor. Regardless of the SM configurations or modulation methods employed, each arm is modeled as a DC voltage source in series with an AC voltage source, as illustrated in Figure 11.2.b.

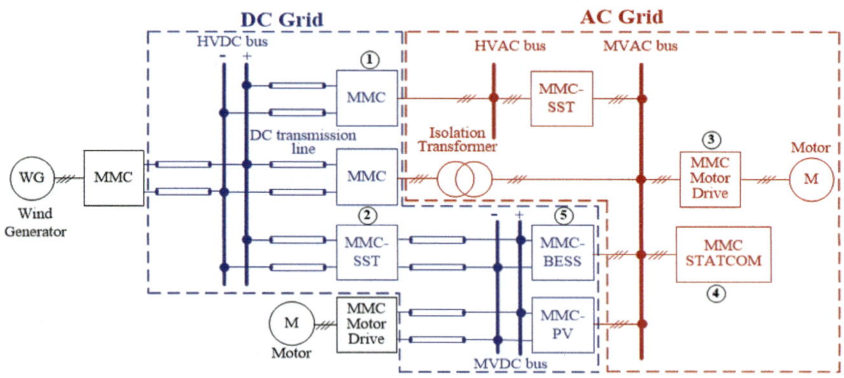

Figure 11.1 Typical applications of MMCs.

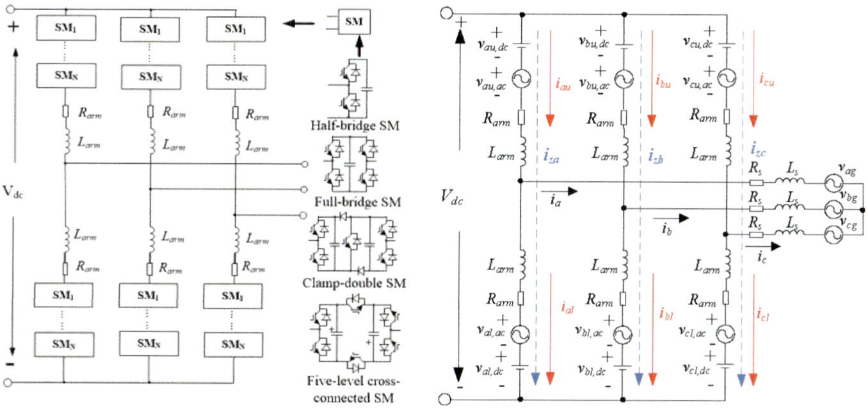
(a) (b)

Figure 11.2 Schematic and equivalent circuit of MMC.

This chapter briefly overviews the operational principles of MMCs. Next, it analyzes the control strategy for an MMC-based hybrid energy storage system (MMC-HESS). When employing the MMCs into isolated DC-DC converters to form multilevel DC-DC converters (IM2DC), their characteristics significantly differ from those of conventional MMCs. Finally, the chapter discusses innovative modulation and voltage balancing techniques aimed at reducing capacitor voltage ripple and optimizing IM2DC performance.

11.2 Operational Principles of Modular Multilevel Converters

The steady-state operational principles of MMCs are analyzed in references [2, 3] and briefly summarized in this section. According to the superposition principle, the equivalent circuit of the MMC is separated into two distinct equivalent circuits: one for DC current and another for AC current.

In various applications, such as MMC-based ESS and photovoltaic (PV) systems, power must be appropriately distributed among the SMs, arms, and phases:

(1) *Submodule power:* In each arm, all SMs share the same arm current. Using phase-shifted carrier-based (PSC) PWM allows each SM to be independently controlled by adjusting its terminal voltage [4, 5]. With nearest-level control (NLC) modulation, the charging or discharging

power depends on the switching states and the direction of the arm current [6].

(2) *Arm power*: eqn (11.1) defines the active power of an arm, assuming that the voltage drop and power losses in the arm inductor are negligible.

$$\begin{cases} p_{\mathrm{xu}} = -\frac{V_{\mathrm{dc}}}{2}\left(\frac{i_{\mathrm{dc}}}{3}+i_{\mathrm{zx,dc}}\right) + v_{\mathrm{xg}}\left(\frac{i_{\mathrm{x}}}{2}+i_{\mathrm{zx,h1}}\right) + p_{\mathrm{xu}}^{\mathrm{ripple}}, \\ p_{\mathrm{xl}} = -\frac{V_{\mathrm{dc}}}{2}\left(\frac{i_{\mathrm{dc}}}{3}+i_{\mathrm{zx,dc}}\right) + v_{\mathrm{xg}}\left(\frac{i_{\mathrm{x}}}{2}-i_{\mathrm{zx,h1}}\right) + p_{\mathrm{xl}}^{\mathrm{ripple}}, \end{cases} \tag{11.1}$$

where p_{xu} and p_{xl} are arm instantaneous power in phase x.

(3) *Phase power*: the phase-leg power is derived from eqn (11.1), and the resulting expression is provided in eqn (11.2).

$$p_{\mathrm{x}} = -V_{\mathrm{dc}}\left(\frac{i_{\mathrm{dc}}}{3}+i_{\mathrm{zx,dc}}\right) + v_{\mathrm{xg}} i_{\mathrm{x}} + p_{\mathrm{x}}^{\mathrm{ripple}}. \tag{11.2}$$

According to eqn (11.1) and (11.2), the circulating current is essential for distributing power among arms and phase legs without impacting the AC-side currents.

11.3 Hierarchical Control of MMC-based Hybrid Energy Storage System

Since weather conditions frequently change, ESSs must compensate for short-term high-power fluctuations, leading to accelerated degradation of storage components [7]. Supercapacitors (SCs), with their high power density, can complement batteries within an HESS to leverage their benefits [8].

Conventional HESSs typically use multistage converters [9, 10], which pose limitations in high-power applications. The MMC has been incorporated into battery energy storage systems (BESSs) [4, 11, 12], where batteries are directly integrated into submodules (SMs). In some systems, both batteries and SCs are combined through MMC to form an HESS [13], with batteries connecting to the DC link while SCs distributed in the SMs. However, managing such systems is challenging, requiring additional equalizer circuits to balance the state of charge (SOC), as discussed in [14], which increases system complexity.

To simplify control strategies and energy management, a novel configuration, illustrated in Figure 11.3, distributes batteries and SCs into different arms [15, 16]. This setup integrates multiple energy resources without

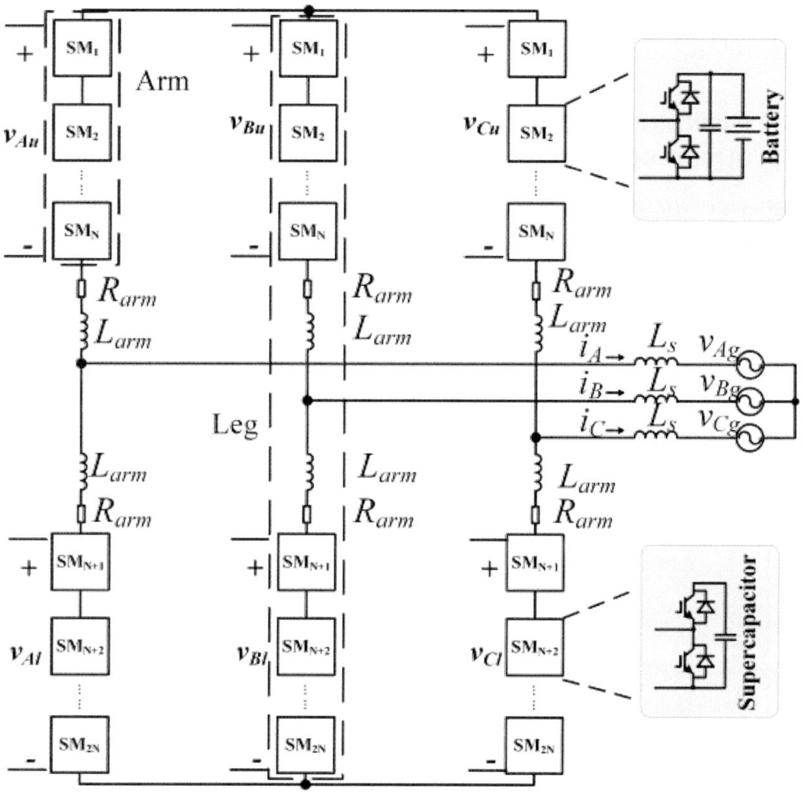

Figure 11.3 Configuration of MMC-based HESS.

requiring additional semiconductor devices or passive components, providing greater flexibility than the HESS proposed in [13]. In this MMC-HESS, decoupled power control assigns long-term and short-term power fluctuations to batteries and SCs. Furthermore, the system enables decoupling of power, which is not achievable in HESSs based on other topologies.

The MMC-HESS control system includes AC-side power control, decoupled power control among arms, and energy management of the storage components. As discussed in Section 11.2, decoupled power control focuses on regulating the internal differential current without affecting the external power, allowing the conventional power control strategy to be applied to the proposed MMC-HESS. Since this control method is well known, it is not covered in detail in this chapter. Key energy management objectives for ensuring the smooth operation of MMC-based HESS are supercapacitor (SC) voltage regulation and battery state of charge (SOC) balancing. The SM duty cycle combines outputs from all control modules to achieve these goals.

11.3.1 Power decoupling control of batteries and supercapacitors

The arm power is analyzed in eqn (11.1). By transforming the essential variables from the *abc* reference frame to the *dq* synchronous reference frame and substituting them into eqn (11.1), the power of arms is described by eqn (11.3) and (11.4), respectively.

$$\begin{cases} P_u = -\frac{V_{dc}}{2} i_{dc} + V_g^d \underbrace{\left(\frac{I_g^d}{2} + I_z^d \right)}_{I_u^d} = \frac{P_g}{2} - \frac{P_{dc}}{2} + V_g^d I_z^d, \\ P_l = -\frac{V_{dc}}{2} i_{dc} + V_g^d \underbrace{\left(\frac{I_g^d}{2} - I_z^d \right)}_{I_l^d} = \frac{P_g}{2} - \frac{P_{dc}}{2} - V_g^d I_z^d. \end{cases} \tag{11.3}$$

$$\begin{cases} Q_{\mathrm{u}} = -V_{\mathrm{g}}^{\mathrm{d}} I_{\mathrm{u}}^{\mathrm{q}} = -V_{\mathrm{g}}^{\mathrm{d}} \left(\frac{I_{\mathrm{g}}^{\mathrm{q}}}{2} + I_{\mathrm{z}}^{\mathrm{q}} \right) . \\ Q_{\mathrm{l}} = -V_{\mathrm{g}}^{\mathrm{d}} I_{\mathrm{l}}^{\mathrm{q}} = -V_{\mathrm{g}}^{\mathrm{d}} \left(\frac{I_{\mathrm{g}}^{\mathrm{q}}}{2} - I_{\mathrm{z}}^{\mathrm{q}} \right) , \end{cases} \tag{11.4}$$

where d and q represent variables under a synchronous reference frame; P_g denotes the AC-side power; while P_{dc} denotes the DC-grid power.

According to eqn (11.3) and (11.4), the internal differential current is responsible for distributing power among the different arms. Typically, due to the energy capacity, batteries are tasked with compensating for long-term power fluctuations. The SCs are distributed in the lower arm to handle short-term fluctuations. Hence, the power fluctuations compensated by the batteries and SCs are linked to the power difference between the upper and lower arms, as described by the corresponding equations.

$$\begin{cases} P_{\mathrm{u}} = (P_{\mathrm{g}} - P_{\mathrm{dc}}) G_{\mathrm{lpf}}(s) = \left(V_{\mathrm{g}}^{\mathrm{d}} I_{\mathrm{g}}^{\mathrm{d}} - V_{\mathrm{dc}} i_{\mathrm{dc}} \right) G_{\mathrm{lpf}}(s) , \\ P_{\mathrm{l}} = (P_{\mathrm{g}} - P_{\mathrm{dc}}) G_{\mathrm{hpf}}(s) = \left(V_{\mathrm{g}}^{\mathrm{d}} I_{\mathrm{g}}^{\mathrm{d}} - V_{\mathrm{dc}} i_{\mathrm{dc}} \right) G_{\mathrm{hpf}}(s) , \end{cases} \tag{11.5}$$

where $G_{\mathrm{lpf}}(\mathrm{s})$ and $G_{\mathrm{hpf}}(\mathrm{s})$ are transfer functions of filters, which are used to separate power fluctuations based on different timescales. The cutoff frequencies of these filters can be designed following the guidelines in reference [9], and it is determined by the capacities of the batteries and supercapacitors (SCs). Batteries are tasked with compensating for long-term fluctuations, so the low-frequency gain of the LPF should be 1. Conversely, SCs handle short-term fluctuations, and the high-frequency gain of $G_{\text{hpf}}(\text{s})$ is 1.

When combining eqn (11.3) and (11.5), the d-axis reference of circulating current is derived as follows:

$$I_{\text{z}}^{\text{d}} = \left(I_{\text{g}}^{\text{d}} - \frac{V_{\text{dc}}}{V_{\text{g}}^{\text{d}}} i_{\text{dc}} \right) \frac{G_{\text{lpf}}(s) - G_{\text{hpf}}(s)}{2}. \tag{11.6}$$

The SCs compensate for reactive power to reduce power losses induced by the internal resistance of batteries and extend their lifespan. As a result, the reactive power is entirely provided by SCs, ensuring that the batteries are primarily used for active power compensation, thereby optimizing the overall system performance and longevity.

$$\begin{cases} Q_{\text{u}} = 0\ , \\ Q_{\text{l}} = Q_{\text{g}} = -V_{\text{g}}^{\text{d}} I_{\text{g}}^{\text{q}}\ . \end{cases} \tag{11.7}$$

Similarly, combining eqn (11.4) and (11.7), the q-axis reference of circulating current is expressed as follows:

$$I_{\text{z}}^{\text{q}} = -\frac{I_{\text{g}}^{\text{q}}}{2}. \tag{11.8}$$

Finally, the I_z^d and I_z^q are transformed back to the *abc* reference frame. The process for synthesizing the circulating current references is illustrated in Figure 11.4.

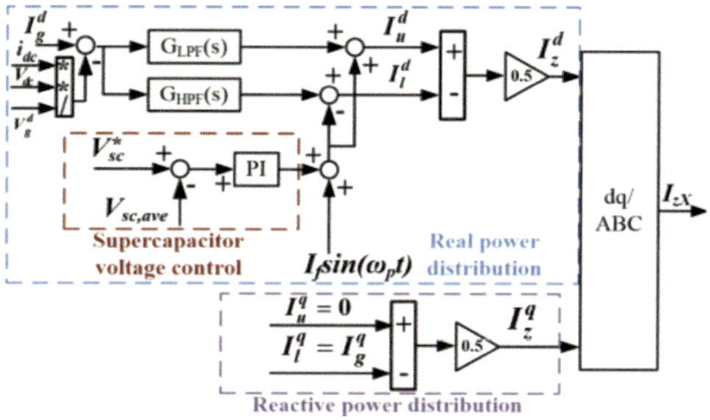

Figure 11.4 MMC-HESS power control reference.

11.3.2 SOC balancing control of BESS

The SOC balancing is crucial for maximizing the storage capacity of batteries. Common SOC estimation methods are outlined in [17]. In an MMC-HESS, it is important to balance the SOC not only across different phase legs but also within the submodules (SMs) of each arm. A widely used SOC regulation method has been proposed in [5] for BESSs based on a cascaded multilevel converter (CHB). When connecting batteries and SCs into different arms, they are similar to the CHB structure, the existing SOC balancing strategies can still be applied to MMC-HESS systems [4, 5]. This strategy is briefly summarized in this section.

To facilitate the rest analysis, some variables are predefined as follows:

$$\begin{cases} \mathrm{SOC_x} = \frac{1}{N}\sum\limits_{i=1}^{N} \mathrm{SOC_{xi}} \\ \mathrm{SOC_{ave}} = \frac{1}{N}\left(\mathrm{SOC_a} + \mathrm{SOC_b} + \mathrm{SOC_c}\right) \end{cases}, \tag{11.9}$$

where $\mathrm{SOC_{xi}}$ represents the SOC of the batteries connected to the ith SM; $\mathrm{SOC_x}$ is the average SOC of the batteries in phase x ($x = a, b,$ and c); $\mathrm{SOC_{ave}}$ denotes the average SOC of all batteries.

To regulate the SOCs of SMs, each SM controls the charging and discharging power of its associated battery pack by adjusting its terminal voltage individually. This adjustment is implemented as shown in Figure 11.5. The SM terminal voltage is expressed as

$$v_{\mathrm{SMi}} = \frac{V_{\mathrm{bat,xi}}}{2}\left(1 + m\sin(\omega t) + d_{\mathrm{a,xi}}\right), \tag{11.10}$$

where $V_{\mathrm{bat,xi}}$ is the battery voltage of the ith SM; $d_{\mathrm{a,xi}}$ is the add-on duty ratio of adjusting SM terminal voltage.

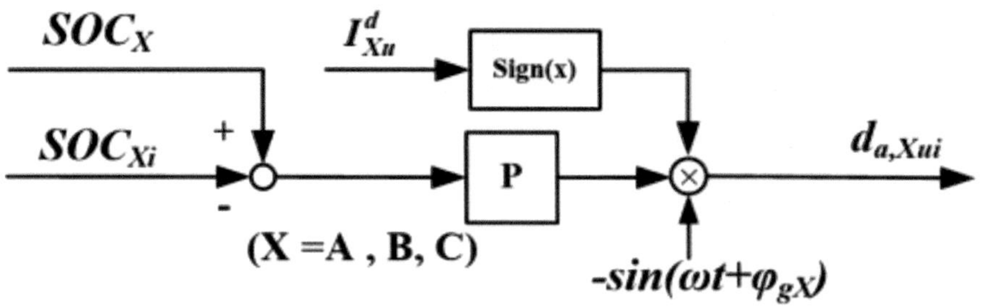

Figure 11.5 Block diagram of SOC balancing within each arm.

To balance the SOCs among three phases, the zero-sequence voltage is injected to redistribute the power. The voltages and the associated power of the upper arms are described by eqn (11.11).

$$\begin{bmatrix} V_{\mathrm{Au,ac}} \\ V_{\mathrm{Bu,ac}} \\ V_{\mathrm{Cu,ac}} \end{bmatrix} = -\begin{bmatrix} V_{\mathrm{Ag}} \\ V_{\mathrm{Bg}} \\ V_{\mathrm{Cg}} \end{bmatrix} + \begin{bmatrix} v_0 \\ v_0 \\ v_0 \end{bmatrix} \begin{bmatrix} P_{\mathrm{A0}} \\ P_{\mathrm{B0}} \\ P_{\mathrm{C0}} \end{bmatrix} = V_0 I \begin{bmatrix} \cos(\theta_0 - \theta_{\mathrm{I}}) \\ \cos(\theta_0 - \theta_{\mathrm{I}} + \frac{2\pi}{3}) \\ \cos(\theta_0 - \theta_{\mathrm{I}} - \frac{2\pi}{3}) \end{bmatrix}, \tag{11.11}$$

where θ_{I} is the phase angle of AC-side current, θ_0 is the phase angle of zero-sequence voltage.

11.3.3 Supercapacitor voltage control

The nominal voltage of the SCs must be kept within a secure range to ensure reliability while providing sufficient capacity for compensating high-frequency power fluctuations. A lower limit is set for the SC voltage to prevent overmodulation as described by eqn (11.12).

$$N \times V_{\mathrm{SC,ave}}^{\mathrm{Low}} \geq 2V_{\mathrm{xg,peak}}, \tag{11.12}$$

where $V_{\mathrm{xg,peak}}$ denotes the amplitude of grid voltage; $V_{\mathrm{SC,ave}}^{\mathrm{Low}}$ represents the lower limit of SC voltage. The nominal voltage must ensure sufficient energy to compensate the transient fluctuations. After compensating for transient power fluctuations, all SCs must be restored to the nominal voltages, as illustrated in Figure 11.4. The SC capacity should be determined by system requirements and the required energy capacity, as explained in [18].

In addition to maintaining the SC voltage, it is essential to balance the voltages of the SCs to prevent abnormal operating conditions, similar to the SOC balancing for batteries, as shown in Figure 11.6(a). The block diagram of SC voltage control has been shown in Fig. 11.6(b). This method ensures that SC voltages are evenly distributed across the SMs, preventing voltage imbalances that could negatively affect the characteristics of the MMC-HESS system.

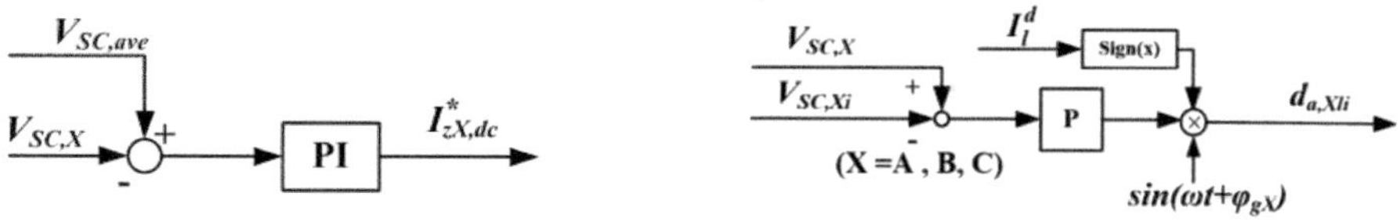

Figure 11.6 Block diagram of SC voltage control: (a) among phases, and (b) in an arm.

11.3.4 Simulation and Experimental Verifications

The configuration of MMC-HESS simulation is shown in Figure 11.3. The main parameters are provided in Table 11.1. The LPF cutoff frequency is set to 0.2 Hz. Figure 11.7 shows the system's transient response when switching from active power compensation to reactive power compensation. Figure 11.8 shows a scaled-down experimental prototype with key parameters listed in Table 11.1.

Figure 11.9 displays the experiment results of transitions from active power compensation to reactive power compensation. These results also validate the SC voltage restoration control, demonstrating that SC voltages can rapidly return to their nominal values following the transient process. This capability underscores the effectiveness of control mechanism within the MMC-HESS, ensuring stability and reliable operation after dynamic shifts in power compensation.

Table 11.1 Parameters of MMC-HESS

	Simulation	*Experiment*
Configuration	Three phases	Single phase
Grid line-line RMS voltage	10 kV	40 kV
Battery unit	332 V, 5 A· h	48 V, 7 A· h
SM per arm	64	3
Arm inductance, L_{arm}	2 mH	5 mH
AC inductance, L_s	1 mH	5 mH
DC capacitor, *C*	470 μF	470 μF
Supercapacitor	30,000 μF	4700 μF
DC load resistor, R_{load}	35 Ω	200 Ω
PWM carrier frequency	200 Hz	2000 Hz

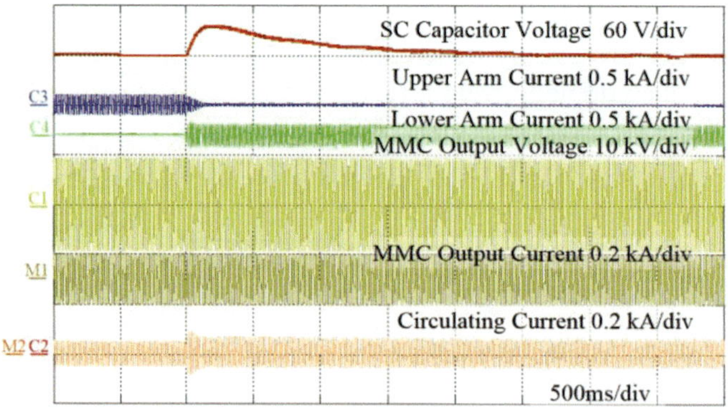

Figure 11.7 Simulation results of dynamic response (active power to reactive power).

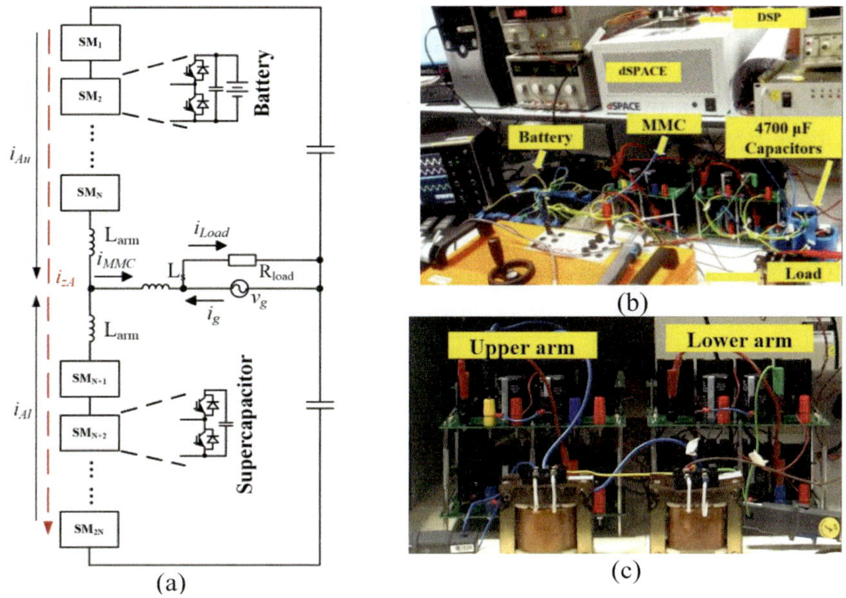

Figure 11.8 Schematic and photo of experimental prototype.

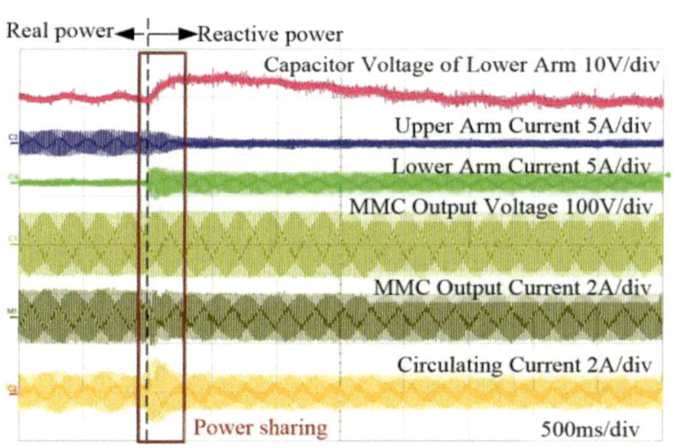

Figure 11.9 Experimental results of dynamic response (active power to reactive power).

11.4 Control Strategy and Optimal Design of DC–DC Solid State Transformer

The solid-state transformer (SST) attracted attention for its potential to optimize the power density and efficiency as comparing to traditional power transformers [19, 20]. This is particularly advantageous for wind farms where dc-dc converters are employed. As shown in Fig. 11.10, the IM2DC generates arbitrary voltage for ac-link transformer, providing more control flexibilities over the SST while mitigating harmful *dv/dt* stress [21].

To produce a multilevel voltage, the phase-shift (PS) modulation methods have been explored in [22]. The trapezoidal current modulation induce a lower switching frequency [23]. This method simultaneously inserts or bypasses multiple submodules (SMs) increasing *dv/dt* stress of transformer.

One of the key challenges in using the MMC for SST applications under medium-or high-frequency operations is managing the balancing of capacitor voltages. Research has explored various strategies to address this:

(1) *Alternating Balancing Algorithm:* This method involves alternately rotating of gating signals, which rotates the switching sequence by one step in each period, as discussed in [22]. A similar approach using

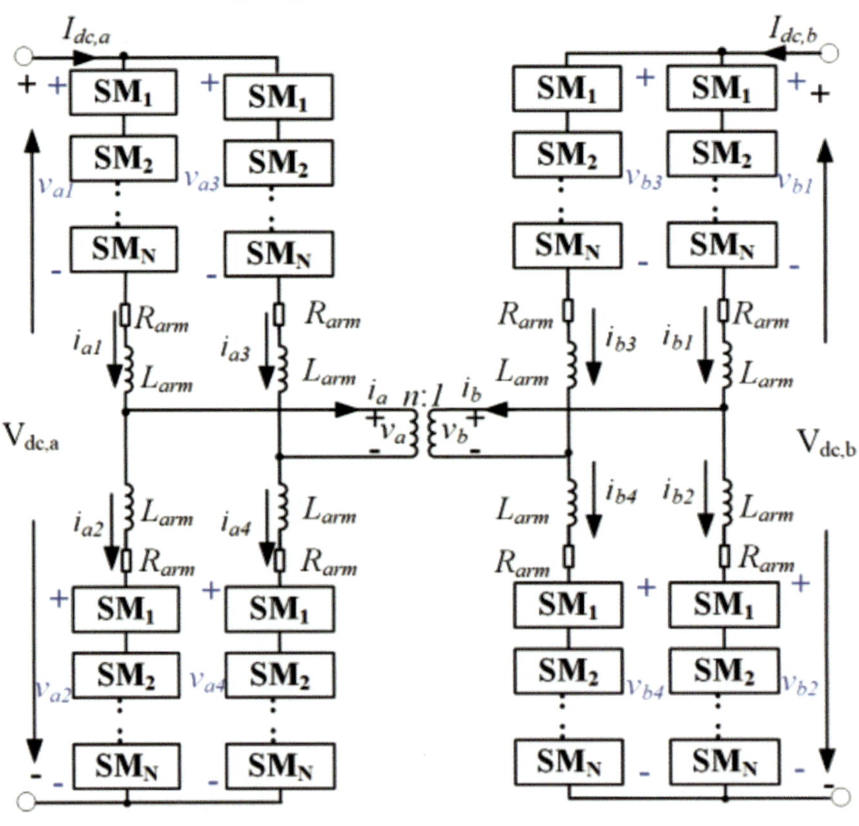

Figure 11.10 Circuit diagram of the IM2DC.

trapezoidal current modulation was examined in [23], where the limitations, including low-frequency capacitor voltage ripple and instability, were analyzed.

(2) *Sorting Algorithm with PS Modulation:* Used in [24], this approach balances the capacitor voltages by leveraging both the balancing methods and PS modulation.

(3) *Dynamic Switching Sequence:* Proposed in [25], this voltage-balancing algorithm uses a dynamic sequence depending on deviations of capacitor voltages from one switching period to the next, leading to an increased computational burden.

The specific charging/discharging characteristics of capacitor have not been comprehensively analyzed for various operating conditions. This section aims to analyze the operating characteristics of capacitor based on various voltage-balancing methods, incorporating both PS and NLC modulation methods. This analysis could offer insights into optimizing the MMC for more stable and efficient medium-or high-frequency operations in SST applications. The more detailed analysis are presented in references [26, 27].

11.4.1 Operational Principles of IM2DC

The PS and NLC modulations are widely utilized for the conventional MMC to produce the multilevel voltages by changing the number of inserted SMs (N_{insert}) according to the voltage reference. Subsequently the switching states are determined by the sorting algorithms.

For the IM2DC of Fig. 11.10, the arm voltages and currents are derived from ac-link and dc-side voltages and currents, which are described by equation (11.13).

$$\begin{cases} v_{a1} = v_{a4} = \frac{V_{dc,a}}{2} - \frac{v_a}{2} \\ v_{a2} = v_{a3} = \frac{V_{dc,a}}{2} + \frac{v_a}{2} \end{cases} \quad \begin{cases} i_{a1} = i_{a4} = \frac{I_{dc,a}}{2} + \frac{i_a}{2} \\ i_{a2} = i_{a3} = \frac{I_{dc,a}}{2} - \frac{i_a}{2} \end{cases}. \tag{11.13}$$

When reflecting secondary voltages to primary side, they are expressed as

$$v_a(\varphi) = \begin{cases} \frac{2V_{dc,a}}{\theta_R}\varphi - V_{dc,a}, & 0 < \varphi < \theta_R \\ V_{dc,a}, & \theta_R < \varphi < \pi \\ -v_a(\varphi - \pi), & \pi < \varphi < 2\pi \end{cases}. \tag{11.14}$$

$$v_b'(\varphi) = nv_b(\varphi) = \frac{nV_{dc,b}}{V_{dc,a}} v_a(\varphi - \Phi), \tag{11.15}$$

where $v_b'(\varphi)$ is the reflected voltage.

Its operational behaviour is similar to DAB converter. Power transfer is regulated by phase-shift angle (Φ) between their respective voltages. The phase-shift angle must remain within $|\Phi| \leq \pi/2$. The ac-link current (i_a) is described by equation (11.16) for $\Phi>0$, and by equation (11.17) for $\Phi<0$. Additionally, θ_R varies between 0 to $|\Phi|$. Given the rated power, the dc-side current can be derived from equation (11.18), and the arm current is determined by equations (11.16) and (11.18). The theoretical waveforms of currents and voltages are derived from analysis as depicted in Figures 11.11(a) and (b).

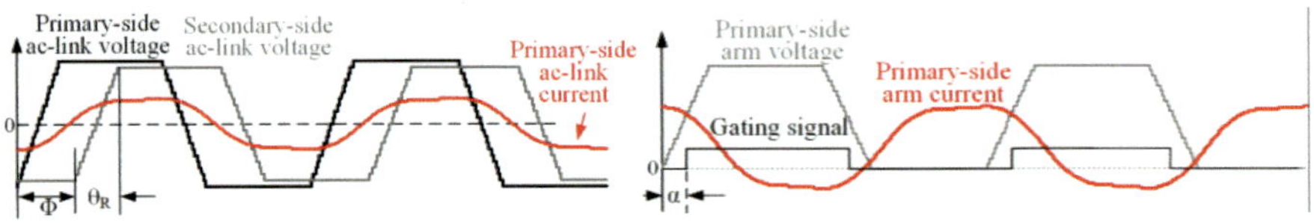

(a) (b)

Figure 11.11 Theoretical waveforms of voltages and currents.

$$
i_{\mathrm{a}}(\varphi) =
\begin{cases}
\frac{1}{\omega L}\left[V_{\mathrm{dc,a}}\left(\frac{\varphi^2}{\theta_{\mathrm{R}}} - \varphi + \frac{\theta_{\mathrm{R}}}{2} - \frac{\pi}{2}\right) + nV_{\mathrm{dc,b}}\left(\varphi - \Phi - \frac{\theta_{\mathrm{R}}}{2} + \frac{\pi}{2}\right)\right], & 0 < \varphi < \theta_{\mathrm{R}}, \\
\frac{1}{\omega L}\left[V_{\mathrm{dc,a}}\left(\varphi - \frac{\theta_{\mathrm{R}}}{2} - \frac{\pi}{2}\right) + nV_{\mathrm{dc,b}}\left(\varphi - \Phi - \frac{\theta_{\mathrm{R}}}{2} + \frac{\pi}{2}\right)\right], & \theta_{\mathrm{R}} < \varphi < \Phi, \\
\frac{1}{\omega L}\left[V_{\mathrm{dc,a}}\left(\varphi - \frac{\theta_{\mathrm{R}}}{2} - \frac{\pi}{2}\right) + nV_{\mathrm{dc,b}}\left(-\frac{\varphi^2}{\theta_{\mathrm{R}}} + \frac{2\Phi}{\theta_{\mathrm{R}}}\varphi + \varphi - \frac{\Phi^2}{\theta_{\mathrm{R}}} - \right.\right. & \\
\left.\left. \qquad \Phi - \frac{\theta_{\mathrm{R}}}{2} + \frac{\pi}{2}\right)\right], & \Phi < \varphi < \Phi + \theta_{\mathrm{R}}, \\
\frac{1}{\omega L}\left[V_{\mathrm{dc,a}}\left(\varphi - \frac{\theta_{\mathrm{R}}}{2} - \frac{\pi}{2}\right) + nV_{\mathrm{dc,b}}\left(-\varphi + \Phi + \frac{\theta_{\mathrm{R}}}{2} + \frac{\pi}{2}\right)\right], & \Phi + \theta_{\mathrm{R}} < \varphi < \pi, \\
-i_{\mathrm{a}}(\varphi - \pi), & \pi < \varphi < 2\pi.
\end{cases}
\tag{11.16}
$$

$$
i_{\mathrm{a}}(\varphi) =
\begin{cases}
\frac{1}{\omega L}\left[V_{\mathrm{dc,a}}\left(\frac{\varphi^2}{\theta_{\mathrm{R}}} - \varphi + \frac{\theta_{\mathrm{R}}}{2} - \frac{\pi}{2}\right) + nV_{\mathrm{dc,b}}\left(\varphi - \Phi + \frac{\theta_{\mathrm{R}}}{2} + \frac{\pi}{2}\right)\right], & 0 < \varphi < \theta_{\mathrm{R}}, \\
\frac{1}{\omega L}\left[V_{\mathrm{dc,a}}\left(\varphi - \frac{\theta_{\mathrm{R}}}{2} - \frac{\pi}{2}\right) + nV_{\mathrm{dc,b}}\left(-\varphi + \Phi + \frac{\theta_{\mathrm{R}}}{2} + \frac{\pi}{2}\right)\right], & \theta_{\mathrm{R}} < \varphi < \pi + \Phi, \\
\frac{1}{\omega L}\left[V_{\mathrm{dc,a}}\left(\varphi - \frac{\theta_{\mathrm{R}}}{2} - \frac{\pi}{2}\right) + nV_{\mathrm{dc,b}}\left(\frac{\varphi^2}{\theta_{\mathrm{R}}} - \frac{2(\pi+\Phi)}{\theta_{\mathrm{R}}}\varphi \right.\right. & \\
\left.\left. \qquad -\varphi - \frac{(\pi+\Phi)^2}{\theta_{\mathrm{R}}} + \Phi + \frac{\theta_{\mathrm{R}}}{2} + \frac{\pi}{2}\right)\right], & \pi + \Phi < \varphi < \pi + \Phi + \theta_{\mathrm{R}}, \\
\frac{1}{\omega L}\left[V_{\mathrm{dc,a}}\left(\varphi - \frac{\theta_{\mathrm{R}}}{2} - \frac{\pi}{2}\right) + nV_{\mathrm{dc,b}}\left(\varphi + \Phi - \Phi - \frac{\theta_{\mathrm{R}}}{2} - \frac{3\pi}{2}\right)\right], & \pi + \Phi + \theta_{\mathrm{R}} < \varphi < \pi, \\
-i_{\mathrm{a}}(\varphi - \pi), & \pi < \varphi < 2\pi.
\end{cases}
\tag{11.17}
$$

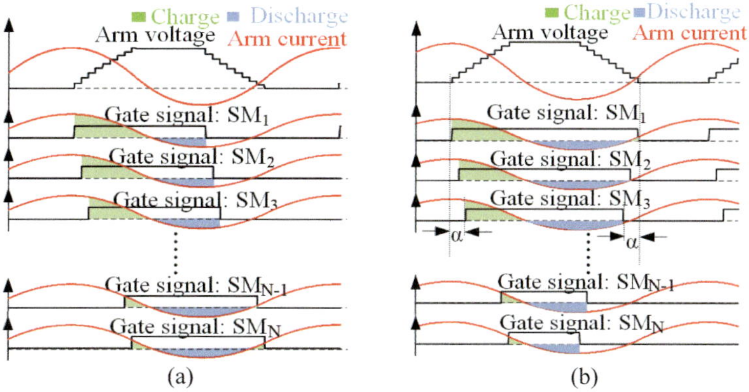

Figure 11.12 Currents, voltages, and gating signals based on (a) the PS and (b) the NLC modulation methods (Φ>0).

$$I_{\mathrm{dc,a}} = \frac{\int_0^{2\pi} v_{\mathrm{a}} i_{\mathrm{a}} d\varphi}{2\pi V_{\mathrm{dc,a}}} = \begin{cases} -\frac{nV_{\mathrm{dc,b}}}{\pi\omega L}\left(\frac{\theta_{\mathrm{R}}^2}{6} + \Phi^2 - \pi\Phi\right), & \Phi > 0, \\ \frac{nV_{\mathrm{dc,b}}}{\pi\omega L}\left(\frac{\theta_{\mathrm{R}}^2}{6} + \Phi^2 + \pi\Phi\right), & \Phi < 0. \end{cases}, \quad (11.18)$$

The operating behaviour of the SM capacitor depends on the modulation method used. Figures 11.12(a) and (b) illustrate currents and voltages of SMs in MMC-SST. In a period, the accumulated capacitor energy is influenced by the initial switching angle (α) and duty ratio, as depicted in figures. These parameters, the switching angle and duty ration, vary depending on the modulation technique. For the PS and NLC modulations, the switching angle of the ith SM is $\alpha_i = \frac{(i-1)\theta_R}{N}$, while the turning-on intervals are π. and (π+θ_R−2α_i, respectively. The capacitor charge can be determined using equation (11.19).

$$\begin{cases} Q = \int_{\alpha}^{\pi+\alpha} i_{arm}(\varphi), & PS \\ Q = \int_{\alpha}^{\pi+\theta_R-\alpha} i_{arm}(\varphi), & NLC \end{cases}. \quad (11.19)$$

where Q represents the accumulated charge during a single switching period. i_{arm} refers to the arm current. Consequently, the capacitor charge as a function of the switching angle α is expressed by equation (11.20) for PS modulation, and by equation (11.21) for NLC method.

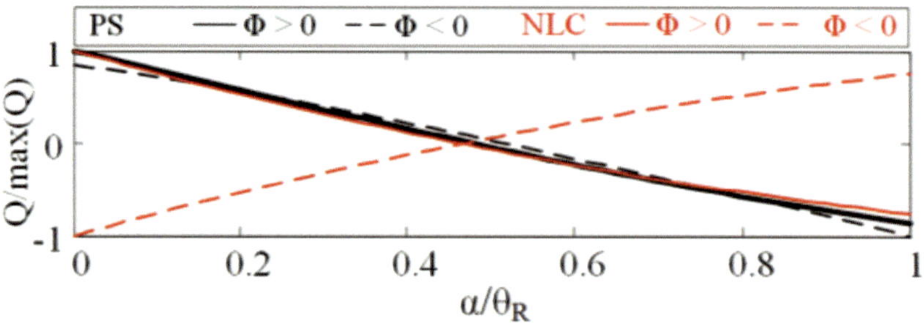

Figure 11.13 SM capacitor charge based on PS and NLC modulations ($\theta_R = \pi/5$).

$$Q(\alpha) = \begin{cases} \frac{1}{\omega^2 L}\Big[V_{\mathrm{dc,a}}\left(-\frac{\theta_{\mathrm{R}}^2}{6} + \frac{\pi\theta_{\mathrm{R}}}{4} + \frac{\alpha^3}{3\theta_{\mathrm{R}}} - \frac{\alpha^2}{2} + \frac{\theta_{\mathrm{R}}-\pi}{2}\alpha\right) \\ \quad +nV_{\mathrm{dc,b}}\left(\frac{\theta_{\mathrm{R}}^2}{12} + \frac{2\Phi-\pi}{4}\theta_{\mathrm{R}} + \frac{\alpha^2}{2} + \frac{\pi-2\Phi-\theta_{\mathrm{R}}}{2}\alpha\right)\Big] & \Phi \geq 0 \\ \frac{1}{\omega^2 L}\Big[V_{\mathrm{dc,a}}\left(-\frac{\theta_{\mathrm{R}}^2}{6} + \frac{\pi\theta_{\mathrm{R}}}{4} + \frac{\alpha^3}{3\theta_{\mathrm{R}}} - \frac{\alpha^2}{2} + \frac{\theta_{\mathrm{R}}-\pi}{2}\alpha\right) \\ \quad +nV_{\mathrm{dc,b}}\left(-\frac{\theta_{\mathrm{R}}^2}{12} - \frac{2\Phi+\pi}{4}\theta_{\mathrm{R}} - \frac{\alpha^2}{2} + \frac{\pi+2\Phi+\theta_{\mathrm{R}}}{2}\alpha\right)\Big] & \Phi < 0. \end{cases} \tag{11.20}$$

$$Q(\alpha) = \begin{cases} \frac{nV_{\mathrm{dc,b}}}{\omega^2 L}\left(-\frac{\theta_{\mathrm{R}}^3}{12\pi} + \frac{\theta_{\mathrm{R}}^2}{12} - \frac{\Phi^2\theta_{\mathrm{R}}}{2\pi} + \frac{\Phi\theta_{\mathrm{R}}}{2} + \frac{\alpha^2}{2} + \frac{6\Phi^2+\theta_{\mathrm{R}}^2}{6\pi}\alpha - \frac{2\Phi+\theta_{\mathrm{R}}}{2}\alpha\right) & \Phi \geq 0 \\ \frac{nV_{\mathrm{dc,b}}}{\omega^2 L}\left(\frac{\theta_{\mathrm{R}}^3}{12\pi} - \frac{\theta_{\mathrm{R}}^2}{12} + \frac{\Phi^2\theta_{\mathrm{R}}}{2\pi} + \frac{\Phi\theta_{\mathrm{R}}}{2} - \frac{\alpha^2}{2} - \frac{6\Phi^2+\theta_{\mathrm{R}}^2}{6\pi}\alpha - \frac{2\Phi-\theta_{\mathrm{R}}}{2}\alpha\right) & \Phi < 0. \end{cases} \tag{11.21}$$

The accumulated capacitor charge depends on the initial switching angle, as illustrated in Fig. 11.13. When using PS modulation, the SM capacitor transit from charging to discharging as the increasing of switching angles, irrespective of the power flow direction. However, for the NLC modulation, when $\Phi<0$ the charging and discharging behaviour is reversed.

11.4.2 Analytical Evaluations of Balancing Algorithms and Modulation Methods

This section provides a detailed analysis and comparison of the operating characteristic of various balancing algorithms when applied with various modulation methods. Based on this analysis, the capacitor voltage ripple is derived for various operating conditions.

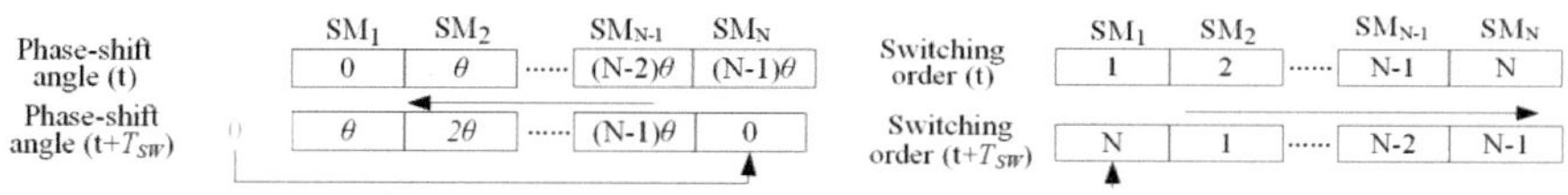

Figure 11.14 Switching sequence of the alternating balancing algorithm: (a) PS, and (b) NLC modulation methods.

In reference [22], the single-step alternating balancing method was explored, where the initial switching angle (α) of SM is incrementally rotates from 0 to $\frac{(N-1)\theta_R}{N}$. This process can similarly be implemented with NLC modulation, which rotates the switching index, as depicted in Fig. 11.14. The index of each SM changes by one step during each period, resulting in a complete charging–discharging cycle that spans *N* switching periods.

This stepwise alternation of the switching angles or the switching order helps distribute the capacitor charging and discharging more evenly over time, contributing to a more balanced voltage profile across all SM capacitors. By rotating the switching order in NLC modulation or adjusting the switching angle in PS modulation, the voltage ripple across the capacitors can be minimized, enhancing the overall stability and performance of the system.

Consequently, the total charging–discharging cycle spans *N* times the switching period T_{SW}. During half of this cycle, the capacitor charges, while in the second half, it discharges, resulting in a low-frequency ripple in capacitor voltage, with a ripple frequency of f_{SW}/N, or equivalently, a period of NT_{SW}.

During a switching period, the relationship between accumulated charge and voltage deviation is Q_k=CΔv_{ck}, where *C* is the capacitance and Δv_{ck} is the voltage deviation. The peak-to-peak low-frequency capacitor voltage ripple, $\Delta v_{c,pp}$, is given by equation (11.22) and shown in Fig. 11.15.

$$\Delta v_{c,pp} = \frac{\sum_{k=1}^{N} |Q(\alpha_k)|}{2C}. \tag{11.22}$$

When *N* increases, the capacitor voltage ripple increases linearly. This increase in ripple highlights the trade-off between adding more SMs for higher efficiency and the need for managing voltage ripple to maintain stability and performance in the system.

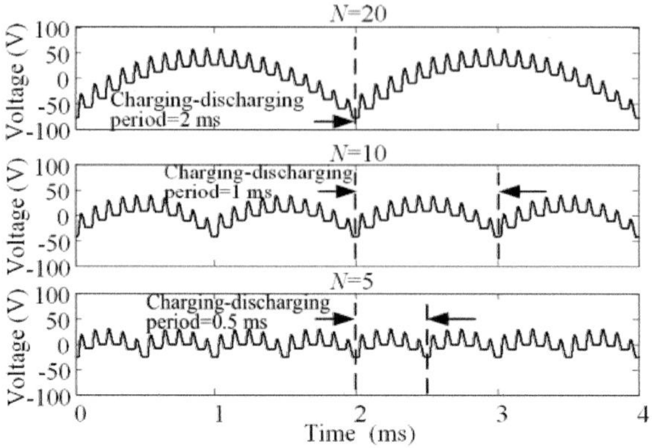

Figure 11.15 Capacitor voltage ripples based on the alternating balancing algorithm ($\theta_R = \pi/5$).

11.4.3 Low-Frequency Ripple Mitigation

According to the above analysis, the existing balancing method results low-frequency ripple, potentially affecting system performance and stability. To mitigate this ripple without inducing extra switching actions, two advanced algorithms have been proposed in [26]. These algorithms aim to reduce the ripple while maintaining efficient operation.

- *Multistep alternating algorithm:* The multistep alternating algorithm, introduced in [26] and shown in Figure 11.16, is designed to alternately charge and discharge capacitors by modifying the switching angles. In this method, the gating signals rotate by N_{rot} at the beginning of each switching period. The switching sequence depends on their index numbers. For instance, SM_1 is initially inserted first, with an index of 1. In the following period, the switching vector rotates by N_{rot} steps making SM_1 the $(N_{rot}+1)^{th}$ in the order. This process continues, and after N switching cycles, SM_1 returns to its original position.
  One possible methods of determining N_{rot} is described by equation (11.23). The N_{rot} must be guarantee balance of charging and discharging energy of all SMs over the alternating cycle, which is essential for maintaining the capacitor voltage [22].
  The switching vector rotates by N_{rot} steps in a period. After k periods, the total steps reach kN_{rot}. When the remainder of $\frac{kN_{rot}}{N}$ is zero, the switching vector return to its initial status. Only when k=N, it ensures all SMs experiencing all switching angles. By using this method to

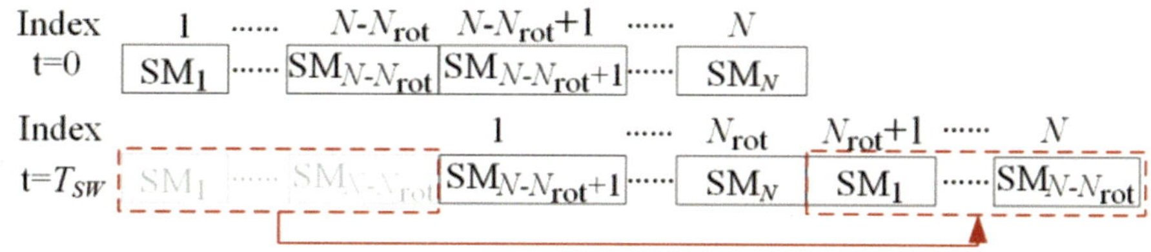

Figure 11.16 Schematic of multistep alternating balancing algorithm.

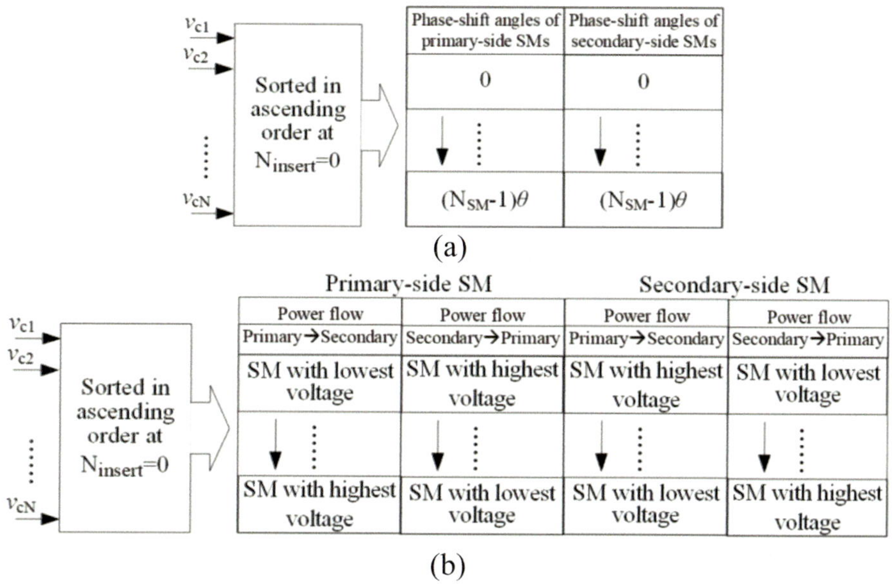

Figure 11.17 Schematic of currentless balancing method: (a) PS, and (b) NLC modulation methods.

select N_{rot}, the SM capacitor alternately charges or discharges, avoiding continuous charging or discharging. This helps to suppress the voltage ripple and improve system performance.

$$N_{rot} = \begin{cases} \frac{N-1}{2}, & N \text{ is odd} \\ \frac{N}{2}+2, & N \text{ is even and } \frac{N}{2} \text{ is odd} \\ \frac{N}{2}+1, & N \text{ is even and } \frac{N}{2} \text{ is even} \end{cases}. \tag{11.23}$$

- *Currentless sorting algorithm:* he currentless sorting algorithm, introduced in [26], offers an innovative approach to balancing SM capacitor voltages. This method sorts the SM capacitor voltages when $N_{\text{insert}} = 0$.

It then assigns smaller switching angles to SMs with lower voltages and larger switching angles to SMs with higher voltages, as shown in Figure 11.17. By doing so, the capacitors are alternately charged and discharged, effectively reducing the capacitor voltage ripple and enhancing voltage balance across the SMs.

11.4.4 Simulation and Experimental Verifications

To verify the effectiveness of the proposed analysis and advanced modulation methods, the IM2DC is modeled in the PSCAD simulation environment. Additionally, an MMC-based SST was constructed to demonstrate the effectiveness of the proposed methods. The relevant system parameters are provided in Table 11.2, and the overall system schematic is illustrated in Figure 11.18.

Figure 11.19 illustrates the gating signals and capacitor charges of MMC, comparing the single-step algorithm with the multistep algorithm. When the multistep voltage-balancing algorithm is applied, the SM capacitors experience alternating charging and discharging cycles. This alternating process helps balance the voltage across the capacitors more effectively, avoiding continuously charging and minimizing low-frequency voltage ripple, thereby improving overall system stability and performance.

Figure 11.20 displays the capacitor voltages corresponding to different balancing methods. In Figure 11.21, the single-step balancing method results in noticeable voltage ripple. However, when using the multistep alternating balancing method, this low-frequency ripple is significantly reduced. This improvement highlights the effectiveness of the multistep algorithm in minimizing voltage fluctuations, leading to better system performance and stability.

Table 11.2 Parameters of IM2DC

	Simulation	*Experiment*
Configuration	Two phases	Single phase
Rated power, P	1 MW	2.5 kW
Primary-side voltage, $V_{dc,\mathbf{a}}$	10 kV	350 V
Secondary-side voltage, $V_{dc,b}$	10 kV	350 V
Switching frequency, f_{SW}	10 kHz	20 kHz
Turn ratio, n	1:1	1:1
Primary-side number of SMs per arm, N	9	6
SM capacitance, C	50 μF	500 μF
Arm inductance, L_{arm}	50 μH	20 μH
Transformer leakage inductance, R_{load}	1100 μH	64 μH

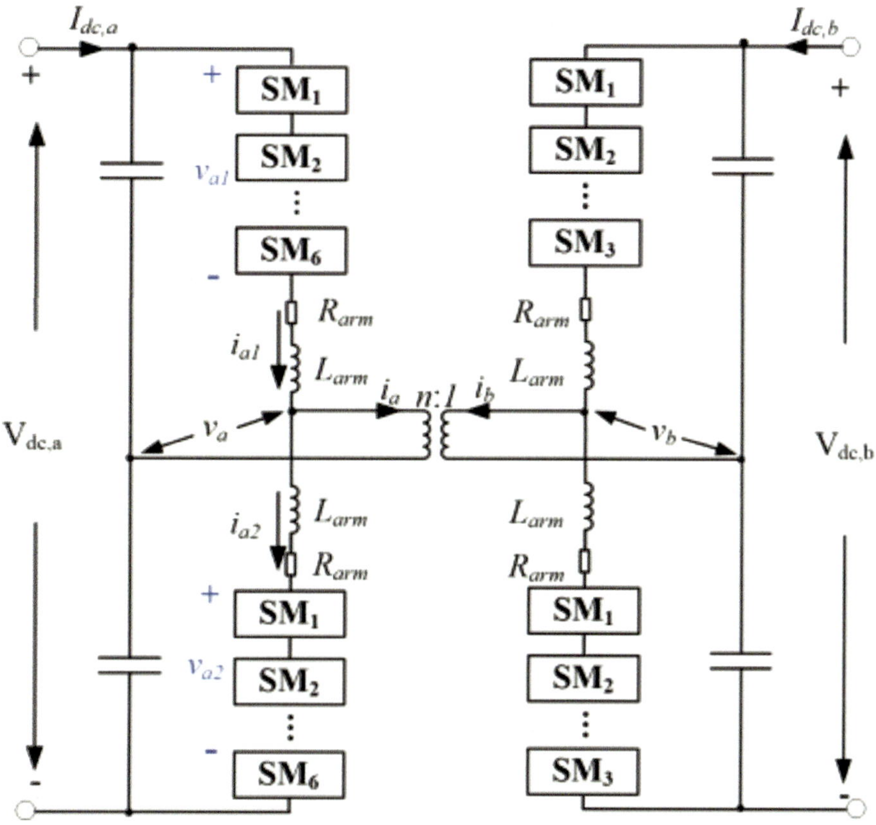

Figure 11.18 Schematic of the experiment prototype.

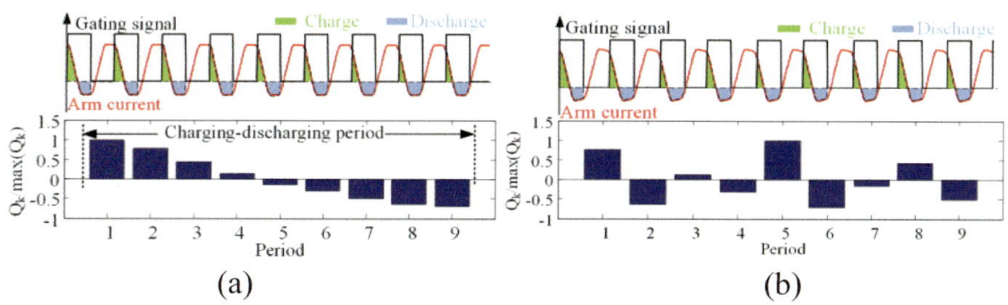

Figure 11.19 Charging and discharging: (a) single-step, and (b) multistep algorithms Φ>0.

Figure 11.21 illustrates voltages and currents, where the phase-shift angle Φ is $\pi/2$. In Figure 11.22, the arm and SM capacitor voltages are using both the single-step and multistep balancing algorithms. The single-step algorithm results in noticeable low-frequency ripple in the capacitor voltage, while the multistep algorithm effectively reduces this ripple, enhancing voltage stability across the capacitors.

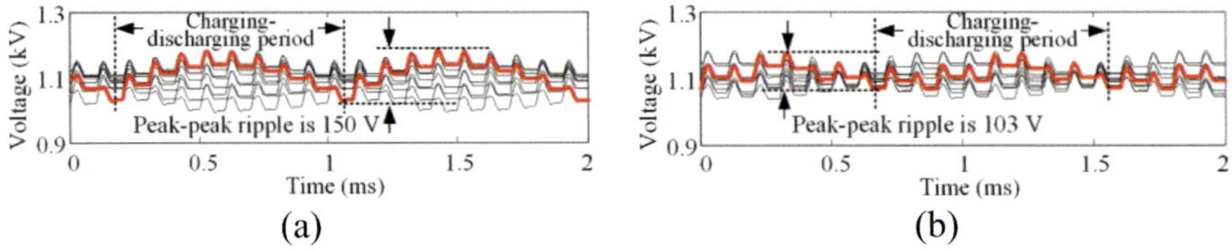

Figure 11.20 Capacitor voltages: (a) single-step and (b) multistep alternating balancing algorithms.

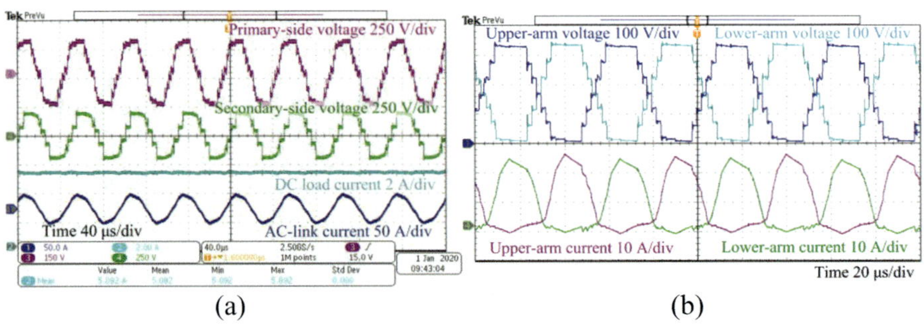

Figure 11.21 Experimental results: (a) ac-link voltages and current, and (b) arm voltages and current.

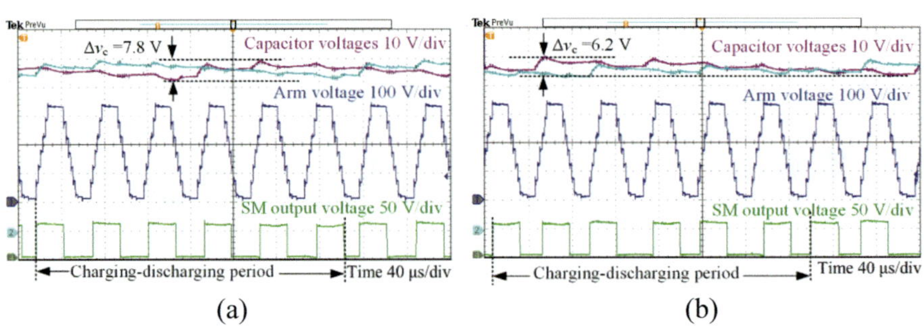

Figure 11.22 Experimental results: (a) single-step, and (b) multistep balancing algorithms.

11.5 Conclusion

This chapter introduces s novel MMC-based configuration for hybrid energy storage, with an optimized control strategy managing diverse components to enhance system performance. These optimized algorithms address low-frequency ripple in submodule capacitor voltage, improving MMC reliability and longevity, as demonstrated by detailed simulations and experimental validations. A comprehensive analysis of various voltage-balancing algorithms and modulation techniques is provides for MMCs in SSTs, proposing two improved algorithms, which are comprehensively verified.

References

[1] Lesnicar, A., & Marquardt, R. (2003, June). An innovative modular multilevel converter topology suitable for a wide power range. In 2003 IEEE Bologna Power Tech Conference Proceedings, (Vol. 3, pp. 6-pp). IEEE.

[2] Zhang, L., Zou, Y., Yu, J., Qin, J., Vittal, V., Karady, G. G., & Wang, Z. (2017). Modeling, control, and protection of modular multilevel converter-based multi-terminal HVDC systems: A review. CSEE Journal of Power and Energy Systems, 3(4), 340–352.

[3] Ilves, K., Harnefors, L., Norrga, S., & Nee, H. P. (2014). Analysis and operation of modular multilevel converters with phase-shifted carrier PWM. IEEE Transactions on Power Electronics, 30(1), 268–283.

[4] Vasiladiotis, M., & Rufer, A. (2014). Analysis and control of modular multilevel converters with integrated battery energy storage. IEEE Transactions on Power Electronics, 30(1), 163–175.

[5] Maharjan, L., Yamagishi, T., & Akagi, H. (2010). Active-power control of individual converter cells for a battery energy storage system based on a multilevel cascade PWM converter. IEEE Transactions on Power Electronics, 27(3), 1099–1107.

[6] Deng, F., & Chen, Z. (2015). Voltage-balancing method for modular multilevel converters under phase-shifted carrier-based pulsewidth modulation. IEEE Transactions on Industrial Electronics, 62(7), 4158–4169.

[7] Lam, L., & Bauer, P. (2012). Practical capacity fading model for Li-ion battery cells in electric vehicles. IEEE transactions on power electronics, 28(12), 5910–5918.

[8] Gee, A. M., Robinson, F. V., & Dunn, R. W. (2013). Analysis of battery lifetime extension in a small-scale wind-energy system using supercapacitors. IEEE transactions on energy conversion, 28(1), 24–33.

[9] Tummuru, N . R., Mishra, M. K., & Srinivas, S. (2015). Dynamic energy management of renewable grid integrated hybrid energy storage system. IEEE Transactions on Industrial Electronics, 62(12), 7728–7737.

[10] Choi, M. E., Kim, S. W., & Seo, S. W. (2011). Energy management optimization in a battery/supercapacitor hybrid energy storage system. IEEE Transactions on Smart grid, 3(1), 463–472.

[11] L. Zhang, F. Gao, N. Li, Q. Zhang and C. Wang, "Interlinking modular multilevel converter of hybrid AC-DC distribution system with integrated battery energy storage," *2015 IEEE Energy Conversion Congress and Exposition (ECCE)*, Montreal, QC, Canada, 2015, pp. 70–77.

[12] F. Gao, L. Zhang, Q. Zhou, M. Chen, T. Xu and S. Hu, "State-of-charge balancing control strategy of battery energy storage system based on modular multilevel converter," *2014 IEEE Energy Conversion Congress and Exposition (ECCE)*, Pittsburgh, PA, USA, 2014, pp. 2567–2574.

[13] F. Guo and R. Sharma, "A modular multilevel converter with half-bridge submodules for hybrid energy storage systems integrating battery and Ultra Capacitor," *2015 IEEE Applied Power Electronics Conference and Exposition (APEC)*, Charlotte, NC, USA, 2015, pp. 3025–3030.

[14] Y. Izadi and R. Beiranvand, "A Comprehensive Review of Battery and Supercapacitor Cells Voltage-Equalizer Circuits," *IEEE Trans. Power Electron.,* vol. 38, no. 12, pp. 15671–15692, Dec. 2023.

[15] L. Zhang, Yi Tang, Shunfeng Yang and F. Gao, "A modular multilevel converter-based grid-tied battery-supercapacitor hybrid energy storage system with decoupled power control," *2016 IEEE 8th International Power Electronics and Motion Control Conference (IPEMC-ECCE Asia)*, Hefei, China, 2016, pp. 2964–2971.

[16] L. Zhang, Y. Tang, S. Yang and F. Gao, "Decoupled Power Control for a Modular-Multilevel-Converter-Based Hybrid AC–DC Grid Integrated With Hybrid Energy Storage," *IEEE Trans. Ind. Electron.,* vol. 66, no. 4, pp. 2926–2934, Apr. 2019.

[17] S. Piller, M. Perrin and A. Jossen, "Methods for state-of-charge determination and their applications," *J. Power Sources,* vol. 96, no. 1, pp. 113–120, 2001.

[18] D. B. W. Abeywardana, B. Hredzak, V. G. Agelidis and G. D. Demetriades, "Supercapacitor sizing method for energy-controlled filter-based

hybrid energy storage systems," *IEEE Trans. Power Electron.,* vol. 32, no. 2, pp. 1626–1637, Feb. 2017.

[19] A. Q. Huang, "Medium-voltage solid-state transformer: Technology for a smarter and resilient grid," *IEEE Ind. Electron. Mag.,* vol. 10, no. 3, pp. 29–42, Sep. 2016.

[20] J. E. Huber and J. W. Kolar, "Solid-state transformers: On the origins and evolution of key concepts," *IEEE Ind. Electron. Mag.,* vol. 10, no. 3, pp. 19–28, Sep. 2016.

[21] S. Kenzelmann,, A. Rufer, D. Dujic, F. Canales and Y. R. d. Novaes, "Isolated DC/DC structure based on modular multilevel converter," *IEEE Trans. Power Electron.,* vol. 30, no. 1, pp. 89–98, Jan. 2015.

[22] B. Zhao, Q. Song, J. Li, Y. Wang and W. Liu, "High-frequency-link modulation methodology of DC-DC transformer based on modular multilevel converter for HVDC application: Comprehensive analysis and experimental verification," *IEEE Trans. Power Electron.,* vol. 32, no. 5, pp. 3413–3424, May 2017.

[23] Y. Qiao, X. Zhang, X. Xiang, X. Yang and T. C. Green, "Trapezoidal current modulation for bidirectional high step-ratio modular DC-DC converters," *IEEE Trans. Power Electron.,* vol. 35, no. 4, pp. 3402–3415, Apr. 2020.

[24] C. Sun, J. Zhang, X. Cai and G. Shi, "Voltage balancing control of isolated modular multilevel DC-DC converter for use in DC grids with zero voltage switching," *IET Power Electron.,* vol. 9, no. 2, pp. 270–280, 2016.

[25] S. Shao, M. Jiang, J. Zhang and X. Wu, "A capacitor voltage balancing method for a modular multilevel DC transformer for DC distribution system," *IEEE Trans. Power Electron.,* vol. 33, no. 4, pp. 3002–3011, Apr. 2018.

[26] L. Zhang, J. Qin, Y. Zou, Q. Duan and W. Sheng, "Analysis of Capacitor Charging Characteristics and Low-Frequency Ripple Mitigation by Two New Voltage-Balancing Strategies for MMC-Based Solid-State Transformers," in *IEEE Transactions on Power Electronics*, vol. 36, no. 1, pp. 1004–1017, Jan. 2021.

[27] L. Zhang, J. Qin, Y. Qiu, K. Wang, Z. Li and Y. Tang, "Analysis and Evaluation of Modular Multilevel Converters for DC–DC Solid-State Transformer Based on Different Operating Conditions," in *IEEE Transactions on Power Electronics*, vol. 38, no. 12, pp. 14898–14917, Dec. 2023.

12

Aspects of MLI Integration in EV Powertrain and Performance Evaluation Using Drive Cycle Analysis

Lairenlakpam Robindro

CSIR-Indian Institute of Petroleum, Automotive Fuels and Lubricants Application Division (AFLAD), Dehradun, India
Email: robindro@iip.res.in; robin.laren@gmail.com

Abstract

A modern EV is equipped with a multilevel inverter (MLI). This chapter discusses the practical aspects and various possible operating conditions of an EV where the MLI is subjected to different operating points and conditions. It discusses the main selection, and integration of electrical subsystems, such as MLI-controller, Li-ion battery, and BMS in an EV. It also discusses the laboratory and real-world operating conditions and performance checks related to the motor and MLI. The MLI used in an EV is always subjected to dynamic operating conditions in the real world. The characteristics of MLI voltage and current versus speed are also plotted and discussed. The wide range of vehicle speed variation puts the MLI and motor in stressful conditions. Therefore, the MLI and motor need to be properly designed and evaluated for the wide range of operating points and vehicle speed. They must withstand the speed and torque variation to overcome real-world driving conditions. Further, the chapter discusses the selection of drivetrain topology–system integration, how it can affect the inverter performance, and aspects of performance checks using driving cycles. Furthermore, the chapter also discusses a case study that explains the inverter operating challenges such as frequent changes in EV speed, torque, battery current, input voltage,

DOI: 10.1201/9781003669845-12

etc. during the different drive modes of standard driving cycles and real-world driving.

Keywords: Electric vehicle (EV), AC motor, MLI, driving cycles, drive modes, laboratory, real-world, performance, speed and torque variation, drive cycle analysis

12.1 Introduction

Global warming and air pollution have had serious impacts on the production and living conditions of the human population across the world. To address the issues of global warming and air quality, COP21 (also known as the Paris Climate Conference), COP26, and COP28 have been organized, and they have targeted to achieve a legally binding and universal agreement on climate, to keep global warming below 2°C [1, 2, 3]. Governments around the world have initiated various electric mobility programs [4, 5, 6] for faster manufacturing and adoption of hybrid electric vehicles (HEVs) and electric vehicles (EVs) for low-carbon footprints.

There are also concerns about health issues and energy resource constraints due to increasing vehicle emissions and fuel consumption. Vehicle tailpipe emissions (CO, NOx, PM, HC, and others) and GHGs can have adverse impacts on health (such as premature mortality and morbidity from cardiopulmonary diseases), lower crop yields, environmental damage, and contribution to global warming. These problems, in turn, can set back the economy. Premature mortality and morbidity can reduce productivity. So, money and resources must be diverted to clean up the environment [7]. A study also reported several thousand premature deaths and a loss of more than 18 million healthy life years because of the ozone and outdoor PM2.5 pollution [8].

Due to zero fuel consumption and low carbon footprints, electric vehicles (EVs) are fast becoming the choice of vehicle owners and the ultimate goal for vehicle manufacturers. At the same time, HEVs are another alternative option in the process of replacing ICE vehicles with EVs [9]. An electric motor generates wheel torque in an EV, and that motor can offer more than 85% efficiency [10], whereas an internal combustion engine (ICE) can turn about 12%–20% of the total fuel consumption into useful wheel torque [11] and have a maximum efficiency of 40% [12].

Every modern EV has an electric powertrain system consisting of subsystems like the battery pack, power converters, electric motor, motor inverter-controller (MLI), and mechanical drivetrain. The modern EVs are equipped with AC electric motors for traction. The MLI is used to control and operation of those EV motors. It is essential to evaluate the powertrain performance with all subsystems working together, especially the battery, MLI, and motor. Therefore, this chapter focuses on the MLI integration, performance, and energy consumption analysis of different driving cycles and drive modes.

12.2 Selection of the Main Electrical Subsystem

A proper selection of the main electrical subsystems is the most critical part of EV development. The specifications of the entire power electronics, such as power converters, MLI (with field-oriented control [FOC] technique), and BMS, are dependent on selecting the right electric motor and its specification in terms of current and power rating [13, 14]. Selection of the right electric motor, MLI, battery, and suitable transmission are significant issues for having an efficient propulsion system for an EV operating in different urban and suburban areas. During the component selection process, one has to consider the limitations of overloading and underloading operation, to achieve longer mileage and safe performance of the drivetrain. The electrical subsystems must match vehicle load characteristics and motor torque-speed characteristics within a safe, efficient operation zone for satisfactory vehicle performance. In a modern EV, motor power is controlled using an MLI

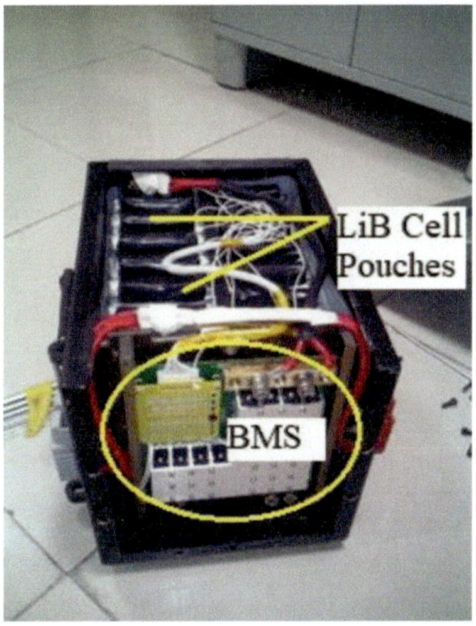

Figure 12.1 LiB cell pouches and BMS.

combined with a control unit (controller). The MLI draws electrical energy from the battery.

In recent times, lithium-ion batteries are generally used as the primary energy source for an EV. The BMS is used to protect the Li battery from overcharge, overdischarge, overcurrent, short circuit, and overtemperature operations. Figure 12.1 shows the Li-ion Batteries (LiB) pack and BMS used for the case study.

Many simulation studies have shown the power consumption analysis of an EV. These studies are conducted considering the variables that relate the energy consumption with its acceleration, speed, and road gradient. Variation of the input power and energy consumption due to the vehicle speed is also evaluated to check energy consumption and efficiency during the vehicle operation [15].

12.3 Drive Train Topologies and MLI Integration

The performance of an EV depends on the powertrain and the energy management strategy used for it. To provide the traction torque, a modern EV powertrain system has a drivetrain consisting of an AC electric motor and transmission, an MLI (converter), DC-DC converters, and a Li-ion battery pack. Several researchers have done many studies on electric drive systems of EVs and HEVs. Despite all these studies, there is still a scope of research work on the drive-cycle analysis and performance evaluation of MLI for different EV categories and powertrain configurations. EV performance is generally subcategorized into dynamic and economic performance. Figure 12.2 shows the criteria for EV performance evaluation. The battery capacity (Ah) influences economic performance, and the drivetrain (incl.MLI and motor)

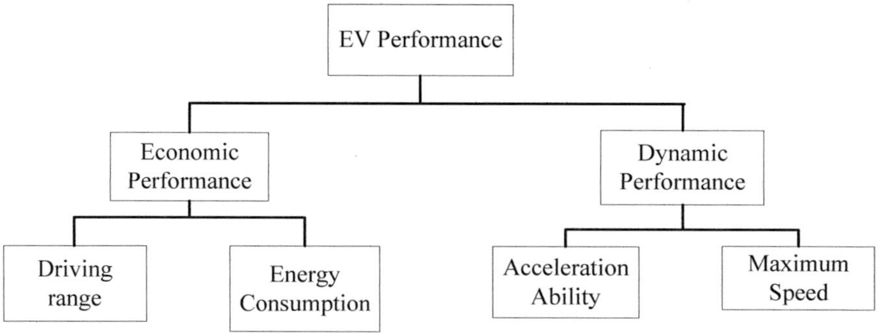

Figure 12.2 Criteria of EV performance evaluation.

influences dynamic performance. Therefore, the appropriate selection and implementation of battery packs and drivetrain is the most critical for an EV. Generally, MLI is mainly associated with the EV dynamic performance.

An EV tends to have flexible drivetrain configurations, with or without any gears/gearbox, as the energy flows from the battery to the wheels via the flexible electrical wires, MLI, and electric motor (EM). Therefore, different propulsion arrangements (drivetrain topology) can be achieved. The drivetrain choice in an EV mainly depends on

(1) modes of propulsion, such as rear-wheel, front-wheel, or all four-wheel drive;
(2) the number of electric machines used in an EV;
(3) type of drive, whether direct or indirect drive;
(4) the number and level of gear, that is, single, dual, or multigears;
(5) variations in electric propulsion and energy sources [16, 17].

Therefore, the selection of suitable MLI depends on the number of motors used. There are six possible drivetrain systems for EVs, and they are shown in Figure 12.3.a–f.

- Figure 12.3.a shows a single EM configuration with a gearbox (GB) and a clutch. It is the conventional type of drivetrain system used in a combustion engine vehicle, where an electric motor replaces the conventional combustion engine for a similar EV configuration. Such configuration does not change the ICE vehicle's drivetrain system. The clutch enables the connection or disconnection of power flow from the motor to the vehicle wheels. The vehicle wheels have high speed, low torque in the higher gears, and high torque, low speed in the lower gears. Due to mechanical simplicity, a single-speed drivetrain can be used and implemented for low-cost EV development.
- Figure 12.3.b shows a single EM configuration without a gearbox and clutch. Such a design has the advantage of a reduced weight of the transmission. However, such a drivetrain configuration demands a complex control of the EM to provide the necessary torque at the wheels [18].
- Figure 12.3.c illustrates the no-transmission type of drivetrain system. The transmission-less type simplifies the conventional type as the transmission is absent.
- The no-transmission type can be further simplified to the differential-less type if the differential gear is eliminated, as shown in Figure 12.3.d. Such topology is known as a cascaded-motor drivetrain (dual-motor), in which two motors are installed to provide functionality equivalent to the

differential. It has a common axle for the motors to transmit power to the wheels. Such topology can also be called the direct-drive configuration.

- In Figure 12.3.e, another simplified version of the transmission-less is illustrated. Two motors are used without a common axle, where each motor is fixed to the individual wheel with reducing gears to drive the wheels. For such a configuration, each motor operates independently, which makes the control system more complex. Two separate MLIs are required for such a configuration, thereby increasing the vehicle cost.
- Figure 12.3.f, the motors are integrated into the hub of the rear wheels themselves to cause rotation directly without gear. Such electric motors are also known as hub motors as the wheel rim itself is the motor. Such a

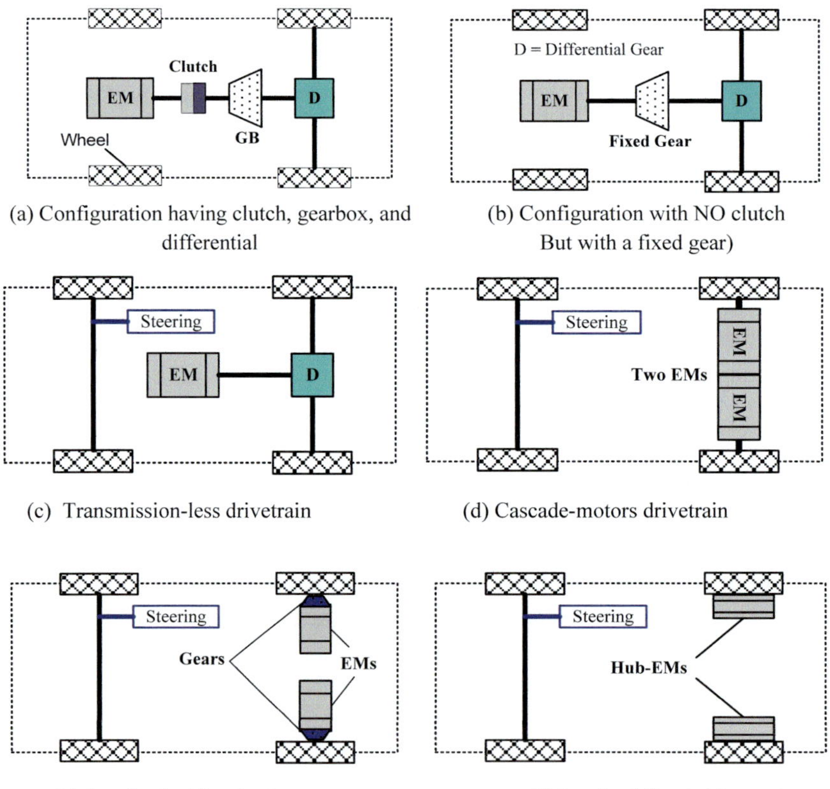

(a) Configuration having clutch, gearbox, and differential

(b) Configuration with NO clutch But with a fixed gear)

(c) Transmission-less drivetrain

(d) Cascade-motors drivetrain

(e) In-wheel with reduction gears

(f) In-wheel direct-drive system

Figure 12.3 Different drivetrain topology.

direct-drive configuration is sometimes known as an in-wheel drivetrain system. A study shows that EVs with a powertrain of the hub-motor drive system with the gear reducer will have lower energy consumption [18].

12.4 Gear Selection and Significance

For an EV drivetrain, the electric motor is the main device that generates the necessary torque for the traction. However, the motor is not uniformly efficient over a wide range of speeds. On the other hand, EVs are required to run at different speeds in various real-world conditions such as highway, city, and hill drive. The electric motor with a suitable gear system is responsible for supplying a wide range of speed and torque to meet the variable speed demand, this situation makes the electric motor run outside its efficient operating region [19]. The MLI is responsible for supplying the electrical power to the motor during those wide ranges of vehicle speed.

There can be either a single-speed or multigear transmission system in an EV. When the number of gears is increased in the mechanical drive system, and the extra mass, the transmission losses will increase, as well as the cost. The multigear drive system causes more energy losses with the gears, clutches, and driveline components. The general practice to search for an optimal gear ratio is to consider the economic performance while constraining the dynamic performance targets. A suitable gear ratio is required, considering the limitations of overloading and underloading operations to achieve longer mileage and safe performance of drive trains [14, 20]. Figure 12.4 illustrates the fundamental relationship between the torque and gear ratios (k_0, k_1) of a vehicle transmission system [21].
where,

T_m = motor torque (Nm);
T_g = transmission torque,
T_d = differential torque
T_w = torque at the wheel
F_w = wheel force (N)
r_w = radius of the wheel
k_0, k_1 = gear-ratios
T_{wl} = torque on the left wheel
T_{wr} = torque on the right wheel
F_{wr}, F_{wl} = wheel forces corresponding to T_{wr} and T_{wl}

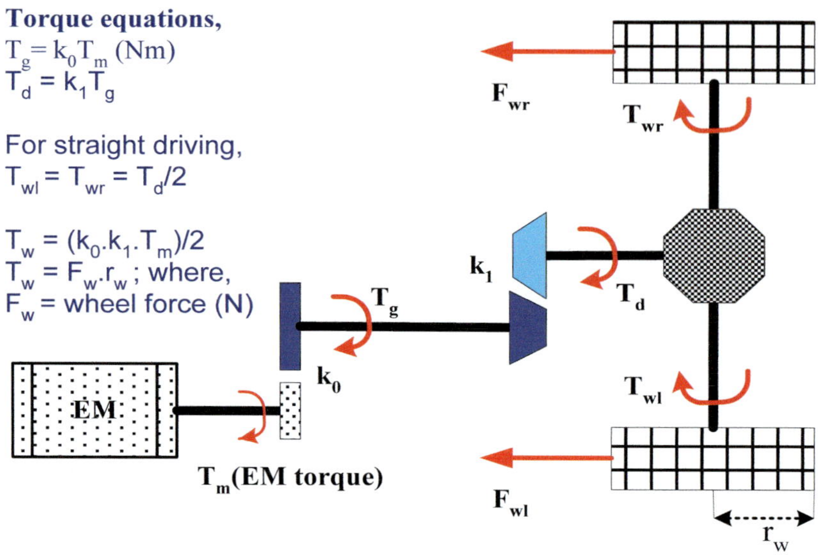

Figure 12.4 General relation of torque and gear ratio.

12.5 EV Data Determination

EVs can be classified into two categories based on their vehicle speed limits. The low-speed EVs are two-wheelers, E-rickshaws, and three-wheeler autos. High-speed EVs are the EV cars, SUVs mini-buses, and passenger buses. One must know their performance with the laboratory driving cycles. There are three driving cycles used for EV development, testing, and study in a chassis laboratory in India. They are namely the Modified Indian Driving Cycle (MIDC) for four-wheelers, the Indian Driving Cycle (IDC) for two, three-wheelers, and the new Worldwide Harmonized Light-duty Test Cycle (WLTC).

There are a couple of measurement methods, namely direct and indirect measurement methods for an EV. These methods are used to generate the vehicle test data of an EV. A suitable data logger can be used to acquire test data of voltage, current, power, and rpm reading through direct measurement for the MLI and motor. A power meter can be used to measure the total electrical energy consumption (kWh) for each battery charge.

The indirect method offers the speed-based indirect energy consumption ($\mathrm{EC_{ind}}$) values and input power of the battery (P_{batt}) for the dynamic speed characteristic observed during testing in the laboratory. The simple formula eqn (12.1) provides the overall energy consumption of each trial. The data acquisition system measures, as well as records, the load-current and battery voltage of the motor drive system during a test. It is necessary to calculate and analyze the energy consumption at every moment for a better understanding of the energy consumption aspect during vehicle operation. For every data point, eqn (12.2) and (12.3) provide the indirect energy consumption in Wh/km and power consumption in kW (kilo-Watt). The chassis dynamometer also provides the mechanical power (kW), tractive force (*F*), and vehicle speed (*v*).

$$M = \frac{N}{D}(\text{Wh/km}), \tag{12.1}$$

where M is the overall energy consumption per kilometer, that is, EC_{tot} (Wh/km), N is the overall electrical energy input in Watt-hour (Wh), and D is the range in kilometer (km).

$$P_{\text{dc}} = VI(\text{W}), \tag{12.2}$$

where P_{dc} = battery input power (PoC), V = batt.voltage in Volts (V), and I= batt.current in Amps (A).

$$\text{EC}_{\text{ind}} = \frac{\text{PoC}}{S}(\text{Wh/km}) \tag{12.3}$$

where EC_{ind} = the indirect EC (Wh/km) derived from speed, referred to as actual energy consumption for the real-world studies in this report; PoC = P_{dc} (W), and S = EV speed (km/h). The above eqn (12.1)–(12.3) are commonly used standard formulas for any EVs.

12.6 A Case Study–Laboratory and Real-world Test of an EV

12.6.1 Physical test setup and testing of an EV in the laboratory

The EV test setup is done on the chassis dyno using the mechanical clamps, as shown in Figure 12.5. The road load simulation test for the EV is conducted on the dynamometer. A driving cycle is loaded in the driver's aid hardware (PC) and utilized for testing the EV. This simulation is also performed to adjust the dynamometer's inertia to correspond to the real vehicle weight.

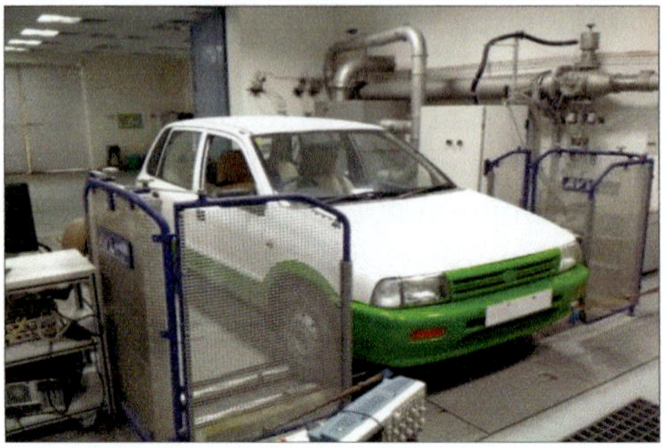

Figure 12.5 Test setup of an experimental EV on a chassis dynamometer.

12.6.2 Performance evaluation of an experimental EV

12.6.2.1 Method: coast-down test

Generally, the coast-down test [21, 22] of a vehicle is carried out to find out the road-load tractive force (F). It is done during the vehicle deceleration by estimating the time elapsed from a speed "v" to zero speed when the gear position is in neutral. IS 14785: 2000 (R2016) specifies the procedure for determining the road-load resistances of a vehicle, such as the aerodynamic and rolling resistance. This resistance data is primarily intended for the road load simulation test on the chassis dyno. The road-load equation is given in eqn (12.4).

$$F = a + bv^2 \tag{12.4}$$

where F = road-load tractive force in Newton (N), "a" and "b" = road load constants, "v" = vehicle speed. However, the fundamental second-degree polynomial equation for the road-load simulation is given in eqn (12.5). The resistance forces F_0, F_1, and F_2 (sometimes known as coefficients) are established using the best curve-fitting method (defined in the IS 14785: 2000 (R2016)).

The coast-down test is also carried out to establish the coefficients of the test vehicle. Figure 12.6 shows the forces of resistance that act on the moving vehicle.

The equation for the road load simulation is as follows [18]:

$$F = F_0 + F_1 v + F_2 v^2, \tag{12.5}$$

where,

F = tractive force (N); g = gravity, 9.81m/s^2

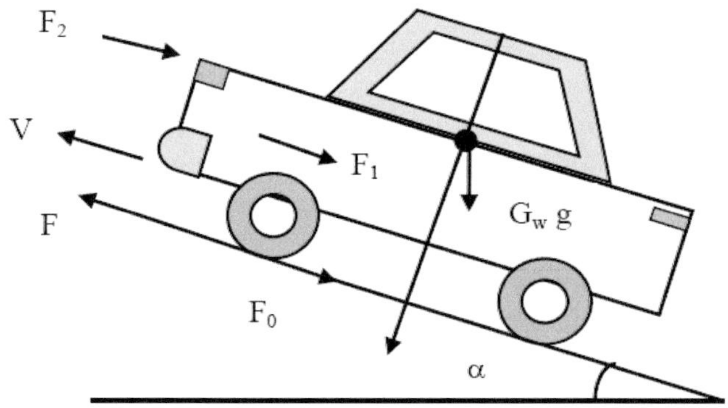

Figure 12.6 Resisting forces acting on a moving vehicle.

G_w = vehicle reference weight (kg).
F_0 = value equal to rolling resistance (N);
F_1 = value equal to losses associated with the chassis dynamometer
F_2 = value equal to the coefficient of aero-dynamic force (N/(km/h)^2)
v = speed of the vehicle (km/h)
α = gradient of the road.

The term $F_1 v$ in the fundamental eqn (12.5) is not considered to get the realistic eqn (12.4) for the road-load of real-world test conditions. It is because all four wheels are rotating during the actual vehicle operation, whereas only two wheels are rotating and the vehicle is not in motion on the chassis dynamometer.

The regulatory driving cycles (WLTC, MIDC, and IDC) are loaded, one after another, in the Driver's Aid hardware. The road load simulation test is conducted on the dynamometer following the cycles.

As a standard practice, the road-load coefficients, F_0, F_1, F_2, and tractive force (F) of the EV are determined as per the procedure (IS 14785: 2000). Later, the values of the coefficients are manually entered in the dynamometer PC for the load and inertia adjustment. The test vehicle EV is tested on the dynamometer following the three driving cycles one after another. The test equipment measures and records the electrical power input, EC, tractive force, vehicle's mechanical power output, and vehicle speed.

The performance evaluation is built on a set of crucial vehicle test parameters (such as vehicle speed, average speed, power input, battery current, mechanical power output, energy consumption, and acceleration test). The top speed and average electric range tests are conducted and observed. The total energy consumption (EC_{tot}) is assessed using the standard formula of eqn (12.1). The basic statistical analysis is performed for the test data which includes the acceleration ability. Acceleration ability is a measurement of the time taken to reach either 60 or 100 km/h from the standstill position, and it is also the overtaking ability. However, the overtaking ability usually measures the time taken to reach a higher speed from a lower speed, for example, the time taken to reach 80 from 60 km/h [19].

A few free acceleration tests are carried out to measure the time (in seconds) to reach 0 to 20, 40, and 60 km/h speed and the corresponding maximum battery currents. Such a sudden acceleration test will subject the MLI to a possible real-world operating condition depending on the driver's behavior. Three sets of readings are recorded for each experiment (e.g.,

Table 12.1 Acceleration test result of the test EV.

Test	Speed (km/h)	Average time (s)	Average battery current (A)/Input current to the MLI
1.	0–20	4.76	245.25
2.	0–40	9.07	296.40
3.	0–60	25.95	301.80

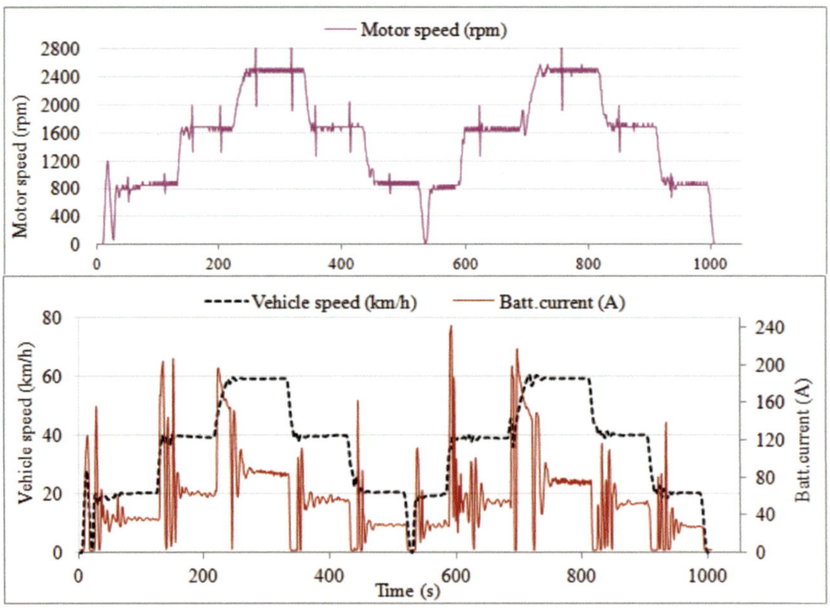

Figure 12.7 Steady speed and MLI current characteristics for the EV.

three readings for the 0–20 km/h test). The average values are provided in Table 12.1. The battery current is noticed to reach a load current of 245 to 301 A momentarily, which indirectly indicates the current drawn by the MLI.

A few constant speed tests of the EV are also carried out at 20, 40, and 60 km/h on the dynamometer. The constant speed traits of the vehicle, battery current, and motor speed are illustrated in Figure 12.7. It shows the range of load current (MLI input current) required to move the vehicle from a standstill condition, as well as the current range required for acceleration of the vehicle from a steady speed to a higher speed.

12.6.2.2 Drive cycle analysis for the EV

Any vehicle test cycle or vehicle driving consists of four modes of drive, namely (1) idling, (2) deceleration, (3) acceleration, and (4) cruising. The test data comprising battery current, power input, mechanical power output, vehicle speed, and energy consumption are assessed and statistically distributed into the four modes. The EV's test parameters are measured and can be analyzed using a computer program in MS Excel to segregate the test data into the four drive- modes. The test data graphs are drawn along with the drive cycles. It is noticed that the battery load current closely follows the vehicle speed in the cruising and acceleration modes, and the current drops to zero as soon as the driver releases the accelerator either in the deceleration mode or during braking. The results indicate that the battery-current directly corresponds to the speed. The maximum batt-current is also noticed in the range of 249 to 260 A during the EV's acceleration mode. Operating the battery and MLI in such a high current condition is undesirable and must be minimized. Such dynamic nature of EV data could be helpful for studying and enhancing the dynamic properties of the Li-ion battery and MLI-based motor controller.

Figures 12.8–12.10 display the change of power input (MLI output power) with the true EV speed along the test cycles. The electrical power input is the product of battery current and voltage, thus the power also closely follows the EV speed whenever the speed and battery current is greater than zero (>0). Such data analysis is crucial as the motor draws power from the MLI. The MLI design parameters need to be considered based on such practical test data.

The electric motor does not propel the vehicle wheels during deceleration, but the wheels keep rotating because of the inertia. Negative mechanical power is observed during the deceleration in each test cycle, which indicates

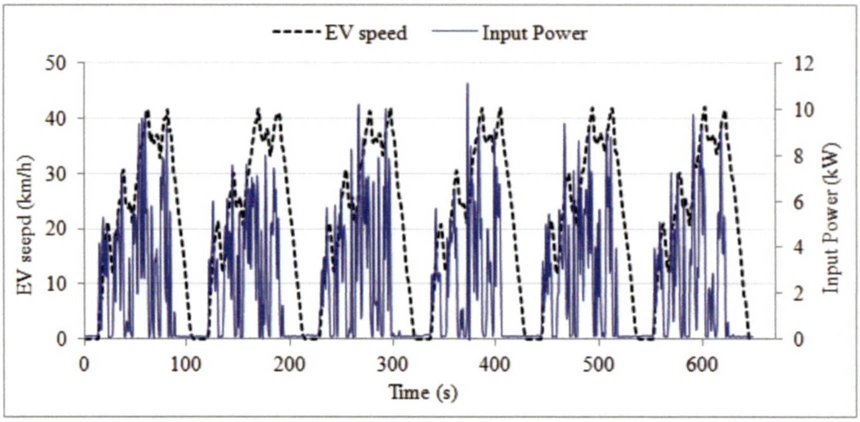

Figure 12.8 Difference of MLI power and EV speed for the IDC.

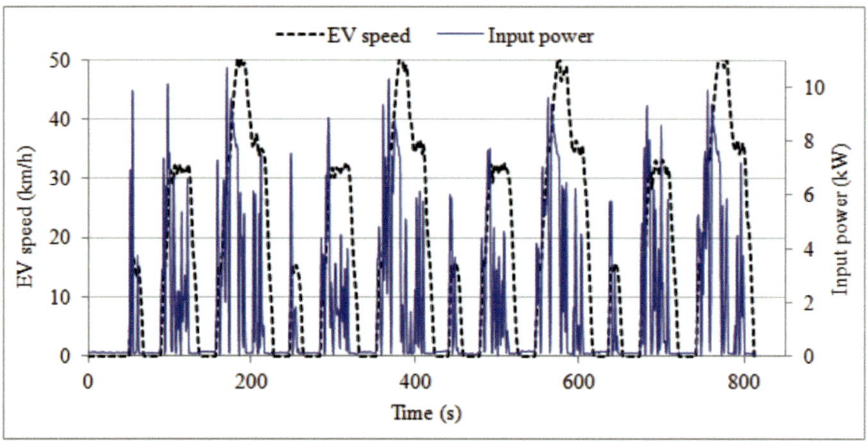

Figure 12.9 Difference of MLI power and EV speed for the MIDC Phase-I.

the regenerative power of the EV. This regenerative power can be recovered and stored in a storage device (such as an ultracapacitor). The motor drive needs more power (i.e., more current) for the high torque requirement during the acceleration mode. Therefore, MLI and motor must be designed to meet maximum operating torque and wide variation of speed.

High torque is normally required at the beginning of every drive attempt. Such high torque at the vehicle start demands a proper and robust design of the MLI and motor.

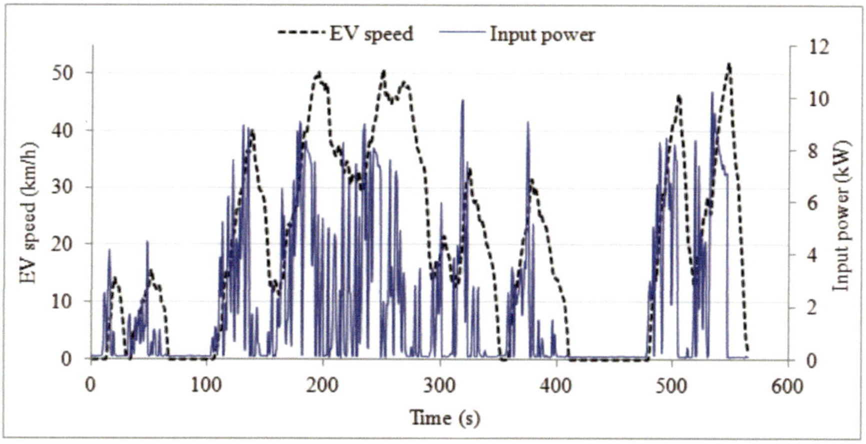

Figure 12.10 Difference of MLI power and EV speed for the WLTC phase-low.

12.6.2.3 Distribution of power input, energy consumption during the individual drive modes

The percentage (%) contribution of each drive mode and corresponding test parameter (i.e., time spent, power input, and EC_{ind} data) are evaluated and analyzed for the three driving cycles. It is generally noticed that most of the power input and EC_{ind} are absorbed in the acceleration mode, this situation can be further investigated for improvement in power consumption [23].

The comprehensive observation shows that the area of improvement for energy consumption is in the cruising and acceleration mode for better vehicle performance and energy efficiency.

The MIDC cycle has two parts, namely Phase I and Phase II. Phase I represents urban driving conditions that have a speed limit of 50 km/h, and Phase II represents highway driving conditions that have a speed limit of 90 km/h. Figure 12.11 indicates how closely the calculated $EC_{\mathbf{ind}}$ follows the path of load current (MLI output current) and EV speed. This speed is consequently dependent on the motor speed (rpm). It is observed that the MLI is subjected to such dynamic operation conditions.

It is generally known that a higher current will cause to having a bigger cable size, and elevated I^2R losses in the MLI and motor. Therefore, it is necessary to improve the motor and MLI design to reduce the load current and the I^2R losses. It is also essential to understand the boundary conditions of motor operation (such as min. and max. current, torque, power, and rpm)

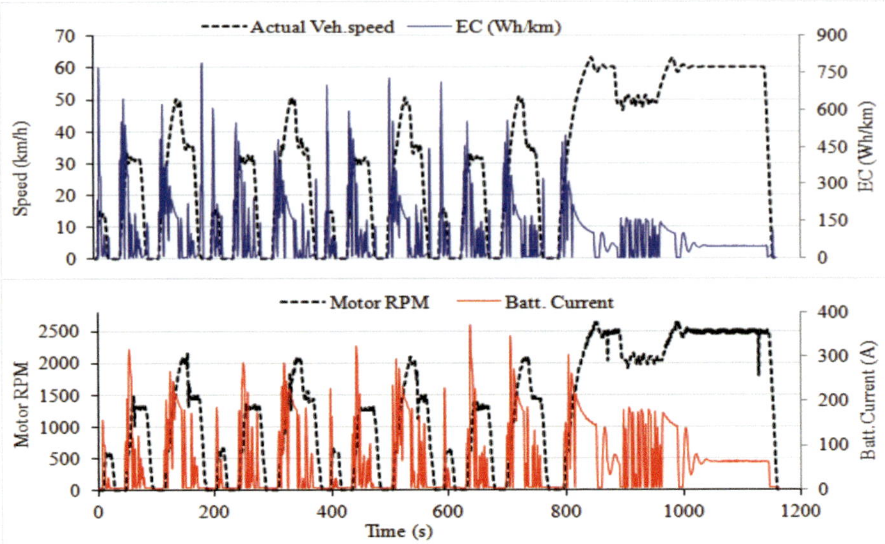

Figure 12.11 Difference of motor rpm, speed, EC_{ind}, and MLI input (load $_{batt}$) current for the MIDC.

to avoid overloading the motor and MLI. Every EV researcher and engineer desires to develop and utilize an energy-efficient motor and MLI for extended electric mileage and life longevity.

12.6.2.4 Real-world EV performance testing

Computer simulation studies have limitations in accurately emulating real-world EV testing. The dynamometer tests in the laboratory are standard, accurate, repetitive, and suitable for better comparisons between EVs under controlled test conditions. However, they also have limitations in emulating EV performance evaluation in real-world operating conditions that are affected by traffic, pedestrians, weather conditions, in-vehicle weights, and changes in road gradients [24]. In real-world conditions, an MLI goes through a wide range of electrical power variations due to continuous changes in vehicle torque and speed. The effect of real-world driving on crucial EV parameters (such as battery current and input power) can be studied for robust MLI design and development. Such a study may provide useful information and data to the researchers who may establish the efficient energy-use strategy and EC factor for the metropolitan and Tier-II cities. The real-world driving simulation may become a faster and more economical testing solution in the future [25]. The MLI performance and durability may be affected by the continuous variation of power in real-world conditions. In a city, the traffic conditions are unforeseeable with pedestrian activities, narrow roads, etc. Figures 12.12 and 12.13 show the dynamic nature of the battery current,

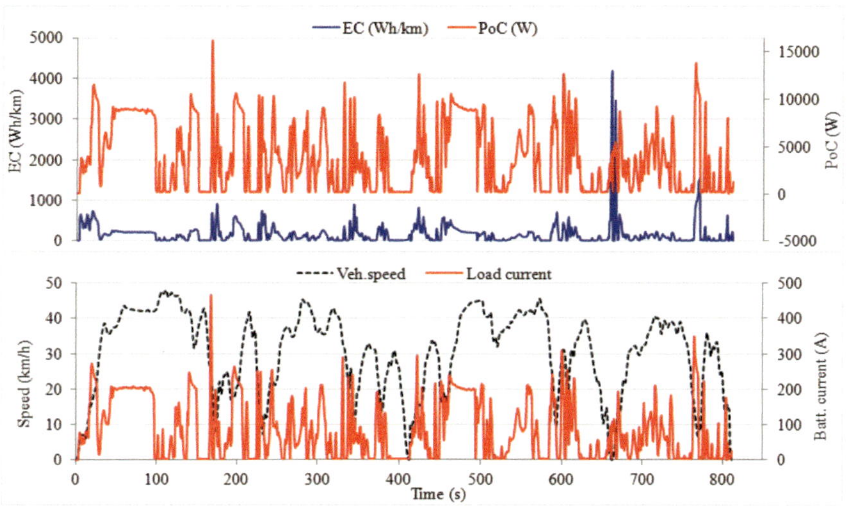

Figure 12.12 Difference of MLI input (load $_{batt}$) current, speed, EC_{ind}, and PoC on a low_traffic route.

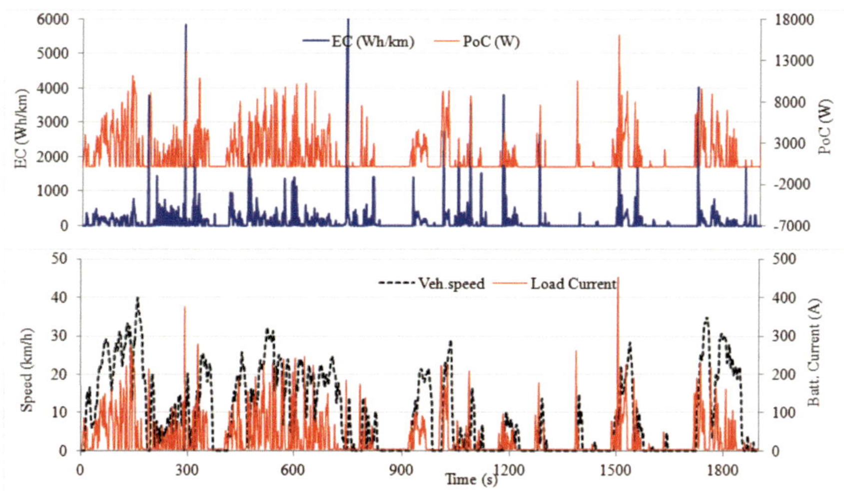

Figure 12.13 Difference of real-world EC_{ind}, PoC, MLI input (load $_{batt}$) current, and vehicle speed for a congested traffic route.

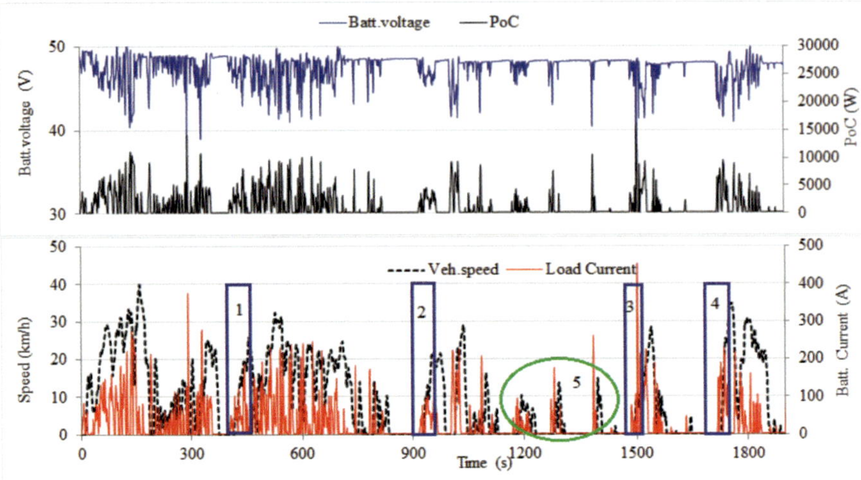

Figure 12.14 Change of MLI input voltage, (load) current, and PoC, from standstill to in-motion state on the congested_traffic route.

EV speed and calculated electrical power, and EC that are obtained from the on-road testing. The test outcome displays that the $EC_{indirect}$ and PoC are closely associated with vehicle acceleration, speed, and repeated stop-go operations.

It is experienced that no idling mode (vehicle stop) happens on the low traffic route leading to the highest average EC_{ind}. Such a situation is shown in Figure 12.12 by the speed's dotted line, where the speed line is not reaching zero speed. Figures 12.12 and 12.13 also indicate the range of vehicle speed, battery current, and power that can be experienced in a similar city.

Figure 12.13 shows longer vehicles idling on the congested_traffic route. It is illustrated by the dotted speed line. It indicates lower average speed and lesser energy consumption on the road and contributes to the lowest average EC_{ind} among the routes.

Figure 12.14 displays the change in voltage, current, and input power when the vehicle acts in motion from a standstill condition. In Figure 12.14, markers 1, 2, 3, 4, and 5 illustrate the swift increase in vehicle speed, load current, and power, indicating the vehicle's movement from the standstill condition. Such an extent of test parameters can be valuable data for enhancing the motor design, MLI, and battery packs. At the start of the EV, the battery voltage is low, however, the battery current is high to meet the torque requirement of "go operation."

12.6.2.5 Analysis of real-driving cycle's drive-modes for EV parameters

As stated earlier, the four drive modes are idling, acceleration, cruising, and deceleration. The load current, energy, and power consumption data are divided into these four drive modes for the on-road test data. It is a standard exercise to follow the limits of deceleration and acceleration as defined in the regulatory "Type Approval Procedure" document [26]. The percentage (%) calculation of the time spent in each drive mode is based on this procedure's restrictions.

Vehicle speed is a key indicator of traffic flow and driving patterns that can influence energy consumption and electric mileage in a city. Many studies reported that EV operation is more energy efficient in congested traffic conditions. The above analysis has also revealed that vehicle operation experiences relatively steady speed with a higher average speed on the highways that causes continuous power consumption from the primary energy source. Least deceleration and braking are observed on the highways, due to which negligible regenerative power is observed. The highest energy consumption is observed during highway driving with a speed of 50 to 70 km/h. The relationship between traffic congestion and EV energy consumption is noted that higher EC happens due to an increased average speed.

It would be a great deal of interest and essential to design a computer model to investigate the effects of the kerb and battery weights, and the resistance forces on the EV performance parameters such as energy consumption, range per charge, maximum acceleration, acceleration intensity, etc. Such a simulation model can provide an interrelationship among the test parameters and compare them with the experimental data for better understanding.

12.7 Conclusion

The case study is conducted to understand better the operating conditions to which the FOC MLI is subjected in an EV. The three driving cycles, that is, IDC, MIDC, and WLTC cycles are generally used for EV performance tests in a chassis laboratory. The case study investigated EV speed variation, motor speed in rpm, torque, battery current, EC, and input/output power during the four drive modes of each test cycle. The performance evaluation using the drive cycle analysis indicates the scope of improvement for the electric powertrain system for better performance, reduction in I^2R losses, and optimization for lesser EC. The experimental study also indicates the importance of the selection of the right MLI, electric motor, and motor drive system. Insightful conclusions can be drawn from such a case study in terms of improvement in efficiency, vehicle speed, acceleration ability, driving range, etc.

As a part of the case study, a real-world experiment is conducted to investigate the limitations of a laboratory study, as well as the gap between the real-world and laboratory data. It is observed that laboratory study has limits due to the inability to emulate actual real-world driving conditions. Many research studies reveal and conclude that laboratory tests may not give a realistic assessment of vehicle performance data of actual on-road conditions. Therefore, real-world tests are necessary for more accurate data interpretation.

The drive cycle analysis used in this case study is established based on the investigation of the standard driving cycle and real driving. The analysis provides the percentage distribution of each drive mode to the load current, EC_{ind}, and power consumption of the vehicle. The EV operation in low_traffic conditions reveals that most of the operating points are within a specific speed range versus torque. Therefore, it is crucial to enhance the motor design with the motor operating points lying within the high_efficiency region for operation in low_traffic conditions. The comparative analysis of the test results revealed the gap between the real-world and laboratory data. The test results provide the following observations:

- The laboratory evaluation of the total electrical EC and PoC (using the MIDC cycle) does not provide the real-world data of the vehicle, and there is a gap between the laboratory and real-world data of a city.
- The EV operation is more energy_efficient on congested traffic routes because of more traffic stops that lead to lower average speed which subsequently results in less energy consumption.
- The load current traits could be useful information and guidance for enhancing the motor design, MLI, and battery pack.
- Higher operating current will make researchers have a bigger cable size and higher I^2R losses in the motor, MLI, and battery.

References

[1] COP21, https://unfccc.int/process-and-meetings/the-paris-agreement, Accesed on 20th April 2024.

[2] COP26, https://www.un.org/en/climatechange/cop26, Accesed on 20th April 2024.

[3] CSE, 2020, 2022 Air Quality and Public Health Report, Centre for Science and Environment (CSE), https://www.cseindia.org/page/air-quality-and-public-health, Accessed on 15th April 2024.
[4] Aayog, N., and R. M. I. (RMI), "India's Electric Mobility Transformation: Progress to date and future opportunities," **NITI Aayog** INDIA and RMI, Tech. Rep., April 2019.
[5] Maya, B.D., "Zero-emission vehicle credits: China program design inputs brief," Tech. Rep., Innovation Center for Energy and Transportation, Beijing, China, 2015.
[6] FAME-II, https://fame2.heavyindustries.gov.in/content/english/32_1_FAMEDepository.aspx
[7] Nemet, G. F., T. Holloway, and P. Meier. "Implications of Incorporating Air-Quality Co-Benefits into Climate Change Policymaking," Environmental Research Letters 5, No. 1, 4007, 2010.
[8] IHME, 2013, The Global Burden of Disease 2010: Generating Evidence and Guiding Policy. Institute for Health Metrics and Evaluation (IHME), Seattle, USA.
[9] Z Li, Amir Khajepour, Jinchun Song, "A comprehensive review of the key technologies for pure electric vehicles," Volume 182, 1 September 2019, Pages 824-839, 2019
[10] Wu, G., Zhang, X., and Dong, Z., "Powertrain Architectures of Electrified Vehicles: Review, Classification and Comparison," Journal of the Franklin Institute, 352(2):425-448, 2014, doi:10.1016/j.jfranklin.2014.04.018.
[11] Lukic, S.M. and Emado, A., "Modeling of Electric Machines for Automotive Applications Using Efficiency Maps," Proceedings of the Electrical Insulation Conference and Electrical Manufacturing & Coil Winding Technology Conference of IEEE, 2003, 543-550, doi:10.1109/EICEMC.2003.1247945.
[12] Williamson, S.S., Lukic, S.M., and Emadi, A.,"Comprehensive Drive Train Efficiency Analysis of Hybrid Electric and Fuel Cell Vehicles Based on Motor-Controller Efficiency Modeling," IEEE Transactions on Power Electronics 21(3):730-740, 2006, doi:10.1109/tpel.2006.872388.
[13] Mustafa Aktasa, et al., "Direct torque control versus indirect field-oriented control of inductionmotors for electric vehicle applications," Engineering Science and Technology, an International Journal 23 (2020), 1134-1143, 2020.
[14] Prasun Mishra, Suman Saha, Ikkurti, H.P., "Selection of propulsion motor and suitable gear ratio for driving an electric vehicle on Indian

city roads," IEEE International Conference on Energy-Efficient Technologies for Sustainability, 692 – 698, April 2013.

[15] Bor Yann Liaw, Matthieu Dubarry, "From driving cycle analysis to understanding battery performance' in real-life electric hybrid vehicle operation," Science Direct, Journal of Power Sources, 174,76–88, 2007.

[16] Xue,X. D., Cheng, K. W. E., and Cheung, N. C., "Selection of electric motor drives for electric vehicles," in Power Engineering Conference, 2008. AUPEC '08, Australasian Universities, 1–6, 2008.

[17] Praveen Kumar, Prof. S. Majhi, "Introduction to Hybrid and Electric Vehicles," of NPTEL, MoHRD, 2009-2012, https://nptel.ac.in/courses/108/103/108103009/, Accessed on 18th October 2020.

[18] Bin Wang, Min Xu, Li Yang, "Study on the economic and environmental benefits of different EV powertrain topologies", Energy Conversion and Management, Vol.86,916–926, 2014.

[19] Ahssan, M., Ektesabi, M., and Gorji, S., "Electric Vehicle with Multi-Speed Transmission: A Review on Performances and Complexities," SAE Int. J. Alt.Power. 7(2):169–181, 2018, doi:10.4271/08-07-02-0011.

[20] Polychronis Spanoudakis, et al., "Efficient Gear Ratio Selection of a Single-Speed Drivetrain for Improved Electric Vehicle Energy Consumption," Sustainability 2020, 12, 9254, 2020, doi:10.3390/su12219254.

[21] Robindro Lairenlakpam, "Experimental Investigation of Energy Consumption and Performance of an Electric Vehicle Powertrain on Different Laboratory and Real-World Driving Cycles and Drive Modes," The PhD thesis submitted for the Degree of DOCTOR OF PHILOSOPHY, IIT Guwahati, 2021.

[22] SAE Vehicle dynamics, https://www.sae.org/images/books/toc_pdfs/R114.pdf, Accessed on 12th April 2024.

[23] Jin Zhang, Zhenpo Wang, Peng Liu, Zhaosheng Zhang, "Energy consumption analysis and prediction of electric vehicles based on real-world driving data," Elsevier Applied Energy, 275 (2020) 115408, 2020.

[24] Ray Galvin, "Energy consumption effects of speed and acceleration in electric vehicles: Laboratory case studies and implications for drivers and policymakers," Transportation Research: Part D, 53 (2017), 234–248, 2017.

[25] Ruoyun Ma, Xiaoyi He, Yali Zheng, Boya Zhou, Sheng Lu, Ye Wu, "Real-world driving cycles and energy consumption informed by large-sized vehicle trajectory data," Journal of Cleaner Production, Vol. 223, 564-574, June 2019.

[26] Automotive Industry Standard (AIS) -39, 40 https://hmr.araiindia.com/api/AISFiles/AIS_039_Rev_1_Corri_1_&_Amd%201_998db079-3c7b-4d36-886e-fdf5127e435a.pdf, Accessed on 15th April 2024.

Index

About the Editors

Dr. Kundan Kumar
Assistant Professor,
Electrical Engineering Dept.,
National Institute of Technology Jamshedpur, India
E-mail: kundan01012020@gmail.com

Kundan Kumar (M' 2013, SM' 2021 IEEE USA) is presently working as an Assistant Professor in the Department of Electrical Engineering, at NIT Jamshedpur. He completed his B. Tech. from the IET MJP Rohilkhand University, Bareilly, in 2008. He has received the M. Tech. degree from NIT Jamshedpur, Jamshedpur, India, and the Ph.D. degree from the University of Padova, Padova, Italy, both in Electrical Engineering, in 2010 and 2016, respectively. Prior to NIT Jamshedpur, he has worked as an Assistant Professor at NIT Manipur and at School of Electrical Engineering, Kalinga Institute of Industrial Technology, deemed to be a university, in Bhubaneswar, India. Dr. Kumar received the Silver Medal for securing the first position during his M. Tech. course. He has also been awarded a University Fellowship for a Ph. D. course at the University of Padova, Italy. He has also been awarded the Best Presentation Recognition award at IECON, held in Japan in 2015, and the Best Paper award in IEEE STPEC-2021 at Bilaspur, India. He has chaired and co-chaired the various Technical Sessions in the prestigious IEEE International Conferences, such as IECON, PEDES, ICPEE, GUCON, and SEFET. He has published more than 50 research articles in reputed International Journals and Conference Proceedings, 7 book chapters, and 5 Indian Patents. He has published one edited book entitled "Electric Vehicle Propulsion Drives and Charging Systems," in Taylor and Francis, CRC Press. He has organized the IEEE International Conference on Power Electronics and Energy (ICPEE-2021) and International Conference on Science, Technology, and Engineering (ICSTE-2023) with the capacity of organizing secretary as well as the 2nd IEEE International Conference on Industrial Electronics: Developments and Applications (ICIDeA-2023) with the capacity of Conference Chair. As well as "2024 1st IEEE International Conference on Green Industrial Electronics and Sustainable Technologies (GIEST-2024)" held at NIT Manipur from 25–26 Oct. 2024. He is a senior member of the IEEE (USA) and a Life member of the Institution of Engineers (IE), India. He is the reviewer of IEEE journals and transactions such as TIE, TPEL, JESTPE, and various reputed conferences such as IECON, PEDES, INDICON, and PEMC. His research interests include Electric Vehicles and their charging infrastructures, Wireless Power Transfer Systems, Dynamic Wireless Charging Systems, Propulsion Drive Systems for EVs, Application of Wide Bandgap Semiconductor Devices (i.e. GaN & SiC), High-Efficiency Power Converters, Soft Switching, Multi-port Converters, Single-Active Bridge (SAB), Dual Active Bridge (DAB) Converters, and Offshore Wind Energy System.

Dr. Ambrish Devanshu,
Assistant Professor,
Electrical Engineering Dept.,
National Institute of Technology Silchar, India,
E-mail: ambrish.devanshu@gmail.com

Ambrish Devanshu earned his B.Tech. in Electrical and Electronics Engineering from Cochin University of Science and Technology, completing his major project at ISRO's Vikram Sarabhai Space Centre in 2011. He worked as a Project Engineer at Wipro Technologies from 2011 to 2014 before pursuing his Ph.D. at Delhi Technological University, which he completed in 2019. Currently an Assistant Professor at the National Institute of Technology (NIT) Silchar, he has previously served at NIT Patna.

Dr. Devanshu has an extensive research portfolio, including numerous journal, conference and book chapter publications, authored books, and patents. He has been recognized with Delhi Technological University's "Premier Research Award" in 2021 and "Commendable Research Award" in 2022 with a cash prize and certificate of appreciation. He is actively engaged in academic and professional activities, having delivered talks at various Institutes of National Importance in India and contributed to IEEE conferences as a technical chair, session chair, publicity chair, and program chair.

In addition to organizing national-level workshops and technical programs, Dr. Devanshu serves as a reviewer for several reputed national and international journals, including prestigious IEEE Transactions, underscoring his commitment to advancing the field of electrical engineering.

Prof. Subrata Banerjee,
Professor
Electrical Engineering Dept.,
National Institute of Technology Durgapur, India
E-mail: bansub2004@yahoo.com

Subrata Banerjee is working as Professor with the Department of Electrical Engineering, National Institute of Technology, Durgapur, India. His research interest includes modeling & control of switch-mode converters and inverters, multilevel inverters & different modulation techniques, electromagnetic levitation, active magnetic bearing, controller design, intelligent control etc. He has successfully completed some research and consultancy projects. He has authored about 240 research papers in national/international journals and conference records. He has filed three Indian patents & one copyright out of which three have been granted. He has guided 13 Ph.D. and 30 M. Tech students and many are pursuing their degree under his guidance. He has written one book (published in Springer) and nine book chapters. He is recipient of several academic awards, including 10 nos. Best Paper Awards , TATA RAO Prize, IETE Bimal Bose Award. He is a Fellow of the Institution of Engineers (India), the Institution of Electronics and Telecommunication Engineers (India), and the Institution of Engineering and Technology (UK), Senior Member IEEE (USA). He has acted as Associate Editor/Editorial Board in IEEE Access (USA), IET Power Electronics (UK)/ IEEE Transportation Electrification Community eNewsletter (USA). Prof. Banerjee has attended & chaired sessions in different IEEE IES/IAS conferences, acted as Member technical committee of many national & International Conferences. He has visited ten (10) countries for different academic assignments. He has been listed among top 2% scientists in the world in 2020, 2021, 2022, 2023, 2024 by survey conducted by a team of scientists at Stanford University.

Dr. Lairenlakpam Robindro
Senior Principal Scientist
CSIR, Indian Institute of Petroleum, Dehradun, India
E-mail: robindro@iip.res.in

Robindro is a Scientist in the areas of e-mobility and energy sector. He obtained B.E. degree with distinction in Electrical & Electronics Engineering from Madurai Kamaraj University in 2000, and holds Ph.D. in Electronics & Electrical Engineering from IIT Guwahati. He has more than 22+ years of research experience in EV powertrain system performance and energy studies, E-vehicle system integration, EV driving and charging energy studies, RE hybrid power systems for Power-to-X, Energy system integration and management, Vehicle dynamics tests. Before joining CSIR-IIP in 2006, he worked for 4 years in a private company that dealt with German companies of vehicle test instruments for vehicle dynamic tests such as brake, acceleration test, lane change test, etc. During the year 2002-2006, he had opportunities to work with the R&D teams of Maruti Suzuki, Yamaha Motors, Tata Motors, TVS Motors, TVS Tires, Kalyani Brakes, and ARAI. During this period, he also gained experience in vehicle dynamic testing on proving grounds at VRDE (Ahmednagar) and WABCO (Chennai) test tracks. He had undergone specialized trainings on various EV and automotive test systems in Belgium, Germany and Japan during 2004-2010. He was also awarded a research trip to FCLAB (University of Franche-Comte, Belfort, France) about Hydrogen Fuel Cell (FC) technology for vehicle and energy solution by French Embassy in India from 19-26 November 2022. He recently developed and evaluated prototype EV kits for M1 category passenger vehicles and a prototype EV charging system powered by renewable hybrid energy sources. He has published several research papers and reviewed many journal & conference papers. He has also done project reviews for some of the Govt. project proposals submitted by various institutes. He has delivered 20+ invited talks on different national and international platforms. He got involved in 25+ research projects in vehicle research area, out of which 10 projects were executed & completed as Principal Investigator. Currently, a few projects on L5 vehicles and hybrid renewable energy systems for green hydrogen are in progress. He is also a committee member of TED-26, TED-27 committees of BIS (Bureau of Indian Standards) and Automotive Industry Standards (AIS) for EVs and HEVs.

Prof. Sheldon S. Williamson
Professor,
Dept. of Electrical, Computer and Software Engg. Ontario Tech Univ., Oshawa, Canada
E-mail: sheldon.williamson@gmail.com

Sheldon S. Williamson received the Ph.D. degree (Hons.) in electrical engineering from the Illinois Institute of Technology, Chicago, IL, USA, in 2006. He is currently a Professor with the Department of Electrical, Computer and Software Engineering, and the Director of the Smart Transportation Electrification and Energy Research (STEER) Group, within the Faculty of Engineering and Applied Sciences, at Ontario Tech University, in Oshawa, Ontario, Canada. His current research interests include advanced power electronics, electric energy storage systems, and motor drives for transportation electrification and autonomous e-mobility. Prof. Williamson is also an NSERC Canada Research Chair (CRC) in electric energy storage systems for transportation electrification and is a Fellow of the IEEE.